Make: More Electronics

Charles Platt

Make: More Electronics

by Charles Platt

Printed in Canada.

Published by Maker Media, Inc., 1005 Gravenstein Highway North, Sebastopol, CA 95472.

Maker Media books may be purchased for educational, business, or sales promotional use. Online editions are also available for most titles (*http://my.safaribooksonline.com*). For more information, contact our corporate/institutional sales department: 800-998-9938 or *corporate@oreilly.com*.

Editor: Brian Jepson
Production Editor: Kara Ebrahim
Proofreader: Amanda Kersey
Cover Designer: Juliann Brown
Interior Designer: David Futato
Illustrator: Charles Platt

May 2014: First Edition

Revision History for the First Edition:

2014-04-25: First release

See *http://oreilly.com/catalog/errata.csp?isbn=9781449344047* for release details.

ISBN: 978-1-449-34404-7

[TI]

In memory of my father, Maurice Platt, who showed me that it is a fine and valuable occupation to be an engineer.

Table of Contents

Acknowledgments

I discovered electronics when I was a teenager, in collaboration with my friends in high school. We were nerds before the word existed. Patrick Fagg, Hugh Levinson, Graham Rogers, and John Witty showed me some of the possibilities. Fifty years later, Graham kindly contributed a schematic to this book.

Several decades after that, Mark Frauenfelder nudged me back into the habit of making things. Gareth Branwyn facilitated *Make: Electronics,* and Brian Jepson enabled its sequel. They are three of the best editors I have known, and they are also three of my favorite people. Most writers are not so fortunate.

I am also grateful to Dale Dougherty for starting something that I never imagined could become so important, and for welcoming me as a participant.

Fredrik Jansson provided advice and corrections while I was working on this project. His patience and good humor have been very valuable to me.

Fact checking was also provided by Philipp Marek. Don't blame Philipp or Fredrik if there are still any errors in this book. Remember that it's much easier for me to make an error than it is for someone else to find it.

Circuits were built and tested by Frank Teng and A. Golin. I appreciate their help. I am also grateful for the conscientious attention of Kara Ebrahim and Kristen Brown in the production department, and proofreader Amanda Kersey.

Preface

This book picks up where my previous introductory guide, *Make: Electronics*, left off. Here you will find topics that I did not explore in detail before, and other topics that were not covered at all because I lacked sufficient space. You will also find that I go a little bit further into technicalities, to enable a deeper understanding of the concepts. At the same time, I have tried to make "Learning by Discovery" as much fun as possible.

A few of the ideas here have been discussed previously in *Make* magazine, in very different forms. I always enjoy writing my regular column for *Make*, but the magazine format imposes strict limits on the wordage and the number of illustrations. I can provide much more comprehensive coverage in this book.

I have chosen not to deal with microcontrollers in much depth, because explaining their setup and programming language(s) in sufficient detail would require too much space. Other books already explain the various microcontroller chip families. I will suggest ways in which you can rebuild or simplify the projects here by using a microcontroller, but I will leave you to pursue this further on your own.

What You Need

Prior knowledge

You need a basic understanding of the topics that I covered in the previous book. These include voltage, current, resistance, and Ohm's law; capacitors, switches, transistors, and timers; soldering and breadboarding; and a beginner's knowledge of logic gates. Of course, you can also learn these topics from other introductory guides. Generally I assume that you have read *Make: Electronics* or a similar book, and you have a general memory of it, although you may have forgotten some specifics. Therefore I will include a few quick reminders without repeating the general principles to any significant extent.

Tools

I'm assuming you already own the following equipment, all of which was described in *Make: Electronics*:

- Multimeter
- 24-gauge multicolored hookup wire (25 feet of each color, in at least four colors)
- Wire strippers
- Pliers
- Soldering iron and solder

- Breadboard (the preferred type is described in the next section of the book.)
- 9V battery, or an AC adapter (with a DC output) that can deliver between 9VDC and 12VDC at 1A

Components
: I have listed the components that you will need to build the projects. See Appendix B. That section also recommends sources for mail-order.

Datasheets
: I discussed datasheets in *Make: Electronics*, but I can't overemphasize how important they are. Please try to make a habit of checking them before you use a component that you haven't encountered before.

If you use any general search engine to find a part number, most likely you'll see half a dozen sites offering to show you the datasheet. These sites are organized for their profit, not for your convenience. You will probably end up clicking repeatedly to see each individual page of the datasheet, because the site owner wants to show you as many ads as possible.

You'll save a lot of time by searching for the part number on the site of a supplier such as *http://www.mouser.com*, at which point you will be able to click an icon to open the entire datasheet as a multipage PDF document. This will be easier to view and print.

How to Use This Book

There are a few differences in style and organization between this book and the previous one. Also, you need to know how to read the arithmetical notation that I have used.

Schematics

The schematics in *Make: Electronics* were drawn in an "old-school" style using semicircular "jumps" wherever one wire crossed another without making a connection. I used this style because it reduced the risk of making errors as a result of misinterpreting a circuit. In this book, I feel my readers have had sufficient practice in reading schematics that it's more important to conform with the more modern style that is most commonly used in the rest of the world. See Figure P-1 for clarification.

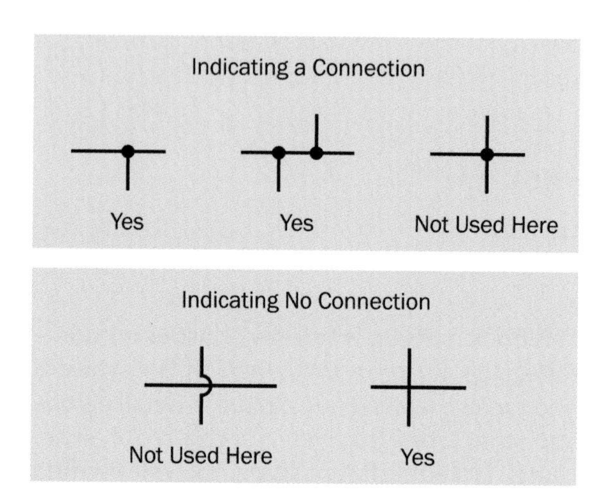

Figure P-1. *Top: In all the schematics in this book, conductors that make an electrical connection are joined with a black dot. However, the configuration at far right is avoided because it looks too similar to a crossover where there is no connection. Bottom: Conductors that cross one another without making a connection were shown in the style at left in Make: Electronics. The style at right is more common, and is used in this book.*

Also in *Make: Electronics* I used the European convention for eliminating decimal points in component values. Thus, values such as 3.3K and 4.7K were expressed as 3K3 and 4K7. I still prefer this style, because decimal points can become hard to discern in a poorly printed schematic. However, some readers were confused by the European notation, so I have discontinued it in this book.

Dimensions

Integrated circuit chips (and many other parts) all used to be equipped with wire legs, properly known as "leads," for insertion into holes in circuit boards. The leads on these "through-hole" components were spaced at intervals of 0.1", and the components were reasonably easy to grasp and position with just your finger and thumb.

This idyllic vision of universal compatibility on a human scale was disrupted initially by an invasion from the metric system. Some manufacturers moved from a pin spacing of 2.54mm (the equivalent of 0.1") to 2mm as the standard, causing frustration for those of us using 0.1" perforated board. Millimeters popped up in other places, too. To take just one example, that most ubiquitous part, the panel-mounted LED, is often 5mm in diameter. This is a fraction too big for a 3/16" hole, but not quite big enough to fit tightly in a 13/64" hole.

Because this book is written and published in the United States, I generally use inches by preference. You will find a conversion table between millimeters and fractions of an inch in *Make: Electronics*.

A much more significant problem is that the entire electronics industry has moved toward surface-mount formats. Instead of a 0.1" pin spacing, there are no pins at all, and a whole component is typically no longer than 0.1". To build a circuit from these parts, you really need tweezers, a microscope, and a special soldering iron. It can be done, but personally I don't find it enjoyable, and you will not find any projects in this book that use surface-mount components.

Math

You won't find a lot of mathematics here, but you do need to understand the simple arithmetic that's included.

I have chosen to use the style that's common in programming languages. The * (asterisk) is used as a multiplication symbol, while the / (forward slash) is used as a division symbol. Where some terms are in parentheses, you deal with them first. Where parentheses are inside parentheses, you deal with the innermost ones first. So, in this example:

```
A = 30 / (7 + (4 * 2) )
```

You would begin by multiplying 4 times 2, to get 8; then add 7, to get 15; then divide that into 30, to get the value for A, which is 2.

Organization

Unlike the previous book, this one has a basically linear structure, mainly because it is more friendly toward handheld devices, which cannot handle the amount of detail and variety scattered around a double-page printed spread. I am hoping that you will progress through the book from beginning to end, instead of dipping into it here and there.

The first project establishes concepts that will be used in the second project, and the second project lays foundations for the third project. If you don't follow this progression, you will run into some problems.

You will find five types of sections identified in subheads:

Experiments
: Hands-on work is the main thread of the book.

Quick Facts
: After I've introduced a new concept, I will often summarize some take-home messages for easy reference later.

Background
: These are short detours from the main thread where I supply additional information that I think is interesting or useful, even though it may not be strictly necessary for building a project. After a brief description, I'll leave you to pursue the topic on your own.

Make Even More
: I don't have space for thorough descriptions of all the possible construction projects, so I am including short summaries of others that I have considered.

Warnings
: Once in a while I will have to mention something that you should try to avoid doing, either for the protection of the components that you are using, or to avoid an inconvenient error, or (rarely) to protect yourself.

If Something Doesn't Work

Usually there is only one way to build a circuit that works, while there are hundreds of ways to make mistakes that will prevent it from working. Therefore the odds are against you, unless you proceed in a really careful and methodical manner. I know how frustrating it is when the components just sit there doing nothing, but if you have a problem, the following steps can usually help you to find the most common errors:

1. Attach the black lead from your meter to the negative side of the power supply, and set the meter to measure volts (DC volts, unless an experiment suggests otherwise). Make sure the power to your circuit is switched on. Now touch the red probe from your meter to various locations in the wiring, looking for erroneous voltages—or no voltage at all.
2. Check very carefully that all the jumper wires and component leads are exactly where they should be on the breadboard.

Two types of breadboarding errors are extremely common: inserting a jumper wire one row higher or one row lower than it should be, and placing two components or connections adjacent to each other on a single row, forgetting that the conductor inside the breadboard will short them together. Figure P-2 illustrates these common problems. Please check that you fully understand them!

In the upper photograph, the leads of the electrolytic capacitor are inserted between rows 13 and 15 of the breadboard, but because they are hidden from this perspective, it's easy to place one end of a blue jumper wire in row 14 by mistake. On the right, pin 5 of the chip is supposed to be grounded through a ceramic capacitor, but because all the holes along each row of the breadboard are connected internally, the capacitor is shorted out, and the chip is connected directly to ground. The lower photograph shows the errors corrected.

Figure P-2. *The two most common types of breadboarding errors are illustrated in the upper photograph, and are shown corrected in the lower photograph.*

If power is being supplied correctly to your circuit, and components and wires are all placed correctly on the breadboard, here are five more possibilities to bear in mind:

Component orientation
: Integrated circuit chips must be pushed down firmly into the board. Verify that no pin has been bent so that it is hidden underneath the chip. Diodes, and capacitors that have polarity, must be the right way around.

Bad connections
: Sometimes (seldom, but it can happen) a component may make a bad connection inside the breadboard. If you have an inexplicable intermittent fault or zero voltage, try relocating some of the components. In my experience this problem is more likely to occur if you buy very cheap breadboards. It is also more likely if you use wire that has a smaller diameter than 24 gauge. (Remember, a higher gauge number means a thinner wire.)

Component values
: Verify that all the resistor and capacitor values are correct. My standard procedure is to check each resistor with a meter before I plug it in. This is time-consuming but can save time in the long run. I'll have more to say about this in the next section of the book.

Damage
: Integrated circuits and transistors can be damaged by incorrect voltages, wrong polarity, or static electricity. Keep spares on hand so that you can make substitutions.

Human burnout
: When all else fails, take a break! Working obsessively for long periods can create tunnel vision that prevents you from seeing what's wrong. If you move your attention to something else for a while, then come back to your problem, the answer may suddenly appear obvious.

Writer-Reader Communication

There are three situations where I may want feedback from you, or you may want feedback from me. They are the following:

- I may want to tell you if the book contains a mistake that will prevent you from building a project successfully. I may also want to tell you if a parts kit, sold in association with the book, has something wrong with it. Naturally, if a problem has been discovered, I'll tell you how to deal with it. This is *me-informing-you* feedback.
- You may want to tell me if you think you found an error in the book, or in a parts kit. This is *you-informing-me* feedback.
- You may be having trouble making something work, and you don't know whether I made a mistake or you made a mistake. You would like some help. This is *you-asking-me* feedback.

I will explain how to deal with each of these situations.

Me Informing You

I can't notify you if there's an error in the book or in a parts kit unless I have your contact information. Therefore I am asking you to send me your email address for the following purposes. Your email will not be used or abused for any other purpose:

- I will notify you if any significant errors are found in this book or in its predecessor, *Make: Electronics*, and I will provide a workaround.
- I will notify you of any errors or problems relating to kits of components sold in association with this book or *Make: Electronics*.
- I will notify you if there is a completely new edition of this book, or of *Make: Electronics*, or of my other book *Encyclopedia of Electronic Components*. These notifications will be very rare, as a new edition may only appear every few years.

We've all seen those warranty cards that promise to enter you for a prize drawing. I'm going to offer you a much better deal. If you submit your email address, which may only be used for the three purposes listed, I will send you an unpublished electronics project with complete construction plans as a multipage PDF. It will be fun, it will be unique, and it will be relatively easy. You won't be able to get this in any other way.

The reason I am encouraging you to participate is that if an error is found, and I have no way to tell you, and you discover it later on your own, you're liable to get annoyed. This will be bad for my reputation and the reputation of my work. It is very much in my interest to avoid a situation where you have a complaint.

- Simply send a blank email (or include some comments in it, if you like) to *make.electronics@gmail.com*. Please put REGISTER in the subject line.

You Informing Me

If you only want to notify me of an error that you have found, it's really better to use the errata system maintained by my publisher. The publisher uses the errata information to fix the error in updates of the book.

If you are sure that you found an error, please visit:

http://oreil.ly/1jJr6DH

This web page will tell you how to submit errata.

You Asking Me

My time is obviously limited, and I can't necessarily solve a problem for you. However, if you attach a photograph of a project that doesn't work, I may have a suggestion. The photograph is essential. Trying to understand why something isn't working, without being able to see it, is generally impossible.

You can use *make.electronics@gmail.com* for this purpose. Please put the word HELP in the subject line.

Before You Write

Before you report an error or tell me that something doesn't work, I have a couple of requests:

- Please rebuild the circuit at least once. Every project here was built by me and by a minimum of one other person before the book went into production, and while it's not very polite for me to tell you that you may have screwed up, the most likely cause of a problem is always a wiring error.

  Bear in mind that I made at least a dozen fatal wiring errors myself while building projects in this book. One error burned out a couple of chips. Another error partially melted a breadboard. Errors do happen, even to me, and even to you.

- Please be aware of the power that you have as a reader, and use it fairly. A single negative review can create a bigger effect than you may realize. It can certainly outweigh half-a-dozen positive reviews. The responses that I received for *Make: Electronics* were generally very positive, but in a couple of cases people became annoyed over small issues such as being unable to find a part that I had recommended. In fact the parts were available, and I was happy to suggest sources, but in the meantime the negative reviews had appeared.

I do read my reviews on Amazon and will always provide a response if necessary.

Of course, if you simply don't like the way in which I have written this book, you should feel free to say so.

Going Further

After you work your way through this book, I think you will be on your way to what I consider an intermediate understanding of electronics. I am not qualified to write an advanced guide, and consequently I don't expect to create a third book with a title such as "Make Even More Electronics."

If you still want to know more, the areas that I have avoided are electronics theory, circuit design, and circuit testing. If you create a circuit yourself, you should know enough theory to understand and predict what's happening in it, and you should have the capability to discover how it is behaving after you have built it. To deal with this, you really need an oscilloscope and circuit simulation software. You will find a list of free software on Wikipedia (*http://bit.ly/1jJrfqX*). Some of these simulators show you the performance of digital circuits, some of them specialize in analog circuits, and some do both. But this topic is beyond the scope of a general book, and probably beyond the scope of most people who view electronics as a hobby rather than a career.

If you want to know more electrical theory, *Practical Electronics for Inventors* by Paul Scherz (McGraw-Hill, 2013) is still the book that I recommend most often. You don't have to be an inventor to find it useful.

For reference, I have always felt that there's a need for an encyclopedia of electronic components. I often wondered why a book of this type did not exist—and so I decided to write one myself.

Volume 1 of my *Encyclopedia of Electronic Components* is now available. There will be three volumes altogether. While *Make: More Electronics* is a hands-on tutorial, the encyclopedia format is designed to enable fast access to information. It is also a little more technical, and is written in a style that is less friendly but gets straight to the point. Personally I think an encyclopedia of components is an invaluable way to refresh your memory about the properties and applications of any parts that you are likely to use.

A Note from the Publisher: Safari® Books Online

Safari Books Online is an on-demand digital library that delivers expert content in both book and video form from the world's leading authors in technology and business.

With a subscription, you can read any page and watch any video from our library online. Read books on your cell phone and mobile devices. Access new titles before they are available for print, get exclusive access to manuscripts in development, and post feedback for the authors. Copy and paste code samples, organize your favorites, download chapters, bookmark key sections, create notes, print out pages, and benefit from tons of other time-saving features.

Maker Media has uploaded this book to the Safari Books Online service. To have full digital access to this book and others on similar topics from *Make* and other publishers, sign up for free at *http://my.safaribooksonline.com*.

Make unites, inspires, informs, and entertains a growing community of resourceful people who undertake amazing projects in their backyards, basements, and garages. *Make* celebrates your right to tweak, hack, and bend any technology to your will. The *Make* audience continues to be a growing culture and community that believes in bettering ourselves, our environment, our educational system—our entire world. This is much more than an audience, it's a worldwide movement that Make is leading—we call it the Maker Movement.

For more information about *Make*, visit us online:

> Make magazine: *http://makezine.com/magazine/*
> Maker Faire: *http://makerfaire.com*
> Makezine.com: *http://makezine.com*
> Maker Shed: *http://makershed.com/*

We have a web page for this book, where we list errata, examples, and any additional information. You can access this page at *http://bit.ly/more-electronics*.

Setup

I made suggestions about a work area, storage of parts, tools, and other basics in *Make: Electronics*. Some of these suggestions should now be revised, while others must be reiterated or elaborated.

Power Source

Most of the circuits in this book can be powered by a 9V battery, which has the advantage of not only being cheap but also supplying a stable current without any spikes or glitches. On the other hand, the voltage from a battery will diminish significantly with use, and will vary from moment to moment, depending on how much current you are drawing from it.

Having a variable power supply capable of delivering 0VDC to 20VDC (or more) is a real pleasure, but may cost more than you are willing to spend. A reasonable compromise is to buy the type of AC adapter that plugs directly into the wall and has switch-selectable voltages, as suggested in my previous book.

Another option is to buy the kind of single-voltage AC adapter designed for laptop computers. Many have an output around 12VDC, which can be passed through a voltage regulator to get the 5VDC or 9VDC that you need for most of the experiments here. Voltage regulators cost less than $1 each, and a laptop power supply shouldn't cost much more than $10, making this an attractive option. The power supply should be capable of delivering up to 1A (1,000mA).

You may be tempted to use a cellular phone charger, especially if you have one lying around after a phone has died. But most chargers deliver only 5VDC, which makes them unsuitable for the 9V projects that I will be describing. Also, because they are designed to function as battery chargers, they may reduce their output voltage, depending on the load.

The bottom line: if you're on a tight budget, and you don't expect to make permanent versions of any of the projects here, a 9V battery will do. Otherwise, look for a 12VDC adapter in your price range.

Regulation

Many of the experiments will require a *regulated* supply of 5VDC. You will need these components:

- LM7805 voltage regulator
- Ceramic capacitors: 0.33µF, 0.1µF
- Resistor: 2.2K
- SPST or SPDT switch of PC-mount type (i.e., its leads will push into the holes in a breadboard)

- Generic LED

Figure S-1 shows how the parts can be squeezed into the top few rows of a breadboard, creating a positive bus on the left and a negative bus on the right, which is the layout I'll be using for many experiments. The photograph shows a 9V battery, but naturally you can use an AC adapter. Make sure it has a DC output of at least 7VDC. To avoid generating excess waste heat, the adapter should not deliver more than 12VDC.

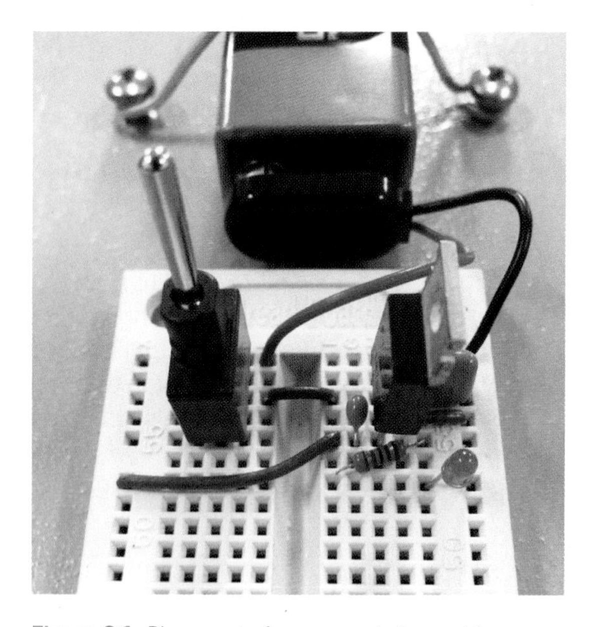

Figure S-1. *Placement of components to provide a regulated 5VDC power supply.*

Figure S-2 shows the same circuit in schematic form. The capacitors should be included even if you are using a battery, because they insure correct behavior of your voltage regulator.

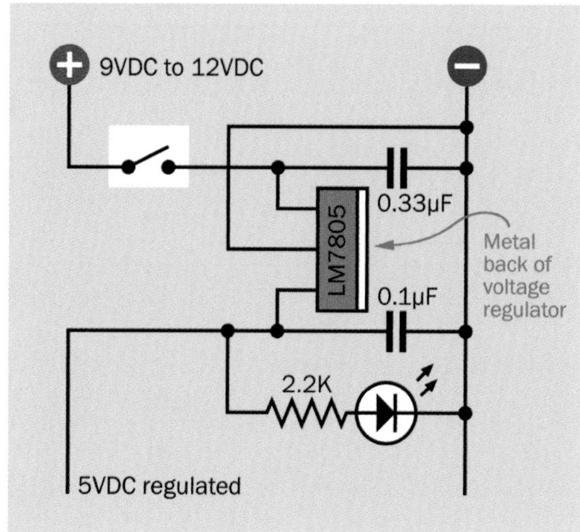

Figure S-2. *Schematic for the 5VDC regulated power supply.*

I'm suggesting that you include a switch and an LED because they're so convenient. When you're wondering why a circuit doesn't work, it's useful to see the LED glowing, confirming that power is reaching the board. And when you're moving wires around to modify a circuit, you'll appreciate being able to switch the power off and on without any hassles. I'm suggesting a relatively high-value 2.2K resistor in series with the LED, to conserve power if you use a battery.

Boarding School

In *Make: Electronics* I used the type of breadboard with a pair of buses down each of its two long edges, so that you had positive and negative power on both sides of the board. In this book I decided to use the simpler type of breadboard, which has only one bus down each long edge, as shown in Figure S-3.

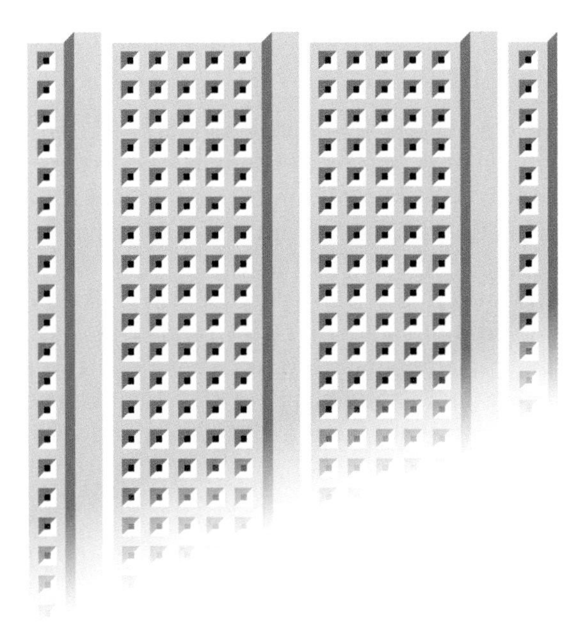

Figure S-3. *The external appearance of a breadboard of the type that has only one bus on each side. All the circuits in this book will be designed for this type of board.*

I have several reasons for making this change:

- Boards of this type are exceptionally affordable, especially if you buy them direct from Asian sources that list their products on eBay. Don't be disconcerted by obscure vendor names such as "herofengstore" or "kunkunh." At the time of writing, you can find breadboards for as little as $2 each, so long as you don't mind waiting ten days or more for international shipping. For more advice about component sources, see Appendix B.

  If you buy several breadboards, you can keep previous circuits on some of them while using a fresh board for each new circuit.

- If you want to make a permanent version of a circuit by soldering components into a printed-circuit (PC) board, the easiest way is to use a PC board where the traces are configured in breadboard format. This type of PC board often has just one bus on each side. (The RadioShack 276-170 is an example.) Transferring the components to it from a breadboard will be much easier if the layout is exactly the same.

- Feedback from readers has shown me that people tend to make mistakes more easily on breadboards where positive and negative buses are paired on both sides. These mistakes can be costly and inconvenient, as some components have very little tolerance for reversed polarity.

It's important that you always have a mental image of the conductors inside a breadboard, so I'm including a version of a diagram that you may remember from my previous book. Figure S-4 shows a cutaway view.

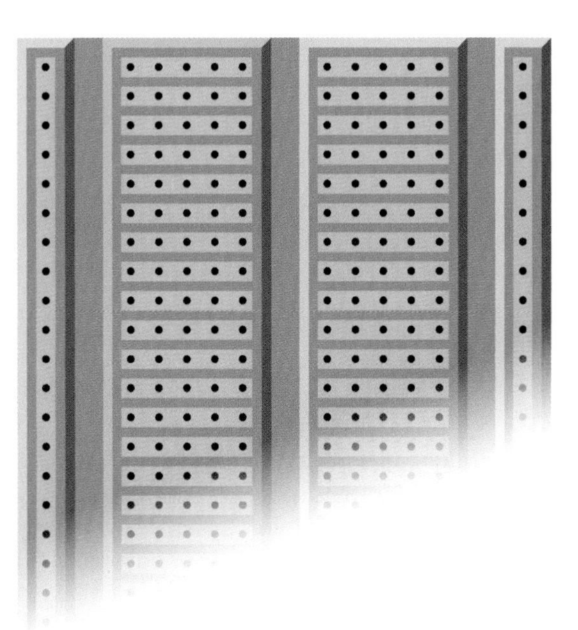

Figure S-4. *A cutaway view showing the conductors inside a breadboard.*

Remember that many breadboards have buses with one or two breaks in them, to allow you to use different power supplies in different sections of the board. I don't expect to make use of that feature, so when you get a new breadboard, you must use a meter to check that each bus is continuous from beginning to end. If it isn't, you need to bridge the gaps in the buses with jumper wires. Forgetting to do this is a common cause of nonfunctional circuits.

Wiring

Once in a while, a reader will send me an email with a photograph of a breadboarded circuit, asking me why it doesn't work. If the reader has used the flexible type of jumper wires with a little plug at each end, my answer is always the same: I cannot offer any advice. Even if I had the circuit in front of me, I still wouldn't be able to offer advice, other than to pull out all the wires and start over.

Breadboard jumper wires are quick and simple to install. I have succumbed to their temptation myself, many times—and have often regretted it, because if you make just one error, you will have extreme difficulty finding it amid the wiring tangle.

In almost most of the photographs in this book, you'll find that I only use plug-type, flexible jumper wires when I need to connect with devices off the breadboard. On the breadboard I use little pieces of solid wire, stripped at each end. They are infinitely easier to deal with when you have to do some troubleshooting.

If you buy ready-cut segments of solid wire in a kit, you'll find that they are color-coded by length. This is not helpful, because I want my breadboard wires to be color-coded according to function. A connection terminating at the positive bus of the breadboard should be red, for example, no matter how long or short it is. Two wires of equal length that run close together should be of contrasting colors, so that I don't confuse one with the other. And so on. This way I can look at a breadboard, quickly assess its function, and find a misplaced wire more easily.

Perhaps you feel that custom-cutting your own color-coded jumper wires is too much of a hassle. If so, I have a suggestion. Figure S-5 shows the system that I used to breadboard all the projects in this book.

First remove an arbitrary amount of insulation (a couple of inches) and discard it. Next, estimate the distance that your jumper should span on the breadboard. I'll call this distance "X." Measure this on the remaining insulation on your wire as shown in Step 2, and apply your wire strippers in the position indicated by the dashed line. Push the insulation down toward the end of the wire, as in Step 3, stopping about 3/8" from the end. Cut on the solid line. Bend the ends, and you're done.

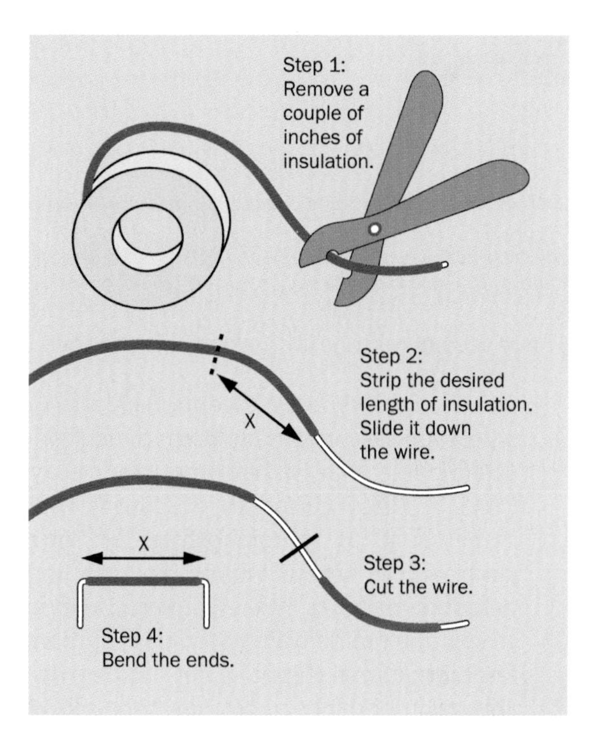

Figure S-5. *A simplified way to create breadboard jumper wires.*

For sorting and storing jumper wires after they have been cut, you can make yourself a wire-length gauge. This is also useful for bending wire ends to the desired length. It consists simply of a triangular-shaped piece of plastic or plywood with steps cut into the diagonal edge, as shown in Figure S-6 and Figure S-7. Because the wire thickness will add slightly to the length of the jumper, your gauge should actually use steps that are about 1/16" less than the length that they represent.

Another way to check the lengths of jumper wires is by comparing them against a piece of plain perforated board (often referred to as "perf board") where the holes are spaced at intervals of 0.1".

Figure S-6. *A homemade wire-length gauge for breadboard jumper wires.*

Figure S-7. *A jumper of 1.1" in length being checked against the length gauge.*

Remember, holes in a breadboard are spaced 0.1" apart, horizontally and vertically, and the channel down the center of a breadboard is 0.3" wide.

As for wire thickness, I think 24 gauge is by far the best choice for breadboarding. If you use 26 gauge, it tends to kink too easily when you're trying to push it into the holes; and after it's inserted, it sits too loosely. On the other hand, 22 gauge is too tight a fit.

You can often find surplus lots of wire on eBay, or from sources such as Bulk Wire (*http://www.bulkwire.com*). Personally I have ten basic wire colors: red, orange, yellow, green, and blue (the spectrum), and black, brown, purple, gray, and white (the shades). If you are systematic, and you assign one color for each purpose on all your breadboards, this will make your life a lot easier.

Lastly, please take another look at Figure P-2 to remind yourself of the two most common breadboard wiring errors. You may think that you would never make such obvious mistakes, but I have certainly made them myself when I've been tired or working under deadline.

Grabbing

In *Make: Electronics* I mentioned "minigrabbers" that you can push onto the probes of a multimeter. These used to be relatively difficult to find, but are now readily available from sources such as RadioShack (catalog part 270-0334, described as "Mini Test Clip Adapters"). Figure S-8 shows a black grabber installed on a meter probe, while the red grabber remains unconnected. I think this is a useful mix. You can hook the black grabber onto any ground wire, then use the red probe to detect voltages around a circuit. The grabber is a very tight push-fit, which I think should add only an ohm or two at most.

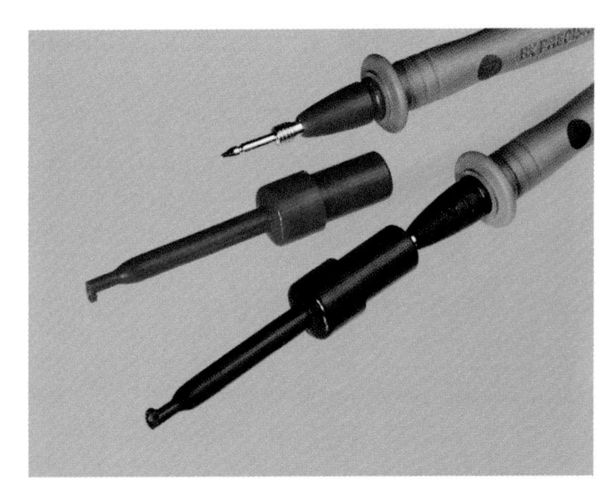

Figure S-8. *Minigrabbers convert one or two probes of your meter so that they will latch onto a wire, freeing you from holding a probe in place.*

The mechanism of the grabber is shown in Figure S-9, where it is in its open state, extended against an internal spring. In Figure S-10, the spring has been released to hold a resistor lead.

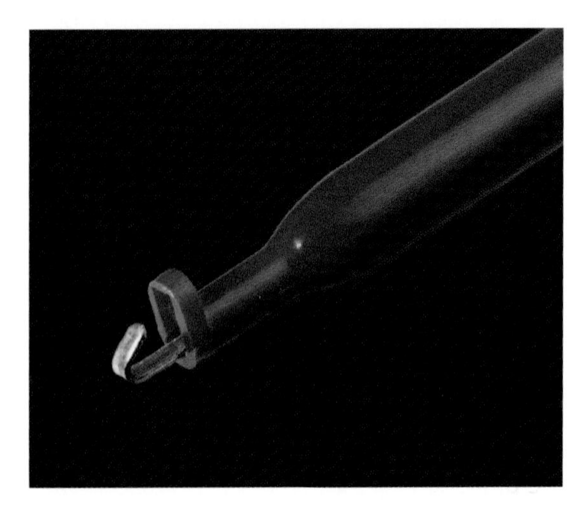

Figure S-9. *A minigrabber with its grabbing clip extended against the force of an internal spring.*

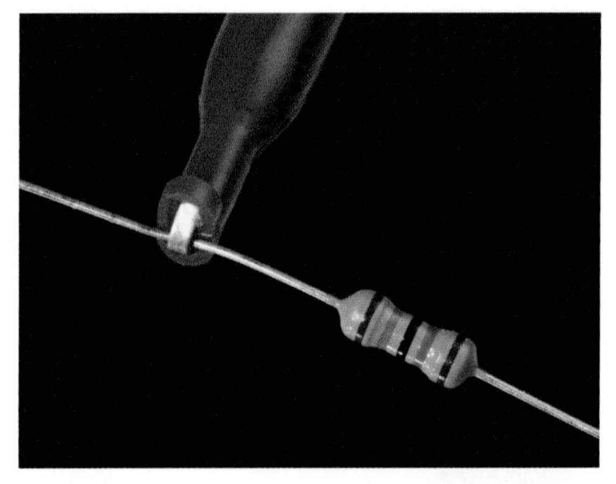

Figure S-10. *When the spring is released, the grabber exercises a firm grip on a thin object, such as a resistor lead.*

Jumper wires with an alligator clip at each end (as shown in Figure S-11) can be used as a substitute, with one alligator gripping a meter probe while the other latches on to a convenient location in the circuit. You'll find that I mention this later in the book where you need a free hand that isn't occupied pressing a probe against a wire. Personally I think grabbers are better, but if you don't want to encumber your meter probe(s) on a semipermanent basis, the double-alligator jumper is an alternative.

Lastly you can buy jumpers that have a micrograbber at each end, as shown in Figure S-12. Here again RadioShack is a source, with part number 278-0016 identified as "Mini-Clip Jumper Wires." The advantage of these jumper wires is that the micrograbber (a size smaller than the minigrabber) can latch onto small parts where an alligator clip would be liable to nudge an adjacent wire and cause a short circuit.

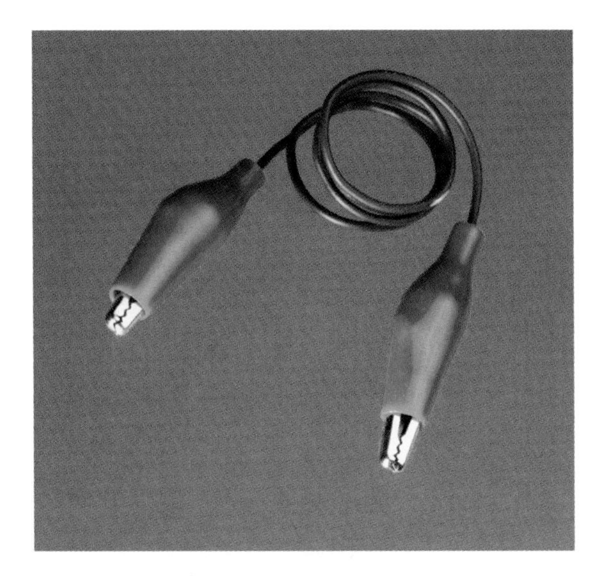

Figure S-11. *This type of jumper wire with an alligator clip at each end can be used as a "grabber substitute," with one alligator gripping a meter probe while the other grips a wire or connection in a circuit that is being tested.*

Figure S-12. *A jumper wire with a micrograbber at each end is useful for locations where a full-size alligator clip would be liable to touch an adjacent conductor.*

Component Storage

For storing capacitors, the reduced size of multilayer ceramics means that my recommendations in *Make: Electronics* are becoming obsolete. Tiny parts are most efficiently kept in tiny containers, and jewelry hobbyists have exactly what we want.

At a crafts store in the United States, such as Michael's, you will find all sorts of clever storage systems for beads. The system I use now for multilayer ceramic capacitors is a bead storage box shown in Figure S-13. Ceramic capacitors fit easily into these little screw-top compartments, which are only 1" in diameter. This enables me to keep an entire range of basic values on my desktop, from 0.01µF (10nF) upward, in a box measuring just 6.5" by 5.5". Moreover, because each container has a screw top, if I accidentally drop the whole box on the floor, the capacitors will remain confined instead of scattering everywhere. This is important because capacitors look so similar, it would be a nightmare trying to separate them by value.

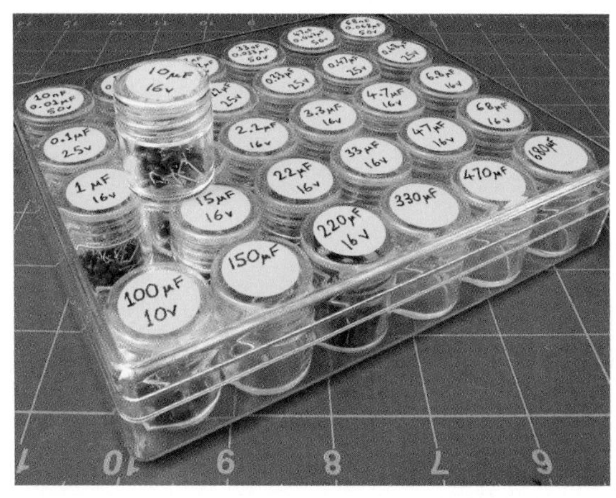

Figure S-13. *Modern multilayer ceramic capacitors are so small, storage containers designed for beads are ideal.*

For resistors, I suggest cropping their leads so that they, too, will fit in smaller containers. We seldom need the full length of a resistor lead—and on the rare occasions when it's useful, an additional piece of insulated wire can be added to the breadboard instead. Figure S-14 shows one option for storing the 30 most commonly used values. Like the storage system for capacitors, this one won't spill any components if you knock it over. Each compartment can hold at least 50 resistors (see Figure S-15).

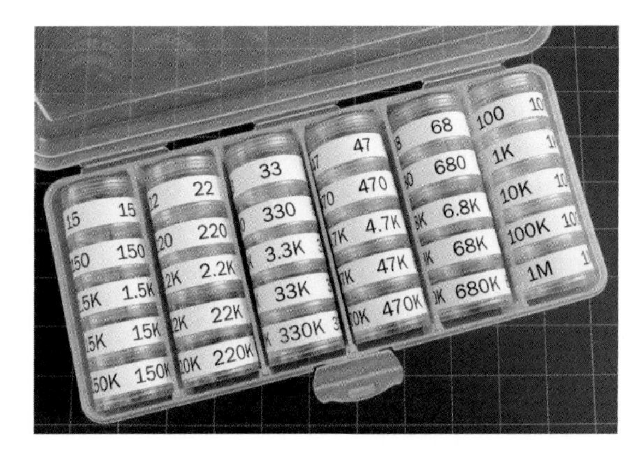

Figure S-14. *Slightly larger jewelry storage containers are good for resistors, if the leads are trimmed.*

Figure S-15. *Fifty resistors can be stored in one of these little containers.*

Verifying

When I'm building a circuit, I try to discipline myself to check the value of each resistor or capacitor before I place it on the breadboard. A 10µF ceramic capacitor looks almost identical to a 0.1µF ceramic capacitor, and resistor values such as 1K and 1M are only one colored band apart. If component values become mixed up, you will find yourself faced with faults that can be truly perplexing.

To simplify the checking process for resistors, I use a mini-breadboard with jumper wires clipped to the probes of an auto-ranging meter, as shown in Figure S-16. All I have to do is push the leads of a resistor into the board, and verification takes about five seconds. The breadboard sockets add a small amount of resistance, but only a few ohms, and I'm usually not concerned with a precise value, anyway. I just want to be sure that I'm not making a significant error. For the same reason, the cheapest possible meter can be used for this task.

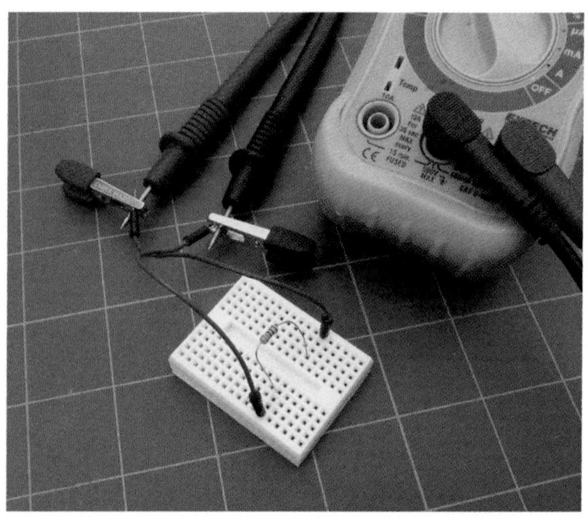

Figure S-16. *A simple system for quickly verifying resistor values before using them in a project.*

So much for the introductory material. Now let's make more electronics!

Experiment 1: Sticky Resistance

1

I want to start with some simple entertainment, because I think electronics should always contain an element of fun.

For this experiment, I'm going to use glue and cardboard. I realize that these materials are not commonly used in electronics books, but they're going to serve two purposes. First, they will remind us that electricity isn't necessarily confined to wires and boards. Second, the experiment will lead to a deepening understanding of that most fundamental and vital component, the bipolar transistor. And third, this experiment will lead into a general conversation about ions, resistance, and resistivity.

I realize that if you read *Make: Electronics* you already learned the basics about transistors, but after I do a small amount of recapitulation, I'm going to move beyond the basics.

- Remember that you will find components for each experiment listed at the back of the book. See Appendix B.

A Glue-Based Amplifier

Figure 1-1 shows the plan. The cardboard will be your foundation for the circuit; you won't be using a breadboard for this project. Begin by pushing the legs of the transistor into the cardboard.

The 2N2222 is sold in two versions, one featuring a little metal cap, the other using a small lump of black plastic. If you happen to be using the metal type, the tab that sticks out should be on the left, viewed from the point of view in the figure. If you have the black plastic type, the 2N2222 or PN2222 will have its flat side on the right—but if you happen to buy the P2N2222 variant (which often pops up as "equivalent" when you search for the other part numbers), the flat side should be on the left. Check the part number with a magnifying glass, and see "Symbology" on page 3 if you are unclear about this.

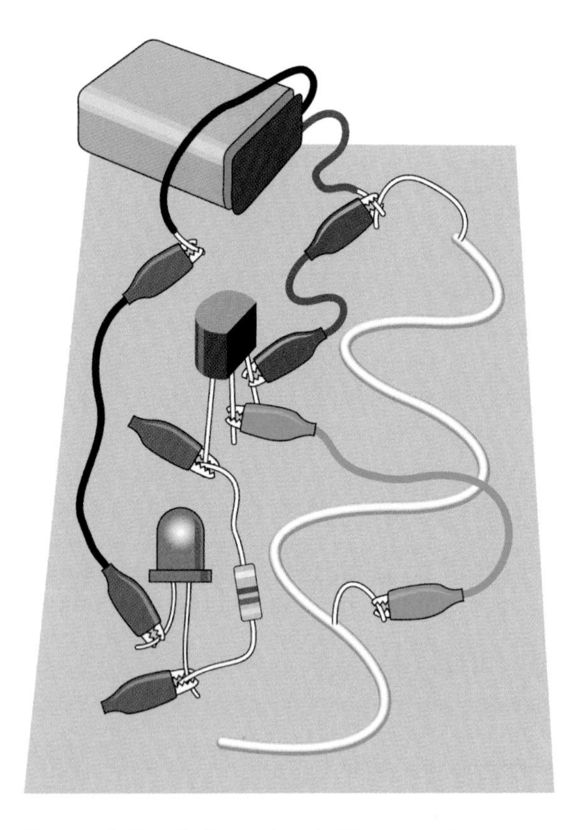

Figure 1-1. *Your first experiment: All you need is a transistor, a 220Ω resistor, a 9V battery, patch cords and pieces of wire, and some white glue and cardboard.*

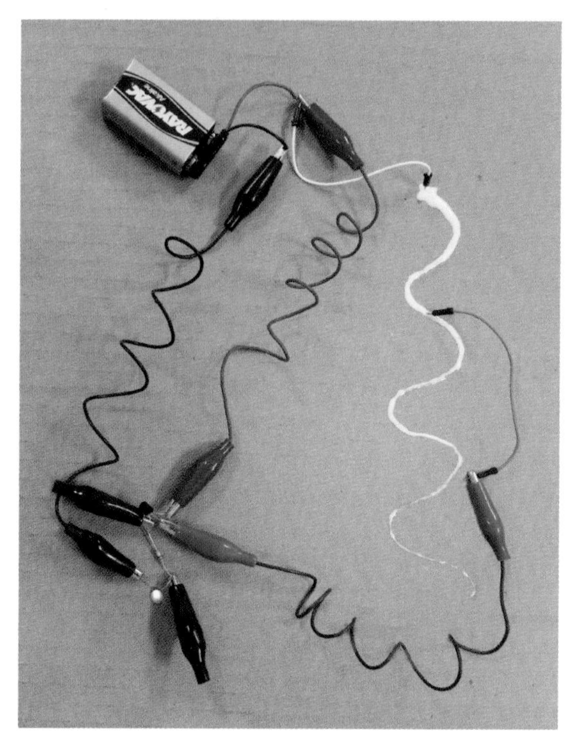

Figure 1-2. *The actual experiment, just in case you were wondering if it really works.*

Connect the components as shown. The long lead of the LED is on the right, and the short lead is on the left. The resistor attached to the long lead of the LED has a value of 470Ω. Don't allow any of the alligator clips to touch each other where they grip the leads of the transistor. Now take your container of Elmer's glue and squeeze out a zigzag path that is about 12" long and less than 1/8" thick. If you can taper it from top to bottom, as shown in Figure 1-2, this is good. Make sure there are no breaks in the trail.

Why am I suggesting Elmer's glue? Because most people have a bottle lying around the house somewhere, and it just happens to have the electrical characteristics that I want. It's not an insulator, but it's not a very good conductor, either.

You have to work fairly fast, before the glue dries. Take the green wire (which connects with the center lead on the transistor) and touch it halfway down the glue trail. The LED should glow quite brightly. Now touch it near the bottom of the glue trail, and the LED should glow less brightly.

If you have read my previous book, you'll know why—but I'm going to tell you anyway.

What's Happening

The path of glue that you squeezed out should have a resistance of about 1 megohm from top to bottom, or 10K per inch. If you want to check this with your meter, use pieces of wire to extend the probes, so that you don't get glue on them.

The transistor acts as an amplifier. It amplifies the current flowing into its base (the center lead). The amplified output emerges from its emitter (the lefthand lead, in Figure 1-1). In the experiment, you restricted the current flowing into the base of the transistor by passing the current through some of the glue, which has a high resistance. The LED is responsive to current, and shows you what's happening by varying its brightness.

To get a visual impression of what the transistor is doing, remove it from the circuit, as in Figure 1-3. The green alligator clip now connects with the series resistor, which connects with the LED, and the LED should remain dark. The resistance of the glue is so high, not enough current gets through to light the LED. If you move the green alligator all the way up to less than a quarter inch from where the positive power supply is connected with the glue, the LED should glow dimly.

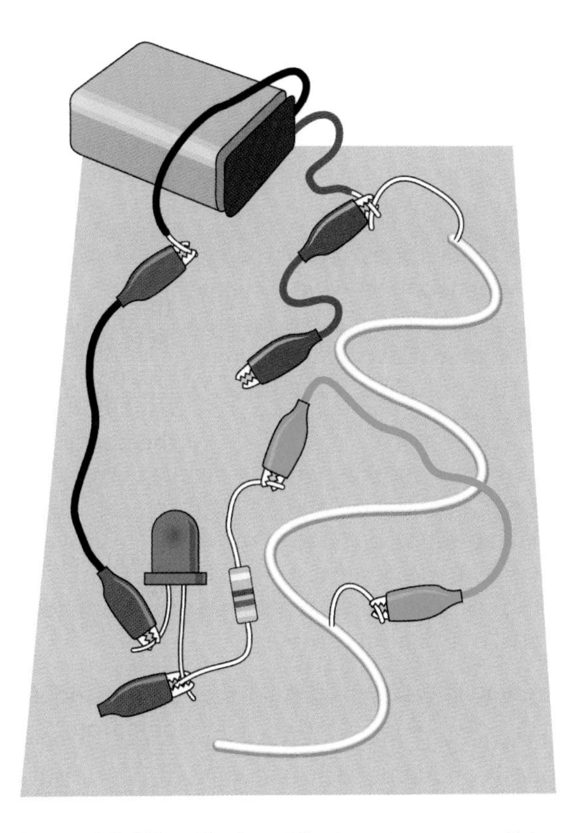

Figure 1-3. *When the transistor is no longer amplifying current to the LED, the resistance of the glue is too high to allow enough current to make the LED light up.*

Symbology

Just in case you have trouble remembering the schematic symbol for an NPN transistor, and the pinouts of actual components, I've included Figure 1-4 to remind you. The tab that sticks out of the metal-can type of transistor may be in either of the orientations shown, or somewhere in between, but it will always be closer to the emitter than to the other leads. As for the schematic symbol, you know that this represents an NPN transistor because the arrow "Never Points iN."

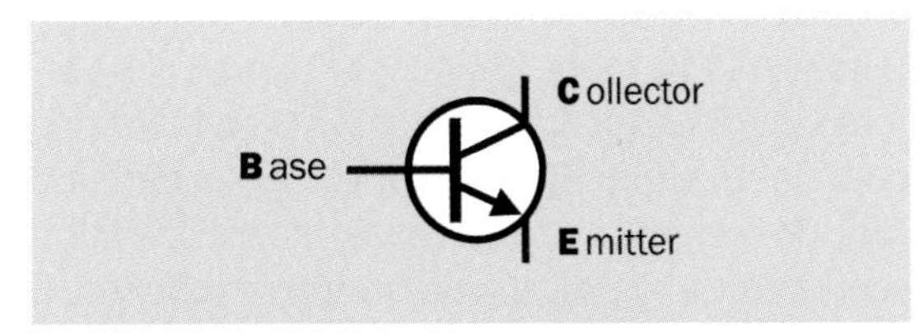

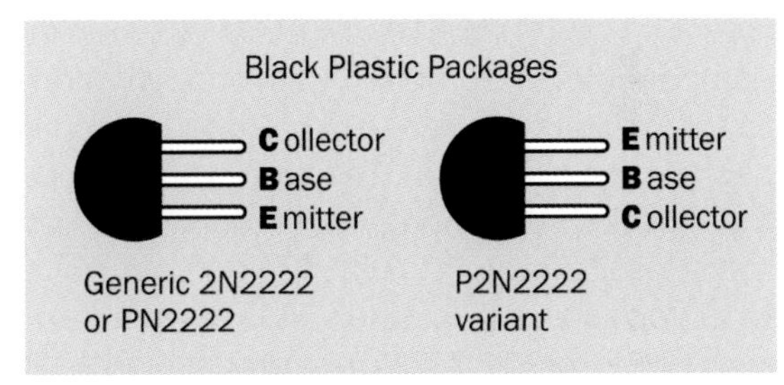

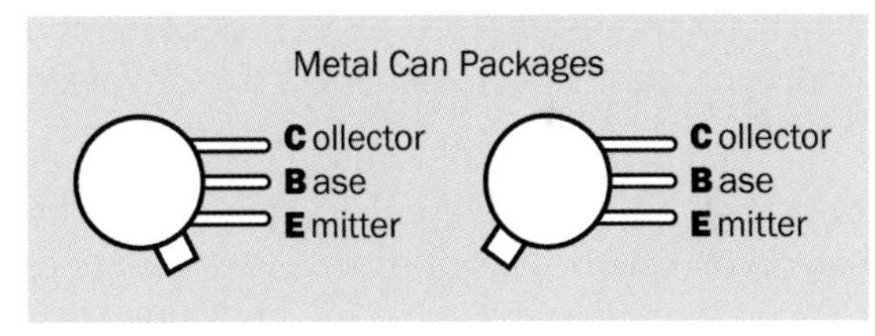

Figure 1-4. *The schematic symbol for an NPN transistor, and simplified views of components seen from above. See the important warning in the text regarding reversed leads on the P2N2222.*

Warning: Nonstandard Leads

For as long as anyone could remember, when you looked at a plastic-packaged 2N2222 transistor from above, and held it with the flat side on the right, the leads were always identified as collector, base, and emitter when reading from top to bottom. Some manufacturers called the transistor the PN2222, but the pinouts were still the same.

For reasons that remain unclear, some time around 2010, a variant with part number P2N2222 was produced by On Semiconductor, Motorola, and possibly some other manufacturers. Its performance was identical to that of the 2N2222 and the PN2222, but the sequence of its leads was reversed.

Suppose you search an online supplier for 2N2222, which is a reasonable thing to do, as 2N2222 is the most generic version of this transistor's part number. Chances are, you will be offered a P2N2222, because your search term is contained within P2N2222. If you go ahead and buy that component because its specification seems to be identical, you are likely to insert it the wrong way around in your circuit.

Compounding the problem, transistors will work to some extent when reversed, although some degradation may occur. Therefore you can use the P2N2222 the wrong way around and get some results from a circuit, although not quite what you expect. If you then discover your error and reverse the P2N2222, quite probably you still won't get the results you expect, because the transistor has been damaged by reversed polarity.

Anyone buying components online should be careful to read the part number and take note of the configurations in Figure 1-4. And, as always, check datasheets carefully!

Background: Conductors and Insulators

You can learn more from your experiment if you wait for the glue to dry. The drier it gets, the weaker the LED response becomes. Why is this? Because some of the water in the glue is evaporating, while the rest of it is absorbed into the cardboard.

As you may recall from *Make: Electronics*, electric current is a flow of electrons. Atoms or molecules which have surplus electrons, or a deficit of electrons, are called ions. I don't know what Elmer's glue is made of, but apparently it contains a chemical that allows ion transfer. The water in the glue helps to enable this, as the ions move through the water.

Water on its own is not a good conductor. To demonstrate this, you need some pure water—not the water that comes out of your faucet, which usually contains mineral impurities. Pure water used to be called distilled water, which was created by boiling water to make steam (leaving the impurities behind), and then condensing the steam. These days people still sometimes talk about distilled water, but it is becoming uncommon because the process of making it is too energy intensive. Instead, you are likely to find "deionized" water, which is usually created by a process such as reverse osmosis. "Deonized" tells you there are no ions in it, right? So, it should be no surprise that the water doesn't conduct electricity very well.

Insert the probes of your meter into a cup of distilled or deonized water, a couple of inches apart. You should find that the resistance is more than 1 megohm. Now dissolve some salt in the water, and the resistance should drop radically, because the salt is a source of ions.

You may wonder where the dividing line is between a conductor and an insulator. To answer that question, you need to know how "resistivity" is measured. It's very simple: if R is the resistance of an object in ohms, A is its area in square meters, and L is its length in meters:

```
Resistivity = (R * A) / L
```

Resistivity is measured in ohm-meters. A very good conductor, such as aluminum, has resistivity of about 0.00000003 ohm-meters. That's 3 divided by 100 million. At the other extreme, a very good insulator, such as glass, has resistivity of about 1,000,000,000,000 (one trillion) ohm-meters.

Somewhere in the middle are semiconductors. Silicon, for instance, has a resistivity of about 640 ohm-meters, although this can be reduced by "doping" the silicon with impurities and biasing it with an electrical potential to encourage electrons to flow through it.

What's the resistivity of Elmer's glue? I'll leave you to figure that out, with the aid of your multimeter. And what about cardboard? Its resistivity is so high, how could you ever measure it? See if you can think of a way.

Make Even More

If you repeat Experiment 1, what happens if you use a trace of glue that is three or four times as wide? What happens if you put two LEDs in parallel—or in series?

Maybe you think you know what the results will be. But it's always good to validate an assumption by testing it.

I mentioned earlier that if you insert a transistor the wrong way around, it will still work to some extent. It can tolerate a small reversed voltage between the base and the emitter (usually less than 6V), but using a 9V battery, you're more likely to cause some damage. Does that actually happen when you try it? And if so, why? If you search for more information about this, you're likely to find yourself learning how the layers in a transistor are structured, and how charges move from one to another. This is good to know.

After you have reversed a transistor in a circuit, it may have sustained some damage and should not be used in other circuits. You can, however, test it as described in the next experiment, and compare its performance with that of a fresh transistor that has not been abused.

Experiment 2: Getting Some Numbers

2

Here's my plan. Looking ahead to the next experiments, I'm going to be showing you some components that were not included in *Make: Electronics*. The first three will be:

- Phototransistor
- Comparator
- Op-amp

These devices will be doing some things to interest and entertain you in Experiments 3 through 14. You'll also be dealing with topics such as circuit design, especially using analog components.

After that, we'll be using digital chips such as:

- Logic gates
- Decoders, encoders, and multiplexers
- Counters and shift registers

Then I'll be discussing randomicity, and sensors—

But first, here and now, I have to make sure that we're all on the same page regarding a few basic concepts. Even if you feel that you are thoroughly familiar with these concepts, well-informed people can still have a few gaps in their knowledge, so please take a few moments to go through this section of the book. You'll need this information to make sense of the sections that come later.

Requirements

- Remember that you will find components for each experiment listed at the back of the book. See Appendix B.

I'm assuming that you already have the 5VDC regulated power supply shown previously in Figure S-2. Any time you see the word "regulated" in a schematic, the basic regulated power supply consisting of an LM7805 and two capacitors will be necessary. In this experiment, because you are going to be making accurate measurements, you need an accurately controlled voltage.

Transistor Behavior

Numbers are unavoidable in electronics. In fact, you can see them as your friends, because they tell you what's going on. Making accurate measurements is also very necessary, because if your measurements aren't accurate, your numbers will mislead you and will be worthless.

Therefore I want to run a version of Experiment 1 with a trimmer potentiometer instead of a trail of Elmer's glue, and a meter instead of an LED, so that you can measure the performance of the circuit. (This will be similar to Experiment 10 in *Make: Electronics*, but I'm going to go further into the topic of amplification.)

Are you good at making accurate measurements? Now's the time to find out.

Step 1

Begin by setting your meter to measure microamps DC. Depending on the type of meter you're using, you may simply need to insert the red lead in the socket reserved for measuring current, and turn the selector knob to amperes. If your meter doesn't do autoranging, you'll have to choose microamps with the selector. Either way, make sure you are measuring DC, not AC, and make sure the red lead of your meter is in its "amperes" socket.

You'll be using your meter in the circuit as shown in Figure 2-1.

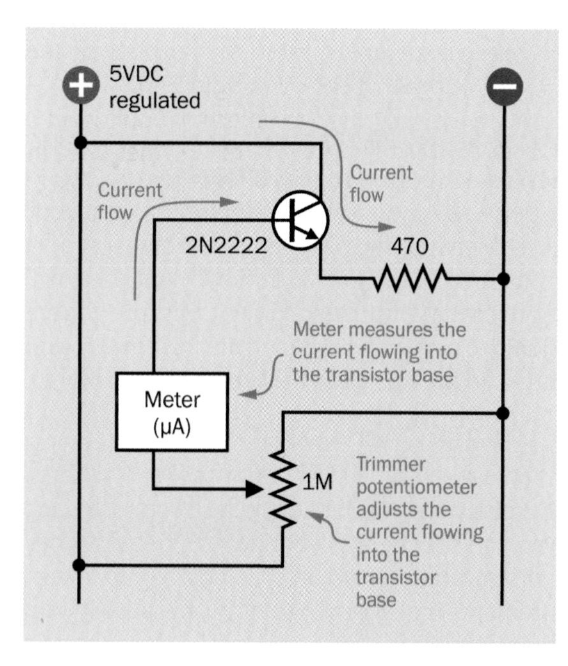

Figure 2-1. *The meter measures current flowing into the base of the transistor.*

If you have any trouble interpreting this schematic, take a look at Figure 2-2. This shows a manual-ranging meter measuring microamps between the wiper of the trimmer potentiometer and the base of the 2N2222 transistor, using flexible jumpers gripped by minigrabbers. The twisted black and red wires entering the picture from the right are supplying 5VDC regulated power to the breadboard. The reading on the meter is arbitrary.

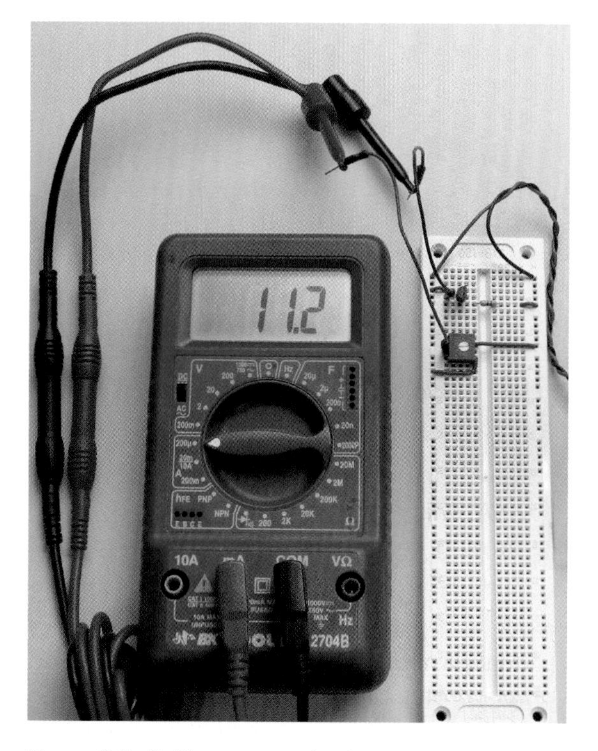

Figure 2-2. *Setting up a meter to measure microamps flowing from the wiper of the trimmer potentiometer to the base of the 2N2222 transistor. See text for more details.*

A closeup of this same breadboard is shown in Figure 2-3. The black and red wires coming in from the left, terminating in round plugs inserted into the breadboard, are from the meter. The trimmer potentiometer is oriented in the same way as in the schematic so that each of its leads is inserted in a separate row of holes in the breadboard. If you turned the trimmer 90 degrees, two of its leads would be in the same row of holes, and it wouldn't work.

Figure 2-3. *A closeup of the breadboard from the previous photograph.*

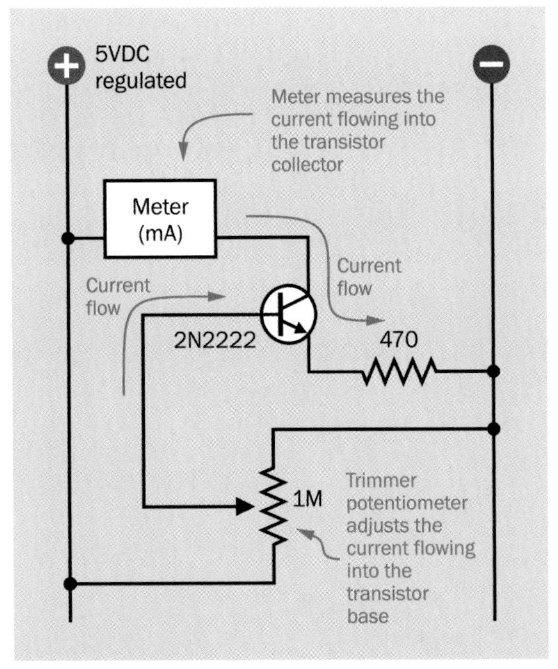

Figure 2-4. *The meter now measures current flowing into the collector of the transistor.*

Adjust the trimmer until your meter shows 5μA. This is the base current—the current flowing through the lefthand side of the trimmer and into the base of the transistor.

Step 2

Make a note of the base current. Maintaining a lab notebook is a really good idea, and you might as well start one now. If you record each experiment on a step-by-step basis, it can be useful to refresh your memory later. *The Maker's Notebook* is a product that makes this convenient.

Step 3

Remove the meter probes from the breadboard and substitute a piece of hookup wire. Change the meter to measure milliamps, if it does not do autoranging, and move it to the position shown in Figure 2-4.

Figure 2-5 shows a photograph of this configuration. The yellow piece of hookup wire has been inserted where the meter wires were before, and the meter now connects the positive bus on the breadboard with the collector of the transistor.

Figure 2-5. *The red and black wires on the left run to the meter, which now measures current flowing into the collector of the transistor.*

Step 4

Alongside the base current that you just wrote down, write the reading that you now see on the meter. This is the collector current.

Step 5

Go back to Step 1, but adjust the trimmer to increase the base current by an additional 5µA. (Don't forget to reset your meter to microamps, if necessary.)

Repeat steps 1 through 5 to make a table in which the left column shows the base current from 5µA to 40µA in steps of 5µA, while the next column shows the collector current at each step. That's eight values altogether—not too much work, even though swapping the meter to and fro is repetitive. The result should look something like the first two columns in Figure 2-6. I measured those numbers myself; are they similar to yours?

Current into Base (I_B in µA)	Current into Collector (I_C in mA)	Current into Collector (I_C in µA)	I_C/I_B Beta Value	Voltage Between Emitter and Ground
5	1.0	1000	200	0.45
10	2.1	2100	210	0.98
15	3.2	3200	213	1.52
20	4.3	4300	215	2.02
25	5.3	5300	212	2.52
30	6.4	6400	213	3.03
35	7.5	7500	214	3.51
40	8.6	8600	215	3.96

Figure 2-6. *Comparing base current with collector current in an NPN transistor.*

Now you need to convert each value for collector current from milliamps to microamps, because there's going to be a division that only makes sense if all the units are the same. There are 1,000 microamps in a milliamp, so you just have to multiply the collector current that you measured in milliamps by 1,000 to get the equivalent in microamps. You can see this in the third column in the table of values that I measured, in Figure 2-6.

Finally, get out a pocket calculator and divide the collector current (in microamps) by the base current (in microamps) for each of your eight pairs of readings. After the first one or two, you should find that the ratio is almost precisely constant. This is shown in the fourth column of my table.

- The current into the collector, divided by the cutting into the base, tells you the amplifying power of the transistor.

Warning: Meter at Risk!

Be careful when measuring current. Excess current can blow the fuse in your meter. It's a good idea to keep spare fuses available. Also, when you stop measuring current and set the meter aside, you may forget to swap the red lead back into the correct socket for measuring voltage. It's a good idea to do this as a matter of habit, because the meter is much less vulnerable in this mode.

Abbreviations and Datasheets

In Figure 2-6, notice the abbreviations I_B and I_C. Remember that letter I is generally used to mean "current." So, if you're thinking that I_B is the current into the base, and I_C is the current into the collector, you're right.

You'll find these abbreviations in almost any transistor datasheet, and they are usually telling you the maximum values that you're allowed to use. This is very helpful information. If you start thinking about building a project of your own, the maximum values for base and collector current will enable you to choose a transistor that won't be overloaded.

Now, what do you think I_E might mean? If you guess that it represents the current flowing out of the emitter, once again, you're right, although this abbreviation is less often used. The reason is that I_E is actually a combination of I_B and I_C. The current that flows in through the base and the collector can only get out of the transistor through its emitter, and therefore:

$$I_E = I_B + I_C$$

Here are some other common abbreviations used with NPN transistors:

- V_{CC} is the power supply voltage. It stands for Voltage at Common Collector, but is used to identify the supply voltage even if there are no bipolar transistors in a circuit.
- V_{CE} is the voltage difference between collector and emitter.
- V_{CB} is the voltage difference between collector and base.
- V_{BE} is the voltage difference between base and emitter.

A datasheet will also usually refer to the "beta value" of a transistor, often using the Greek letter β. This expresses how much a transistor will amplify its base current, and is calculated simply by taking I_C and dividing it by I_B, which is what you did after Step 5. Notice that the fourth column in the table is headed "Beta Value."

The consistency of the beta value in the fourth column of the table in Figure 2-6 tells you that the transistor is a *linear* device. In other words, if you plot a graph of the values, you get a straight line—as shown in Figure 2-7.

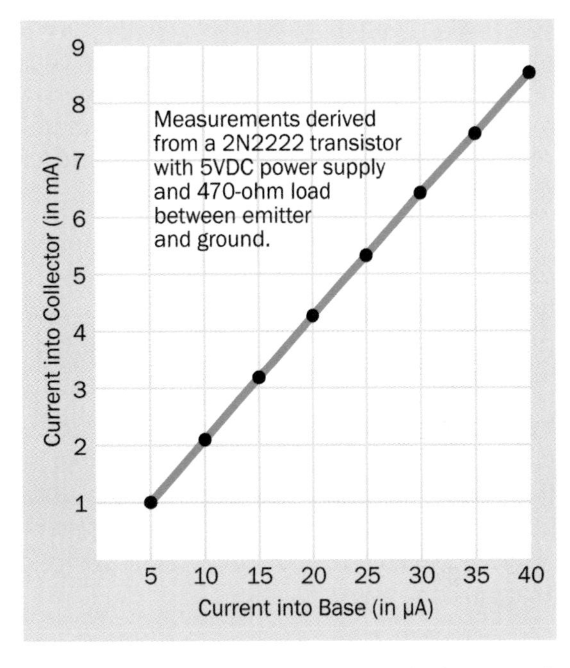

Figure 2-7. *A graph drawn from data in the first two columns in the preceding table.*

You can create a graph from your own numbers, using graphing software (Excel will do the job) or old-style graph paper. *The Maker's Notebook* is full of graph paper, or many sources online will allow you to download PDF files that will generate graph paper on your computer printer. Just search online for "print graph paper."

Why doesn't the beta value that you measure stay absolutely, exactly the same for each pair of values? Because your meter isn't absolutely accurate (especially when measuring very small currents in microamps), and your transistor may have tiny manufacturing imperfections. Still, the amplification ratio is sufficiently stable that a transistor can be used for accurate amplification of sensitive fluctuating signals, such as audio signals. (When we use a transistor as a switch, we don't care so much about this.)

Why is it that the numbers you measured may not be exactly the same as the numbers that I measured? Because there are many *uncontrolled variables*. Your meter and my meter will most likely be from different manufacturers. Your voltage regulator may be slightly different from mine. Meter probes may not make a good connection. The temperature of the transistor can make a small difference. The world is full of uncontrolled variables. We can never get rid of them.

In addition, transistors may have manufacturing differences. A datasheet may indicate a range of beta values that you may find in components of the same type, even if your measuring equipment is highly accurate.

People who write software are accustomed to using values that are absolutely precise—but in the world of hardware, the best we can do is to build circuits that will produce a fairly consistent result in a reasonable range of circumstances. That's just the way it is.

What About the Voltage?

Maybe you remember from *Make: Electronics* that a transistor is a current amplifier. Introductory books always make this statement, and the beta value is a measurement of current amplification. But people often fail to mention that the voltage on the emitter from an NPN transistor also tends to vary when the base current varies, so long as other factors (such as the load on the transistor) remain the same.

Figure 2-8 shows the schematic that will prove this for you. Remember, you most often measure voltage between the point that interests you in a circuit and the negative side of the power supply. Therefore, you do not put the meter in series with the 470Ω resistor in this circuit! And a reminder: don't forget to set your meter to measure voltage rather than amperage, and move the red lead to the appropriate socket on the meter if this is necessary (it usually is).

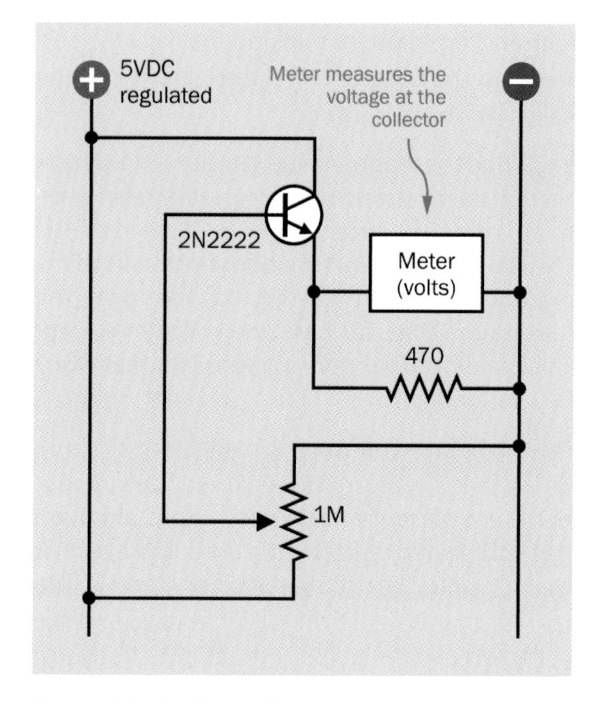

Figure 2-8. *In this configuration, the meter measures voltage between the emitter of the transistor and the negative side of the power supply (so long as you remember to set the meter to measure voltage rather than current).*

In Figure 2-6, the fifth column shows the measurements I made of voltages. I used these numbers to draw another graph, shown in Figure 2-9, comparing base current with emitter voltage—and once again, it's a fairly straight line.

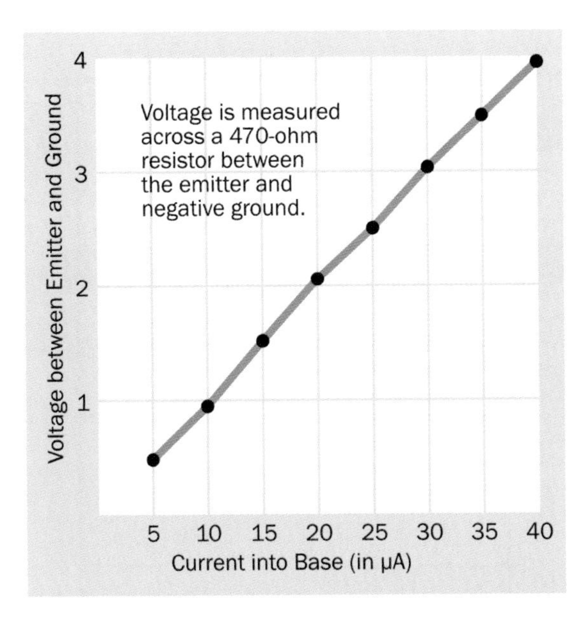

Figure 2-9. *Emitter voltage varies almost linearly with base current, in a 2N2222 transistor. This graph was derived from numbers in the table shown previously.*

If a transistor is a current amplifier, how come it is changing the voltage on its emitter, as well as the current? Well, let's think about what is really happening inside the transistor:

- An increase in the base current causes a reduction in the effective internal resistance of the transistor. This is why the current flowing through the transistor increases.
- But the transistor is in series with the 470-ohm resistor. The two components create a kind of *voltage divider*.

Maybe you remember from *Make: Electronics* that when you have two resistances in series, they divide the voltage drop between them, depending on their resistance relative to each other. If the first resistor has a low value, it doesn't block much voltage, so the second resistor imposes a bigger drop—and vice versa.

Take a look at Figure 2-10. In this schematic, instead of using a transistor in series with a 470Ω resistor, I've used various other resistors. Can you predict the voltages that will be measured at points A, B, C, and D? This is an experiment that you can perform for yourself very quickly. I'll provide the theoretical answers at the end of this section of the book.

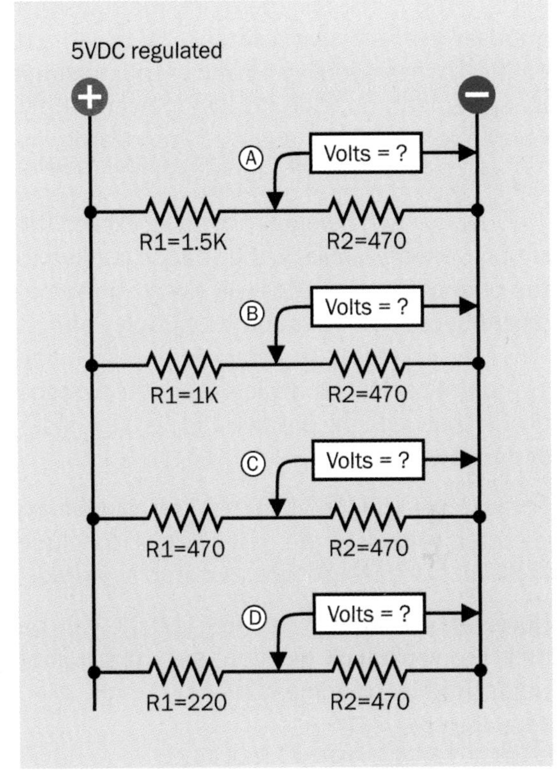

Figure 2-10. *The concept of a voltage divider is fundamental in electronics. Make sure that you understand it clearly.*

I'll remind you of the formula for calculating the voltage at the midpoint between two resistors. In the formula:

- V_M is the voltage at the midpoint.
- V_{CC} is the supply voltage.
- R1 is the value of the resistor (in ohms) on the positive side.
- R2 is the value of the resistor (in ohms) on the negative side (as in Figure 2-10).

The relationship looks like this:

$$V_M = V_{CC} * (R2 / (R1 + R2))$$

Perhaps now you see why the voltage at the emitter of the transistor goes up when the base current increases, in the circuit in Figure 2-8. The base current reduces the effective internal resistance of the transistor. Therefore, the transistor imposes less effective resistance between the emitter (where you are making your voltage measurement) and the positive power supply. Consequently, the voltage that you measure increases. This is shown in Figure 2-11.

The voltage at the emitter can never exceed the supply voltage. Similarly, the voltage applied to the base of the transistor will always be somewhere between 0 volts and the supply voltage. Why? Because the base voltage is taken from the 1M potentiometer, which acts as another voltage divider, between the positive and negative sides of the power supply.

Because the voltage at the emitter cannot exceed the voltage at the base, we can conclude that a bipolar transistor does not amplify voltage.

However, variations in emitter voltage can be useful, as you'll see when you start using a phototransistor in Experiment 3.

Quick Facts About Voltage

So far I have made some assumptions without justifying them. I have assumed:

- The positive voltage in a circuit is a fixed quantity.
- All the points in a circuit wired directly to the negative side of the power supply have a potential of 0 volts.
- Simple arithmetic defines the behavior of a voltage divider.
- We can measure volts or milliamps anywhere we like, without disturbing the circuit.

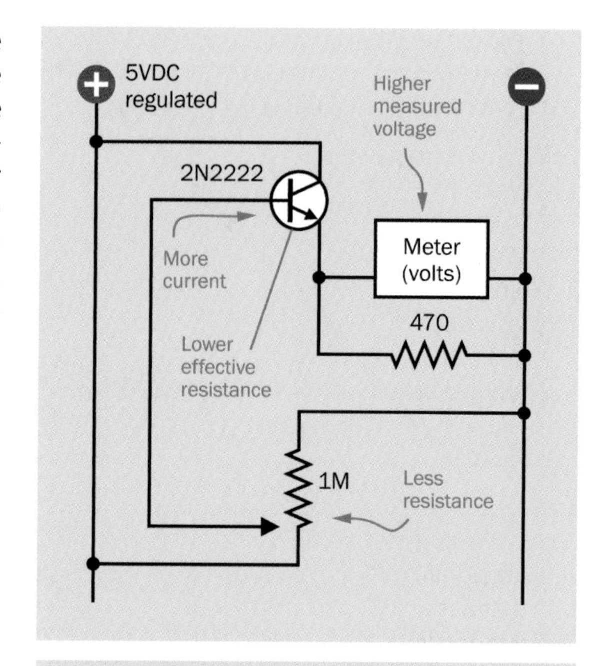

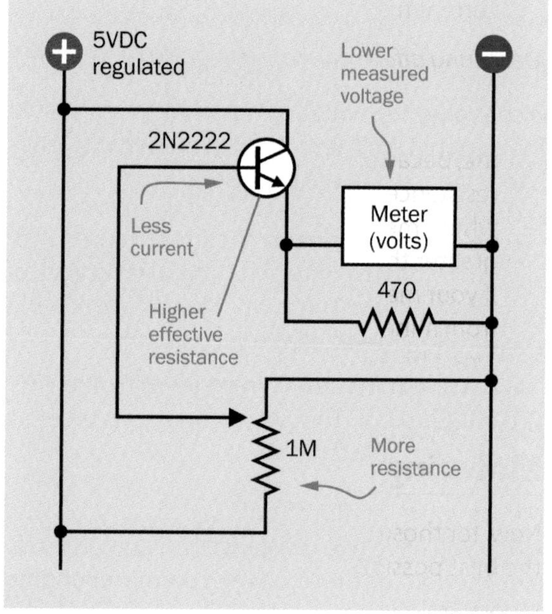

Figure 2-11. *The consequences of more current (upper schematic) and less current (lower schematic) flowing through the base of an NPN transistor.*

In the real world, these assumptions are not quite true!

V_{CC} may be less than V_{CC}

Any power source has its limits. When you load it heavily with a component of low resistance, this can pull down the voltage. The LM7805 regulator does a good job of fighting this tendency, but it is not perfect.

Zero may be greater than zero
: We think of ground as being at 0 volts, but various components are sinking current into the ground connection, and the wire running back to the power source has a small resistance of its own. Depending where you make a connection with the ground wire, the ground potential may not be precisely zero relative to the negative side of the power source.

Voltage dividers are approximate
: The voltage at the midpoint will be thrown off significantly if you attach a component that has a relatively low resistance and sinks current into the voltage divider, or draws current from it.

Measuring affects measurement
: Even the process of measuring voltage (or current) can affect the values that you measure, because your meter itself has an internal resistance. This is very high, but not infinite, when measuring volts. It's very low, but greater than zero, when measuring current. If your meter's internal resistance is different from mine, the values that you and I measure won't be exactly the same.

Make Even More: Old-School Metering

Now, for those who love gadgets (as I do), here's the best possible way to redo Experiment 2.

Long before the invention of multiranging, multifunction meters, you could buy analog meters that did only one thing. They would measure volts or millivolts or amperes or milliamperes, within a fixed range. In fact you can still buy this kind of meter, and you can install a pair of them in your transistor testing circuit, so that you don't have to move your multimeter to and fro.

Through eBay, I was able to order some nice little meters from Hong Kong for only $5 each. (You can also find them on Amazon for slightly more money, but with quicker shipping.) I chose one that measured microamperes on a scale of 0 to 50, while the other measured milliamperes on a scale of 0 to 10. These ranges were exactly what I needed. I set up the meters as shown in Figure 2-12. I enjoyed watching the needles move to and fro, perfectly synchronized, as I adjusted the trimmer potentiometer. Maybe this isn't everyone's idea of fun on a Saturday night (or even on a Monday night), but it makes a nice demo.

Figure 2-12. *Two analog meters can give an immediate display of the current amplification capability of a basic 2N2222 bipolar transistor. The rectangular blue component in the foreground is a trimmer potentiometer with screw adjustment.*

Quick Facts About Transistors

Some people enjoy working with numbers. Some don't. I realize it may be more fun just to throw some components together and see what happens (with or without some Elmer's glue), but the further you go into electronics, the more important it is to know what's really going on, and for this purpose, you need some arithmetic. This is not very onerous, because in a DC circuit, you seldom do anything more difficult than multiplication and division. In an AC circuit, real mathematics is involved—but that's beyond the scope of this book.

Here are your must-remember, take-home messages from this simple demo:

- A bipolar transistor is a *linear* device, meaning that the ratio of current entering the collector to current entering the base is approximately constant, and a graph of these two variables is almost exactly a straight line.
- The *beta value* of a transistor is its amplification factor—the ratio of the current flowing in at the collector, to current flowing in at the base.
- The voltage on the emitter of an NPN bipolar transistor will vary with the current, so long as the emitter has a constant load on it.
- A bipolar transistor is not a voltage amplifier, because the voltage at the emitter cannot exceed the voltage at the base.

And now a few additional facts:

- The *forward bias* applied to an NPN transistor is positive voltage at the base, relative to voltage at the emitter. *Negative bias* would mean that the base has a lower voltage than the emitter. Try to avoid this, because it's bad for the transistor.
- The *cutoff region* is where V_{BE} (the forward bias) is less than about 0.6 volts. In this region, charge carriers inside the transistor are not sufficiently energized, and nothing happens. Only a tiny amount of current, known as *leakage*, passes through the component when it is insufficiently forward-biased. This allows a transistor to be used as a switch.
- The *active region* for a bipolar transistor is the range where it functions as a current amplifier. The upper limit of this region is where the effective internal resistance between collector and emitter drops so low, there is almost no limit to the current flowing between them. This is the *saturation region* in which overheating will occur.

Of course, a transistor may also overheat in the active region if you don't limit the current. Always include some resistance (either a resistor or some other component that has resistance) with the transistor. Never apply the two sides of a power supply directly to the collector and the emitter.

Datasheets may have terms such as $V_{CE(SAT)}$ to tell you where the saturation limits are. Datasheets can be frustrating—for example, when they forget to define a term, or they don't bother to include a schematic to show you how a component is typically used. Still, datasheets are essential if you decided to be creative and modify circuits or build your own. When you use a component for the first time, it's a very good idea to go online, find the datasheet, and print a copy for future reference.

Answers to Voltage Divider Examples

A. `5 * (470 / 1970) = about 1.2V`

B. `5 * (470 / 1470) = about 1.6V`

C. `5 * (470 / 940) = 2.5V`

D. `5 * (470 / 690) = about 3.4V`

So much for the numbers. Now it's time to play with light.

Experiment 3: From Light to Sound

3

In this experiment you'll get acquainted with a phototransistor. Its symbol is shown in Figure 3-1, looking very much like a bipolar NPN transistor. In fact, its collector and emitter serve the same functions. The big difference is that the base is energized by incoming light, indicated by one or two arrows pointing toward it.

Sometimes the circle around the symbol is omitted. Sometimes there may be a single zigzag arrow instead of two straight arrows. These variations don't indicate any difference in operation. However, if a connection is shown protruding from the base, a base connection is available to supplement the effect created by the incoming light. I'm just mentioning this so that you'll recognize it if you see it. I won't be using that type of phototransistor in this book.

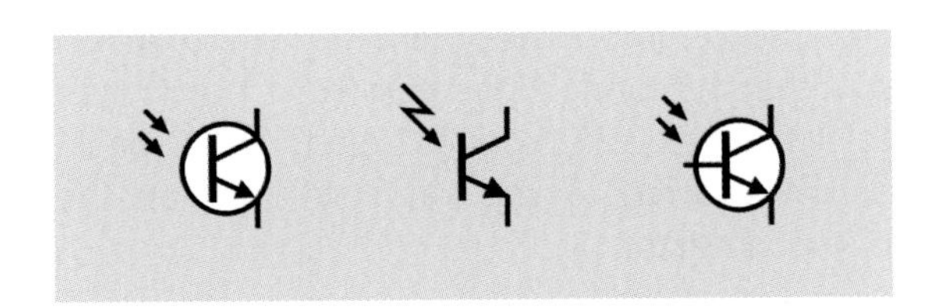

Figure 3-1. *Schematic symbols representing a phototransistor. Those at left and center are functionally identical. The symbol on the right indicates that a connection to the base is available to supplement the voltage induced by incoming light.*

Don't confuse phototransistors with photoresistors! Photoresistors are commonly known as photocells, and I referred to them in *Make: Electronics*. They are convenient to use, because they don't require a separate power supply. They simply vary their resistance in response to light. Because they usually contain cadmium sulfide, which is regarded as hazardous to the environment, they are not widely stocked at large retailers such as *http://www.mouser.com*. You can still find them on eBay, but I avoid specifying them in a circuit because they may be difficult to obtain in the future.

Phototransistors are now widely used as a substitute, with applications ranging from switching streetlights on and off to sensing an infrared signal when you press a button on your TV remote.

Photosensitive Audio Pitch

- Remember that you will find components for each experiment listed at the back of the book. See Appendix B.

Begin by assembling the circuit shown in Figure 3-2. I'm using a 555 timer because it's always useful for doing a demo. The output from the original bipolar version of the 555 can drive an LED, or a relay—or a small loudspeaker, as in this example. The more modern CMOS version, which often has a part number with 7555 in it, cannot deliver so much output power.

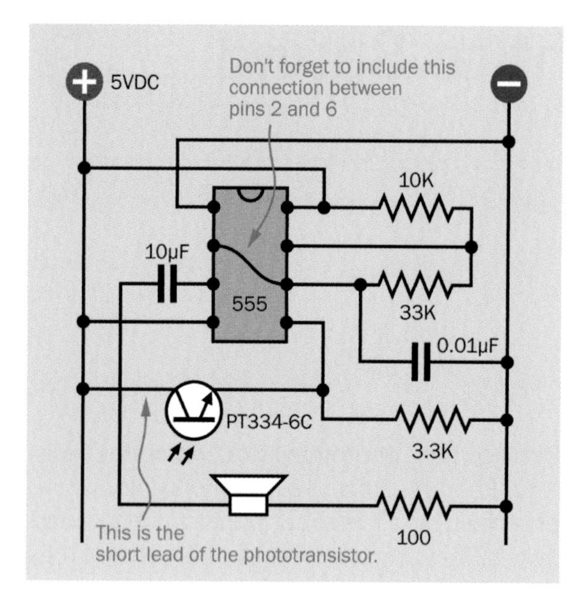

Figure 3-2. *This demo circuit creates an audible indication of the function of a phototransistor.*

You have to be careful to install the phototransistor the right way around. Here's the rule:

- Positive current enters a phototransistor through the *short* lead, and exits through the *long* lead.

Therefore, the short lead should be at the left in this schematic.

This is confusing, because LEDs look identical to phototransistors, and you know that the long lead of an LED always has to be "more positive" than the short one. Phototransistors are the other way around. You can think of a phototransistor as being opposite to an LED, because it absorbs light instead of emitting light. Consequently, its connection is opposite.

Here's another rule. Because LEDs and phototransistors can look virtually identical:

- Make sure you store your phototransistors in a container that is carefully labelled, so you don't get them mixed up with your LEDs!

A photograph of this circuit is shown in Figure 3-3. In this photograph, the 100Ω resistor in series with the loudspeaker is omitted, because I used a loudspeaker with 63Ω impedance. All the other connections in the circuit are the same.

Figure 3-3. *A breadboarded version of the phototransistor test circuit using a 555 timer. The phototransistor is the transparent object at the center near the yellow wires. A loudspeaker is partially visible on the right.*

Remember to include the jumper wire connecting pin 2 of the 555 timer with pin 6. This jumper is a green wire in the photograph, and is shown crossing the chip in the schematic. When you feel sure that the connections are correct, power it up to discover how the sounds from the loudspeaker will change as you vary the light on the phototransistor.

The pinouts of the 555 timer are shown in Figure 3-4. I'll remind you of their functions in more detail in the next experiment.

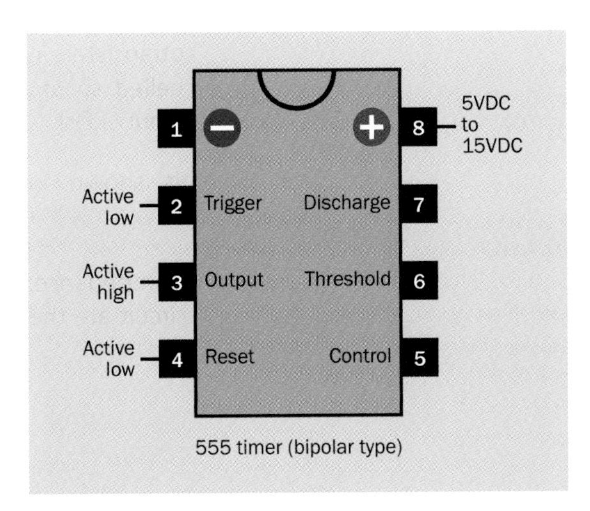

Figure 3-4. *Pin functions of the 555 timer. The supply voltage range applies only to the original TTL, bipolar type of the chip.*

Experiment by substituting different values for the 10K and 33K resistors. Or try a slightly larger or smaller value for the 0.01µF capacitor. Do you recall the formula for determining the frequency of a free-running timer? I'll remind you of the basic info in the next experiment. The key fact here is that pin 5 (the pin at the bottom-right corner of the chip) is the control pin. The voltage applied to this pin adjusts the reference value that the timer users, when it decides that each "on" cycle is complete, and flips into its "off" cycle. Consequently, the control pin adjusts the pitch of the sound that the timer creates when it is running at an audio frequency.

The combination of the phototransistor and the 3.3K resistor works as a voltage divider, in the same way that I described on "What About the Voltage?" on page 12. When light shines on the phototransistor, its effective internal resistance drops, changing the voltage to pin 5 on the timer. But how do we know how much the voltage drops? Let's find out.

Experiment 4: Measuring Light

4

- Remember that you will find components for each experiment listed at the back of the book. See Appendix B.

Check the very simple schematic in Figure 4-1, which looks quite similar to the circuit in Figure 2-8. You can add it as a separate circuit to your breadboard, without dismantling the circuit that you built in Experiment 3. Just move the phototransistor and the resistor farther down your breadboard.

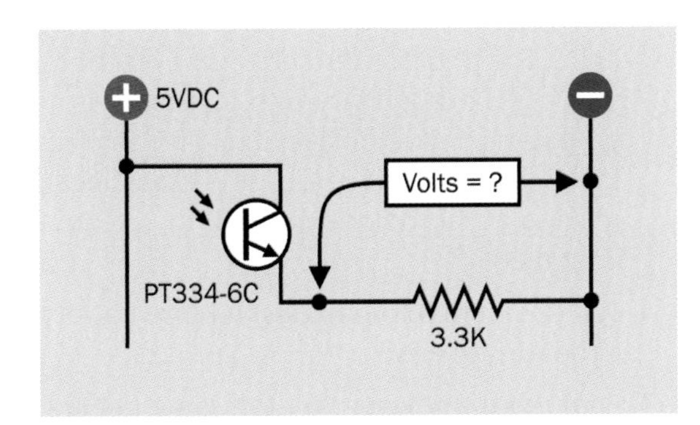

Figure 4-1. *A test circuit for a phototransistor.*

The resistor that I have chosen is 3.3K, because I want to get the widest possible range of voltages from the phototransistor emitter, and I found that 3.3K does the job.

Now, shine some light on the phototransistor, and measure the voltage at the emitter. Try using a desk lamp, a white LED, a flashlight, or a colored LED. The incoming light creates a very small current in the base of the phototransistor, which amplifies a larger current flowing from the collector through to the emitter.

- A *brighter* light *lowers* the effective internal resistance. You can remember this by thinking of the light driving the resistance away.

If you remember what I wrote about a voltage divider at the end of Experiment 2, you'll see that when the phototransistor has a *low* effective internal resistance, the voltage that you measure in the circuit shown in Figure 4-1 will *increase*. The phototransistor provides less of a barrier between the measuring point and the source of positive current. In this particular circuit:

- Bright light increases the voltage at the emitter.

This only holds true if you wire the components as in Figure 4-1.

Now swap them around as in Figure 4-2. The voltage will go down in brighter light, because the phototransistor is now providing less resistance between your measuring point and negative ground.

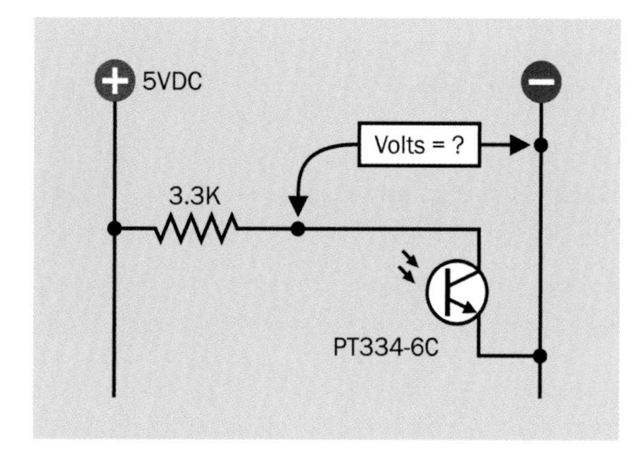

Figure 4-2. *If the phototransistor and the resistor trade places, the voltage at the point between them will be reduced in brighter light.*

Using Phototransistors

Phototransistors are available in many variants. The one I chose is sensitive to a wide range of light frequencies, so you can shine just about any color of light at it and get a response. Many phototransistors are only sensitive to infrared frequencies, because they're intended for activation by infrared-emitting LEDs. When the phototransistor and the LED have the same narrow band of sensitivity, there's less chance of picking up noise or erroneous signals.

Bear in mind that your meter has a high resistance. If you remove it and substitute a component with a relatively low resistance, it will compete with the 3.3K resistor to suck current out of the phototransistor, possibly overloading it. Fortunately for our purposes, logic chips, microcontrollers, and other digital devices have a high input impedance, so we can attach them directly to the phototransistor emitter—so long as we use an appropriate power supply, which will usually be 5VDC.

If you drive the input of a digital chip with the output from an analog device (such as a transistor or phototransistor), you should always be careful to measure the actual voltage going into the chip under all conditions that may exist in the future, to make sure that the voltage stays in an acceptable range. See Figure B-4 on page 332.

Quick Facts About Phototransistors

- Phototransistors are classified according to the wavelengths of light to which they are sensitive, measured in nanometers, abbreviated nm.
- The human eye can perceive light ranging from around 380nm to 750nm.
- Infrared has a wavelength longer than 750nm. Ultraviolet has a wavelength shorter than 380nm. Phototransistors exist that are sensitive only to ultraviolet, but they are uncommon.
- Infrared phototransistors are usually solid black in appearance.

Background: Photons and Electrons

Light is a source of energy, and the phototransistor uses that energy to induce a flow of electrons. Several types of light-conversion components exist:

- *Photodiodes* contain a semiconductor that can be penetrated by photons ("particles" of light). The photons dislodge some electrons that cross a boundary into an adjacent n-type semiconductor layer, building an electrical potential. The response is quite linear, making a photodiode suitable for use in a light meter.
- *Solar cells* are photodiodes that have a very large surface area.
- *Phototransistors* work on the same general principle as a photodiode, except that an external DC power source helps to energize the electron flow, which is now controlled by light instead of being created by light.
- A *photodarlington* is a phototransistor that functions as a two-stage amplifier, like a darlington transistor. It has greater light

sensitivity than a regular phototransistor, but a slower response time.

- *Photoresistors,* also often referred to as "photocells," lower their resistance in response to light.

Quick Facts About the 555

Make: Electronics contained a very thorough section on the 555 timer. I'm just going to refresh your memory by summarizing a few details that are important:

Pin functions
: It's useful to be able to grab a 555 without a lot of thought whenever you want a single pulse or a train of pulses to test a component. Turn back to Figure 3-4 for a reminder of the names of the pins on the timer.

Monostable circuit
: Figure 4-3 may help to remind you of the basic behavior of a timer in monostable mode, also known as one-shot mode. A transition to *low* voltage on the trigger pin creates a *high* pulse from the output pin. The duration of the pulse is determined by the value of resistor R1 and capacitor C1, as the resistor charges the capacitor. If the reset pin of the timer will not be used, it is tied to the positive side of the power supply to prevent it from being activated unintentionally.

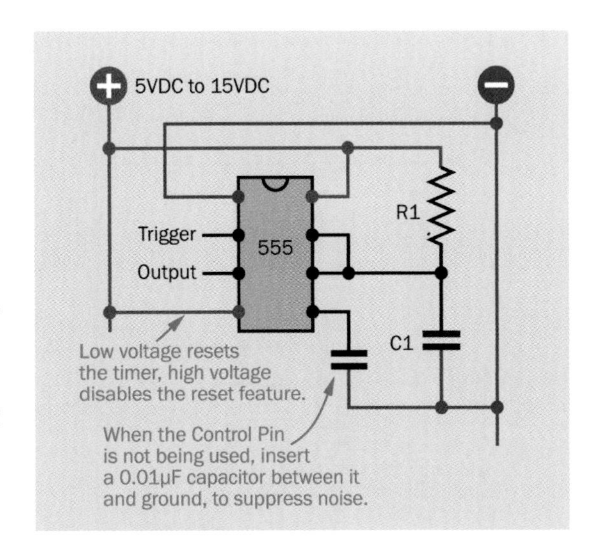

Figure 4-3. *A simplified schematic showing typical connections of a 555 timer in monostable mode.*

Monostable pulse duration
: Figure 4-4 is a quick-reference table of pulse durations, in seconds, for various values of resistor R1 and capacitor C1, when the timer is in monostable mode. You'll find a more detailed table in *Make: Electronics* or manufacturers' datasheets.

Capacitor C1	Monostable pulse duration for values of R1				
	10K	33K	100K	330K	1M
0.01µF	0.00011	0.00036	0.0011	0.0036	0.011
0.1µF	0.0011	0.0036	0.011	0.036	0.11
1µF	0.011	0.036	0.11	0.36	1.1
10µF	0.11	0.36	1.1	3.6	11
100µF	1.1	3.6	11	36	110

Figure 4-4. *Pulse duration, in seconds, of a 555 timer running in monostable mode.*

Astable circuit
: Figure 4-5 reminds you of the basic wiring to make the timer function in astable or free-running mode, creating a stream of pulses. In this configuration, the timer starts itself and continues running so long as (a) power is applied to it and (b) voltage on the reset pin is not allowed to go low.

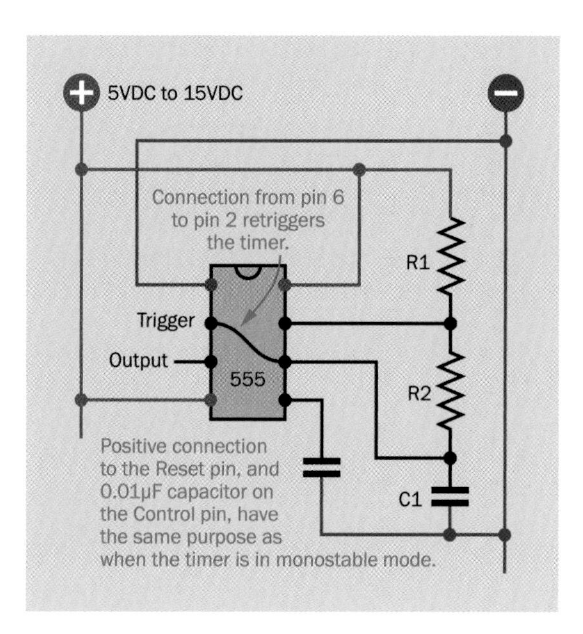

Figure 4-5. *A simplified schematic showing the 555 timer functionality in astable mode.*

Basic astable principle

Figure 4-6 shows the basic principle of operation in astable mode, as the capacitor charges through R1 and R2 and then discharges through R2 into the chip. This explains why the "on" output duration is always longer than the "off" output duration.

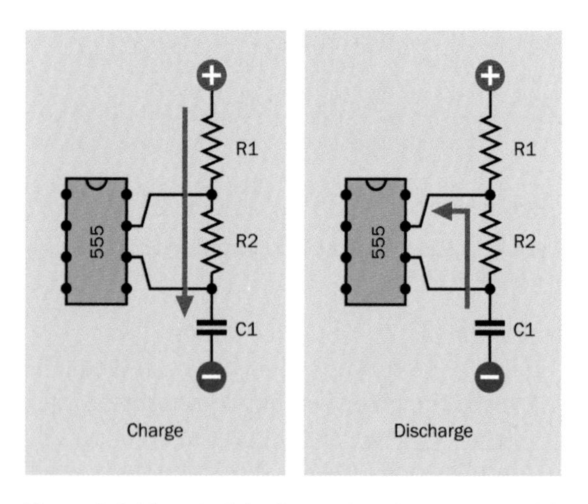

Figure 4-6. *The principle of operation of a timer in astable mode, charging capacitor C1 through R1 + R2 and discharging through R2.*

Astable frequency

Figure 4-7 is a quick-reference table of output frequencies, in Hz, for various values of C1 and R2, assuming 10K as a value for R1, when the timer runs in astable mode. (You can use lower values for R1, but the chip will consume more power.)

Capacitor C1	Astable frequency (Hz) for values of R2 (R1=10K)				
	10K	33K	100K	330K	1M
0.001µF	48,000	19,000	6,900	2,200	720
0.01µF	4,800	1,900	690	220	72
0.1µF	480	190	69	22	7.2
1µF	48	19	6.9	2.2	0.72
10µF	4.8	1.9	0.69	0.22	0.072
100µF	0.48	0.19	0.069	0.022	0.0072

Figure 4-7. *The output frequencies for a 555 timer running in astable mode with a fixed value of 10K for R1 and various values for R2 and C1.*

Total cycle time

The total cycle time of a timer in astable mode is proportional to R1 + R2 + R2, because a full cycle consists of one "on" pulse plus the gap between it and the next pulse. This is shown graphically in Figure 4-8.

Frequency calculation

If you measure R1 and R2 in kilohms, while C1 is measured in microfarads, you can calculate the frequency, F, in Hertz, of a 555 timer running in astable mode, using this formula, where R is R1 + R2 + R2:

```
F = 1440 / (R * C1)
```

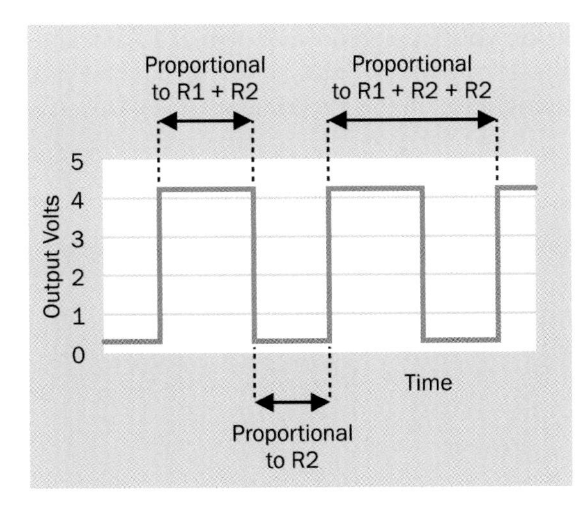

Figure 4-8. *A graphical representation of "on" and "off" durations in a 555 timer (using a 5VDC power supply) running in astable mode, showing why the total time from the beginning of one cycle to the beginning of the next is proportionate to R1 + R2 + R2.*

Large capacitors
: Very high capacitor values (above, say, 470µF) are likely to produce unreliable results, because of leakage. This is the undesirable feature of capacitors (especially electrolytics) whereby imperfections allow some loss of charge. Leakage from a big capacitor can become comparable to the current flowing into it, if you're trying to charge it through a high resistance.

Speed measurement
: If you want to know how fast your timer is running, and the speed is faster than you can log with a stopwatch, and you don't have an oscilloscope, you can substitute a capacitor of 10 times or 100 times the correct value, and the timer duration should increase proportionately. Because capacitors are manufactured within very wide tolerances, and because of the leakage issue mentioned above, substitution will only give you an approximate result.

Power source
: The power supply for a 555 can be as low as 5VDC and as high as 15VDC without affecting the pulse rate significantly.

Output voltage
: The 555 high output voltage is slightly lower than its power supply voltage. If you want the output to drive a logic chip that is fussy about high and low voltage inputs, check the voltage output of the 555 while it is running extremely slowly (e.g., a five-second pulse) to give your meter time to respond. A 10K pullup resistor or pulldown resistor may be used on the input to the logic chip.

If any of these reminders is unclear to you, please go back to *Make: Electronics*, or another introductory book, or a manufacturer's datasheet, for clarification.

Quick Facts About CMOS Versus Bipolar

The original type of 555 timer (still being manufactured) contains bipolar transistors. It is often referred to as a TTL chip, and has these characteristics:

- Not very vulnerable to static electricity
- Accepts a wide range of power-supply voltages
- Sources or sinks up to 200mA
- Creates spikes of noise when it switches on and off
- Uses a relatively large amount of power

The more modern CMOS version has different characteristics:

- Is more vulnerable to static electricity
- Requires a narrower range of voltage
- Cannot source or sink much current (the amount varies, depending on the manufacturer)
- Does not create voltage spikes caused by switching
- Uses very little power

Confusingly, the CMOS and bipolar versions may both be referred to as "555 timers," and their part numbers can be very similar. From Texas Instruments, for instance, the TLC555-Q1 is a CMOS version, while the NE555P is bipolar. Even more confusing, some CMOS versions are 3.3-volt devices, while some require 5VDC, and others will accept a wider voltage range.

If you go shopping on your own, read the datasheets carefully. A CMOS timer will not drive the loudspeaker in this experiment.

Experiment 5: That Whooping Sound

5

Instead of using a phototransistor to adjust the voltage on the control pin of the 555 timer, you can use the output from a second timer running at a much slower speed. This automates the up-and-down shifts in sound frequency.

In Figure 5-1 the previous schematic has been extended downward. (A photograph of a breadboarded version appears in Figure 5-2.) The output from a second timer is connected up to the control pin of the first timer, through a 47µF coupling capacitor. Why is the coupling capacitor there? To create a "whooping" sound. What do I mean by that? Well, you'll recognize it when you hear it.

The second timer uses a 150K resistor to charge your choice of a 1µF or 10µF timing capacitor. Try the 10µF capacitor first. This will cause the timer to run at about one cycle per second. Initially it has has no effect on the first timer, but the output from the second timer slowly charges the 47µF coupling capacitor, causing the pitch of the first timer to rise up gradually. Then the lower timer reaches the end of its "on" cycle and flips into its "off" cycle. At this point, the coupling capacitor discharges, and the frequency of the upper timer drops back down.

I included a circuit of this type in Experiment 17 in *Make: Electronics*, but the sound it produced was different. You can compare that schematic with the following one, and see if you can figure out why.

In Figure 5-1, if you substitute a 1µF timing capacitor for the 10µF timing capacitor, everything happens ten times as fast, and you get the distinctive noise that is typical of burglar alarms. (That's what I meant by a whooping sound.) You can have fun trying different values for the timing resistors and capacitors, and varying the coupling capacitor, to achieve the most absolutely annoying, aggravating noise possible.

The phototransistor adds to the possibilities. Experiment by varying the light that falls on it, and see what happens if you wave your fingers over the phototransistor to vary the light very quickly.

What could you achieve by using two phototransistors to adjust the voltages on the control pins of both timers?

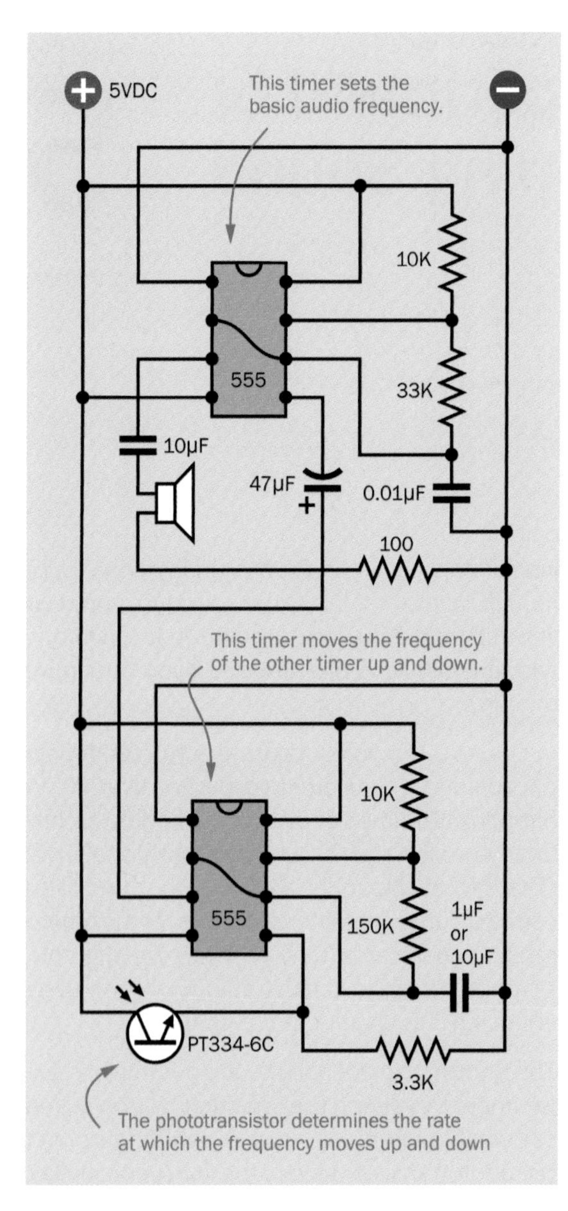

Figure 5-1. *By using a second 555 timer to adjust the voltage on the control pin of the first timer, you can achieve a really annoying "whooping" sound.*

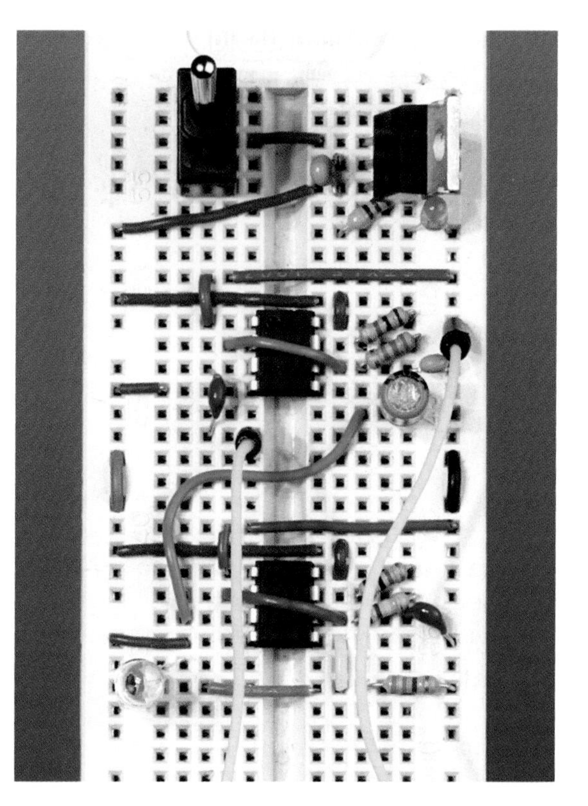

Figure 5-2. *A breadboarded version of the two-timer schematic.*

Make Even More

We have many alternatives to the 555 timer that were designed to circumvent the limitations that I listed at the end of the previous experiment:

- The 7555 is pin-compatible but uses less current, can be run from a voltage as low as 2VDC, and creates less noise in a circuit. The maximum supply voltage and maximum source current vary, depending on the manufacturer.
- The 4047B provides additional features that make it more versatile. One trigger pin responds to a positive-to-negative transition; another responds to a negative-to-positive transition. It also has two complementary outputs, one of which is high when the other is low. The timer runs as a one-shot or in astable multivibrator mode, depending on

the setting of another pin. The power supply range is equivalent to that of the old bipolar 555.

- The 74HC221 is a dual monostable timer chip. In other words, it contains two timers, each of which runs in one-shot mode. When they are wired to trigger each other, the result is a pulse stream where you can set the high duration and low duration independently of each other. The maximum supply voltage is 7VDC, but this is really intended to be used as a 5VDC device.
- The 4528B is a dual monostable timer using a concept that is similar to the 74HC221, but as an old-style CMOS device, it tolerates a wider range of power supply voltage (up to 15VDC).
- Other dual monostable timers include the 74HC123, 74HC423, 74HC4538, and 4098B, all adopting the same general principle but with slightly differing specifications.
- The 556 is a dual timer consisting of two 555s in one chip. They can be used to trigger each other, but have the usual limitations of a classic 555. The 556 chip is not as popular as it used to be, and may eventually become unobtainable. For this reason, and because it is often more convenient to place a single timer exactly where you want it on a circuit board, 556 chips are not used in this book.
- Lastly there is the 74HC5555, which contains a 24-stage counter. This divides the clock frequency by values up to about 16 million, allowing the timed interval to extend for days if you wish. An external crystal oscillator can be used for greater accuracy than a resistor-capacitor combination, but the typical high speed of a crystal oscillator will reduce the maximum pulse duration of the timer—unless you use two or more in a chain, with each triggering the next.

With so many options, all of which boast features lacking in a 555 timer, why is the old bipolar 555 still so popular? Probably because everyone is familiar with it. Like a QWERTY keyboard, it's not ideal, but we all know how to use it. Also, the old original bipolar through-hole version can source more current than any of its successors. This is handy for quick-and-simple circuits.

And it's cheap!

You might consider trying some of the variants listed above. Personally I never get tired of playing with timers, because these simple chips create so many possibilities. But it's time to move on, because we have a new component to deal with: the comparator.

Experiment 6: Easy On, Easy Off

6

You saw in the last two experiments that a phototransistor changes its output gradually in proportion with the amount of light falling on it. This is a useful capability—although not as useful as it could be. For practical purposes, we often want a light-sensitive gadget that has two precisely defined states: "on" and "off." An intrusion alarm, for instance, triggered by someone interrupting a light beam, has to give a clear signal. It cannot function gradually or intermittently.

Is there a way to convert a gradual output from a phototransistor into a clearly defined signal? Most definitely. A comparator is the tool for the job.

Making Comparisons

Put together the schematic in Figure 6-1. The 500K potentiometer is a trimmer that you plug into your breadboard. The phototransistor is in series with a 3.3K resistor as before, except that this time the output from the emitter connects through a 100K resistor to an input on an LM339 chip. This chip contains a comparator—in fact, it contains four comparators, although we will only use one of them right now. During this demo, the unused comparators can be left unconnected.

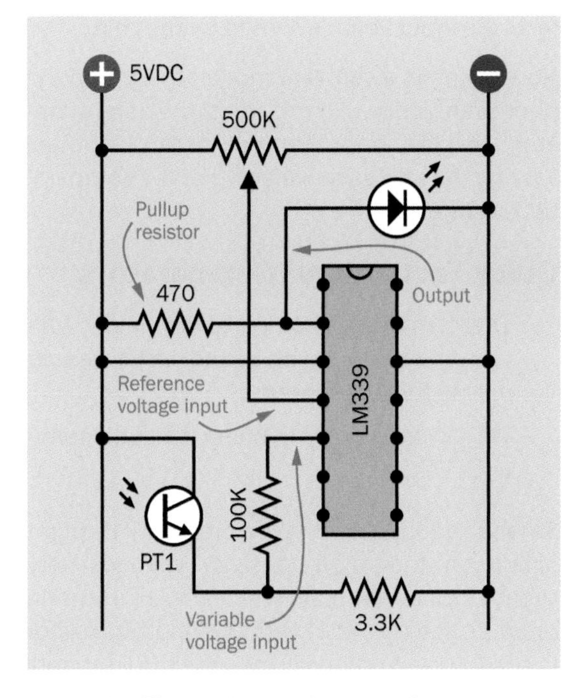

Figure 6-1. *The initial stage of a circuit that uses a comparator to switch an LED in response to light falling on a phototransistor.*

Set the potentiometer around the middle of its range. Begin with the phototransistor covered, so that no light falls on it. Now allow light to intrude, and you should see the LED light up. Dim the light again, and the LED goes off.

The 500K potentiometer sets a *reference voltage* for the comparator. When the wiper of the potentiometer is in the middle of its range, the reference voltage will be about 2.5V, because the potentiometer is functioning as a voltage divider between the positive and negative-ground sides of the power supply.

In dim light, the voltage from the emitter of the phototransistor is lower than 2.5V, so the comparator doesn't respond. As the light brightens, the voltage on the emitter of the phototransistor rises above 2.5V (do you remember why?). The comparator detects the difference and changes its output. (The 100K resistor will be necessary when other components are added in the next step. Because the input impedance of the comparator is extremely high, the resistor hardly affects the input voltage sensed by the chip.)

Now maintain a constant moderate light on the phototransistor while you adjust the potentiometer. The LED goes off and on because you are varying the reference voltage that the comparator is using.

Quick Facts About Comparators

- The comparator compares a *variable voltage* on one input with a fixed, *reference voltage* on the other input.
- We can use a potentiometer to set the reference voltage.

So far, so good. But we have a problem if the light falling on the phototransistor changes very slightly around the point where the LED turns on or off. To see this, start with the phototransistor in shadow, and gradually increase the light until the LED comes on. Now decrease the light just a tiny fraction, and the LED should flicker.

This is shown graphically in Figure 6-2. The flickering is known as "hunting," as the comparator is hunting to and fro, unable to make up its mind whether its output should be in the on state or the off state.

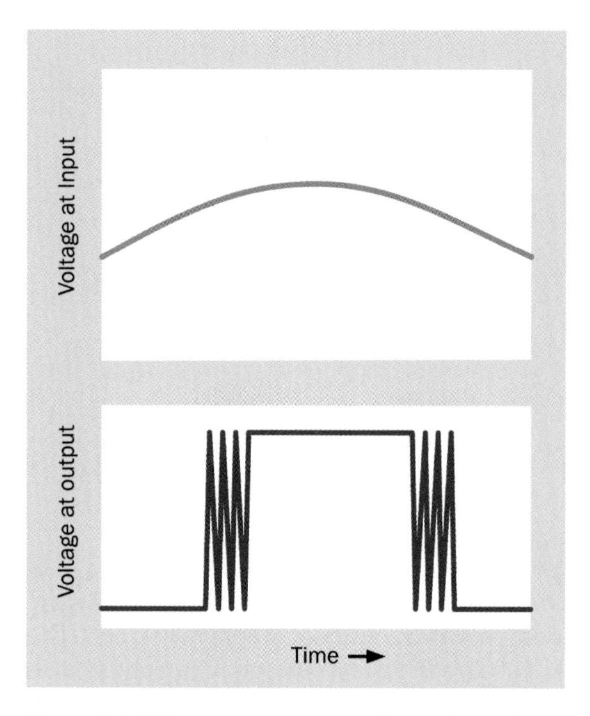

Figure 6-2. *When a comparator receives a slowly changing input (upper graph), its output tends to oscillate unpredictably (lower graph) between "on" and "off."*

How can we prevent this? The answer is to use a very powerful technique known as *positive feedback*.

Feedback

Figure 6-3 shows the same circuit as before, but with an additional potentiometer on the right. Figure 6-4 shows a breadboarded version of the circuit.

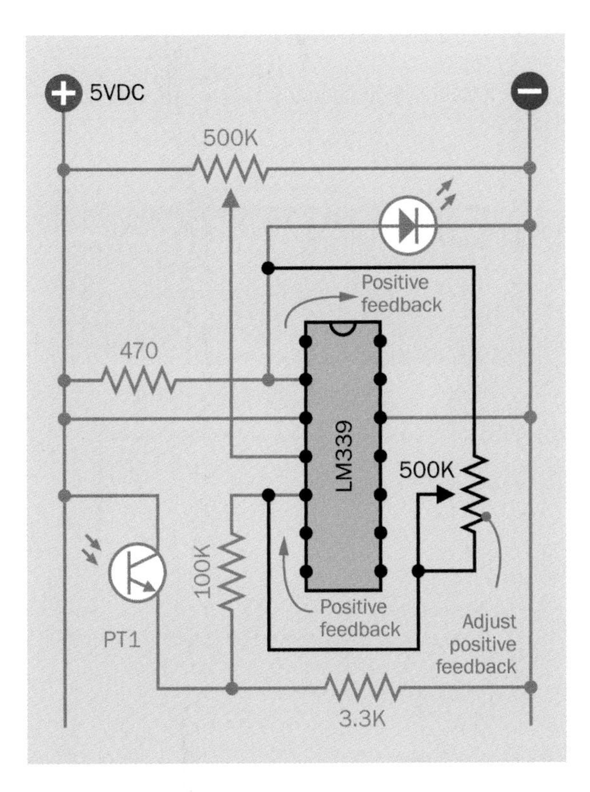

Figure 6-3. *The basic comparator circuit has been modified to clean up its output, by the addition of positive feedback. Previous wiring is gray.*

Figure 6-4. *The positive-feedback comparator schematic breadboarded, with two trimmer potentiometers, a phototransistor at bottom-left, and an LED to display the output.*

The basic idea of this circuit is shown in Figure 6-5. Notice that pin 2 of the comparator chip is connected back to pin 5. Pin 2 is the output (remember, it controls the LED). Pin 5 is the variable input—the phototransistor is connected with it through the 100K resistor. So the second potentiometer in Figure 6-3 is taking some of the voltage from the output and feeding it back to the input. This is the positive feedback.

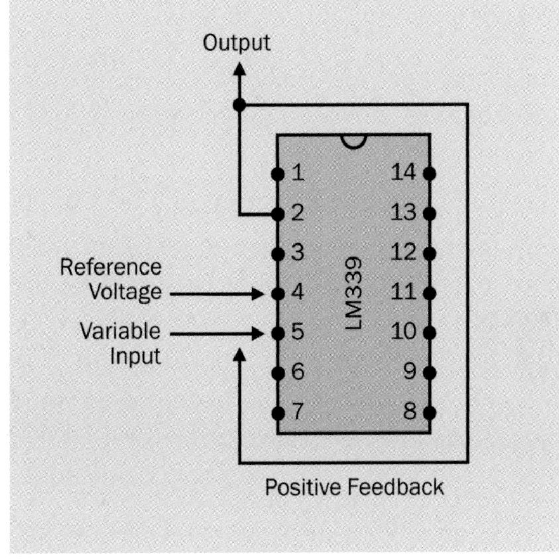

Figure 6-5. *The basic concept of positive feedback.*

With both of the trimmers near the middle of their scales, if you make small adjustments to the light falling on the phototransistor (as you did before), you should find that the LED doesn't flicker anymore. It is either "on" or "off."

Positive feedback works like this:

- When the output becomes more positive, it circulates back through the feedback loop, which adds it to the input.
- The input voltage goes up, which boosts the output.
- The higher output circulates back and reinforces the input some more.

This all happens very, very quickly, so that the LED lights up and stays lit. Now if the light falling on the LED gradually diminishes, at first nothing happens, because there's still enough feedback to sustain the input. But as the light fades, this happens:

- The lower input creates a lower output.
- Feedback from the output doesn't reinforce the input so much anymore.
- Deprived of positive feedback, the input drops suddenly, and the comparator's output goes low.

This all happens so quickly, the LED blinks off instead of flickering or gradually fading out.

Hysteresis

Turn the righthand trimmer to reduce its resistance in the circuit. This will increase the positive feedback so that it becomes more easily seen.

Now adjust the light falling on the phototransistor very, very gradually. If you have it under a desk lamp, hold your hand close to the lamp, so that you cast a shadow with a very soft, blurry edge.

You should find that when the LED lights up, you can reduce the light a little, and the LED will remain illuminated. You can think of the comparator becoming "sticky," because it tends to stick in the "on" state.

When it finally goes out, slowly increase the light, and now the comparator will tend to stick in the "off" state. Figure 6-6 illustrates this.

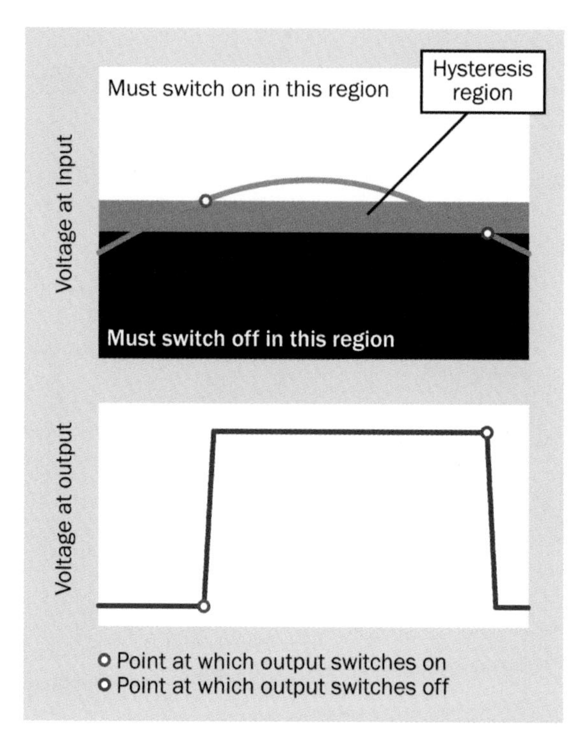

Figure 6-6. *With positive feedback, the output from a comparator tends to stick in its high or low state. The "sticky" zone is the hysteresis region.*

This phenomenon is known as *hysteresis*, and it's very useful. Suppose you are using a phototransistor to switch on a lamp at sunset. The light fluctuates a little as the clouds pass in front of the setting sun. Do you want the lamp to turn on and off with every little variation in the light? No, once the lamp comes on, you want it to stay on, regardless of small variations.

Suppose you have a thermostat that controls a heater. You want the heater to come on when the room temperature drops to around, say, 70 degrees Fahrenheit. Once the heater is on, you don't want it to switch off just because someone walks in front of the thermostat and creates a brief current of slightly warmer air. You want the heater to ignore small fluctuations until the temperature reaches, say, 72 degrees. Then you want the heater to turn off and stay off until the temperature drops back down to 70. In this case, the hysteresis region extends between 70 and 72 degrees.

The amount of hysteresis can be adjusted by increasing or decreasing the amount of positive feedback to a comparator. With a lower resistance allowing more feedback, the comparator will ignore larger fluctuations in its input, to create a simplified output. Figure 6-7 illustrates this.

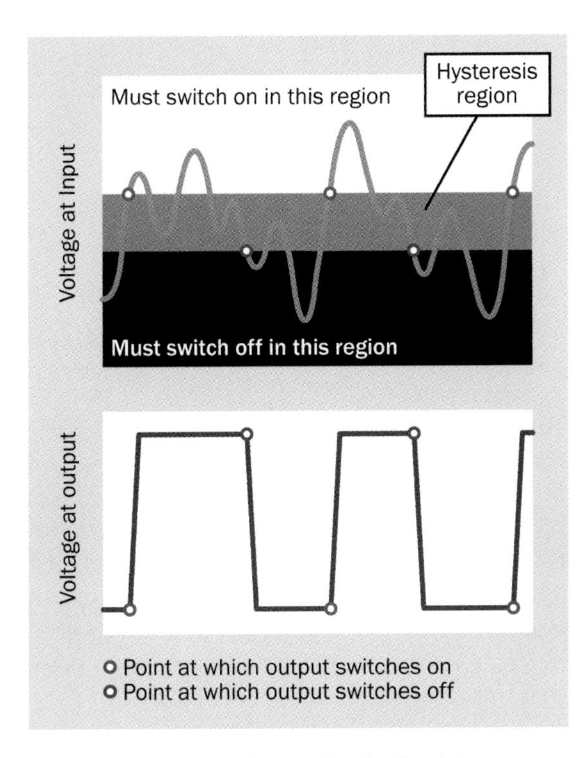

Figure 6-7. *Increasing the positive feedback in a comparator circuit creates more hysteresis, causing the comparator to ignore substantial variations in an erratic signal.*

The lower half of the figure shows the output that we want from the comparator, ignoring all those little wiggles in the input. Basically, the comparator will ignore any changes in the gray area, and will only react when the signal rises through the gray area and emerges into the "must switch on" region, or falls through the gray area and emerges into the "must switch off" region.

You should know that hysteresis is normally shown using a graph that looks like the one in Figure 6-8. This is the graph that you tend to find in most electronics books, but it's a little difficult to understand. The righthand part of the curve shows the output from the comparator (measured on the vertical axis) when the input voltage is smoothly and gradually increasing (measured on the horizontal axis, from left to right). The comparator waits a bit before it allows its output to come on. Then if the input voltage smoothly and gradually starts to go down, the curve to the left shows that the comparator waits a bit before it allows its output to go off.

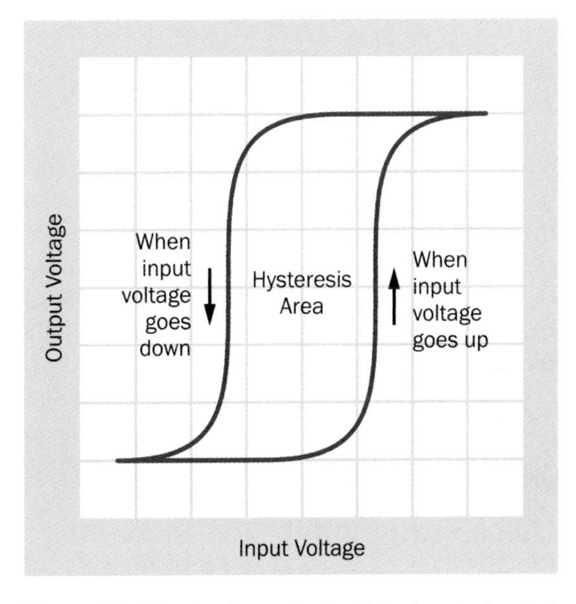

Figure 6-8. *The classic way to depict hysteresis. See text for details.*

The Symbol

Now for some details about the comparator. First, its symbol is shown in Figure 6-9. The comparator requires its own power supply, like a logic chip. I've shown this with the positive and negative signs, but in a schematic, often the power supply to a comparator is omitted. Everyone knows it has to be there, so people who draw schematics may not bother to include it.

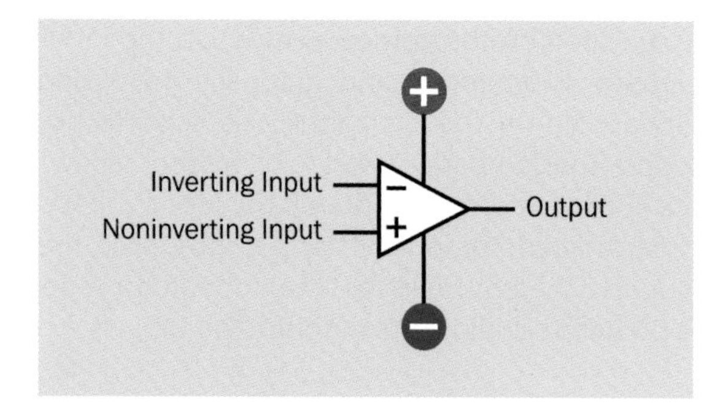

Figure 6-9. *Schematic symbol for a comparator. The power supply is always necessary, but is not always shown in a circuit.*

The reference voltage in your circuit was actually being applied through the "inverting" input of the comparator. You used the "noninverting" input for the variable voltage from the phototransistor. I'll explain why the inputs have these names a bit later. In the schematic symbol, the two inputs have their own plus and minus signs —which are confusing, because they do *not* mean that you are supposed to apply positive or negative voltage to them.

Quick Facts About Plus and Minus

- The comparator switches on when the voltage on the "plus" input in a comparator changes to become more positive than the voltage on the "minus" input. The "plus" input is called the noninverting input.
- Similarly, the comparator switches on when the voltage on the "minus" output in a comparator changes to become more negative than the voltage on the "plus" input. The "minus" input is called the inverting input.

The Output

I've been talking about the output from a comparator, but in fact, many comparators don't create a simple high or low output. They have an *open collector* output. This is shown in Figure 6-10.

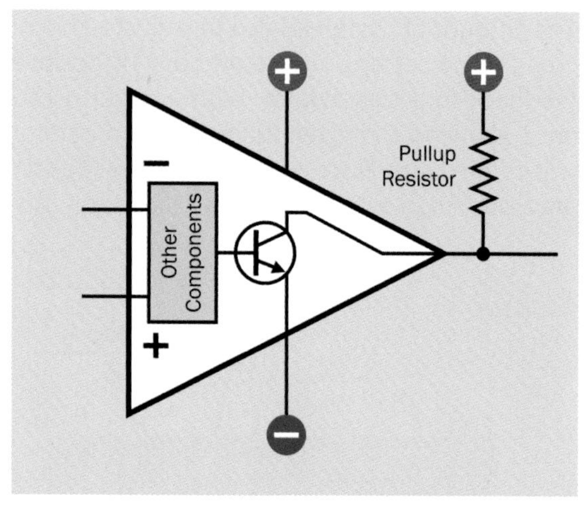

Figure 6-10. *Simplified view of the interior workings of a comparator. The two positive voltages do not have to be the same, so long as they share the same negative ground.*

The comparator actually contains various components, but the one of interest to us here is an output transistor, which is often a bipolar type. When the transistor switches on, it conducts current, so it sucks the current from an external pullup resistor and dumps it into negative ground. It also sucks current from any components connected with the comparator. This makes it seem that the comparator has a *low* output.

When the transistor switches off, it blocks current. The current from the pullup resistor cannot sink through the comparator anymore, so it goes out to any components connected with the output. This makes it seem that the comparator has a *high* output.

Figure 6-11 provides a graphical view of the way that a comparator works.

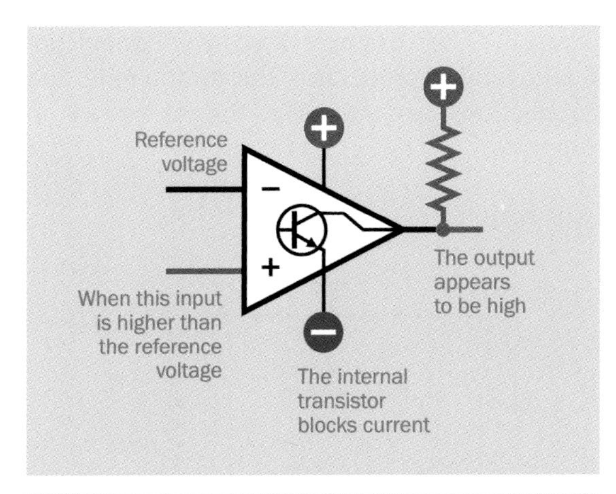

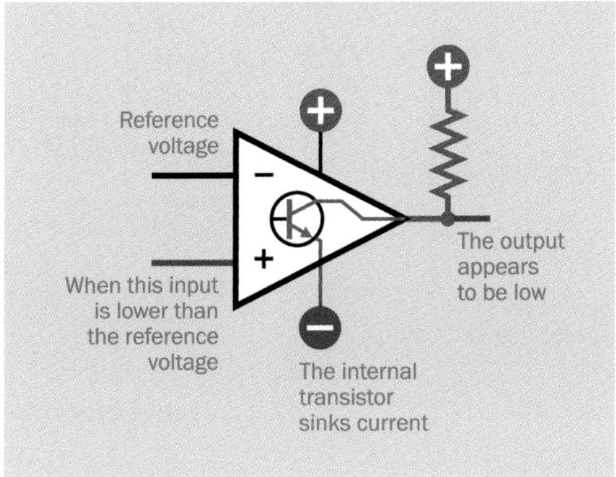

Figure 6-11. *When a comparator is used in noninverting mode (with voltage applied to its "plus," or noninverting, terminal), its output behaves as shown here.*

In practice, you don't need to remember what the output transistor inside the comparator is doing. You just have to remember that the "high" output from the comparator is really being supplied through the pullup resistor, while the "low" output means that current is sinking through the comparator.

In Figure 6-1 you may have wondered why I didn't show the LED with a resistor in series with it, as is the usual practice. This is because the LM339 has an open-collector output, so really the LED is being powered through the 470Ω pullup resistor.

Time, now, to sum up what we know about output.

More Quick Facts About Comparators

- You must include a pullup resistor on the output of a comparator if it has an open collector or open drain. Otherwise, the comparator won't work. Check the datasheet for any comparator that you use.
- If you use a low-value pullup resistor, more current will sink through the comparator, and you can burn it out. When in doubt, use your meter to check the current flowing from the pullup resistor into the comparator's output pin.
- Most comparators should have their outputs connected to devices with high input impedances, such as logic chips. That kind of chip demands very little current, so you can use a relatively high-value pullup resistor, which is typically 5K. The 470Ω resistor in the schematic in Figure 6-3 is relatively low because the comparator had to drive an LED.
- The comparator can't sink much more than 20mA. You can always add an external transistor to the output if you need more current.

Now, this next point is very important, and is a new concept:

- The positive voltage feeding the pullup resistor does not have to be from the same source as the voltage powering the comparator, so long as they share the same negative ground. For instance, you could have a power supply for the comparator that is 5VDC relative to its negative ground, and a pullup resistor powered by 9VDC relative to the same negative ground. Consequently, a comparator can function as a voltage amplifier.

Very interesting—we now know how to amplify current, with a transistor, and voltage, with a comparator. This information can be useful in the future.

Just be careful that you don't sink too much power through that internal transistor. The datasheet for a comparator will tell you what the limits are.

Inside the Chip

I used an LM339 in this experiment because it's one of the oldest types of comparators, but still very widely used—and very cheap! In Figure 6-12 you'll see that it is actually a *quad comparator*, meaning that it contains four comparators, only one of which we are using (so far) in this experiment.

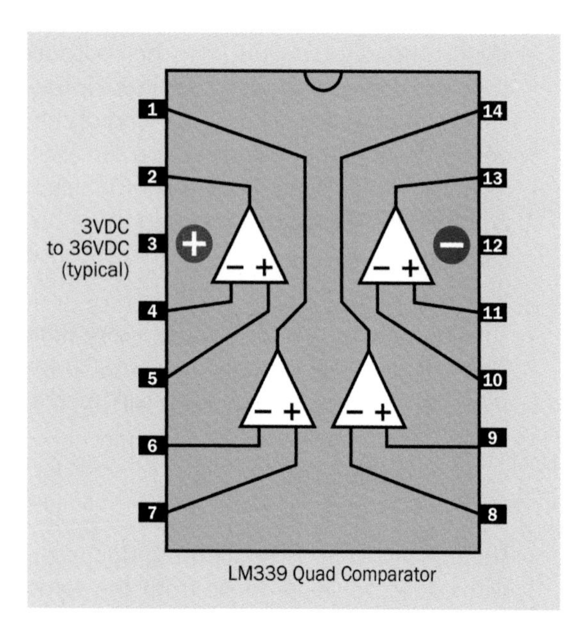

Figure 6-12. *Four comparators are built into the LM339 chip.*

The Circuit Redrawn

The schematic that I showed you in Figure 6-3 was laid out so that it would be as easy as possible for you to transfer to a breadboard. But other people don't make it so easy for you to breadboard a circuit. They tend to put positive voltage at the top, negative ground at the bottom, input on the left, and output on the right. This convention is used because the circuit is easier to understand when you look at it for the first time.

Figure 6-13 is an example of a conventional schematic using a comparator. The components and connections are actually the same as in Figure 6-3.

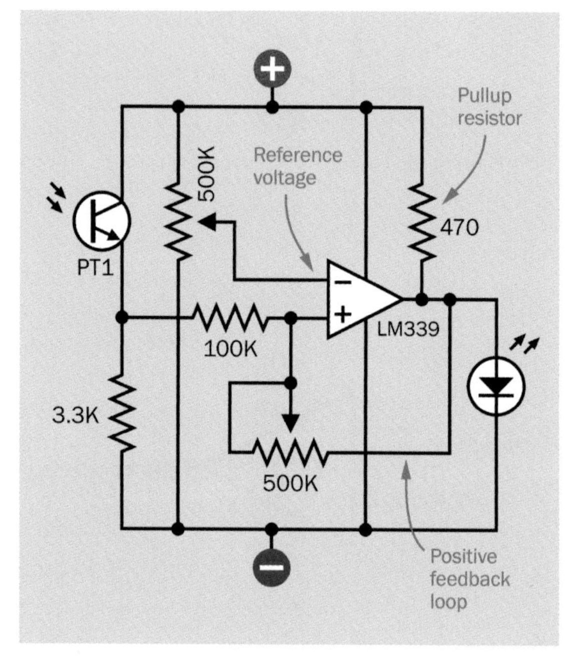

Figure 6-13. *A more typical layout of components in a basic positive-feedback loop using a comparator.*

And the essence of it is shown in Figure 6-14.

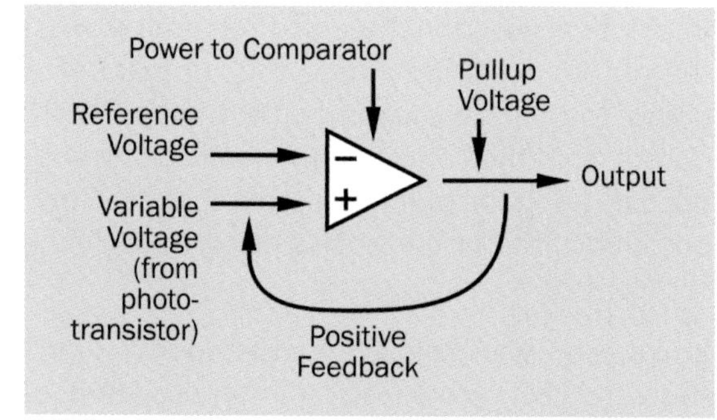

Figure 6-14. *The basic concept of hysteresis in a positive-feedback circuit.*

Warning: Inverted Comparators

In this book, I will always show comparators (and, later, op-amps) with the noninverting, "plus," input below the inverting, "minus," input. In schematics that you may see elsewhere, this is the most common configuration—but it is not universal. Sometimes the person who draws a schematic may feel that it's more convenient to show the noninverting input above the inverting input, because this will allow fewer wiring crossovers, or will enable components to be placed closer to each other.

This can be confusing. You really have to be careful to note the positions of the "plus" and "minus" inputs. If you fail to see that a comparator has been drawn "upside down," your circuit will do the opposite of what you expect.

Comparisons with a Microcontroller

Comparator chips are a bit old-school. People these days tend to reach for a microcontroller when they want to process a variable input and create an on-or-off output.

The hardware in many microcontrollers incorporates one or more analog-digital converters (often abbreviated ADC). Typically each ADC is assigned to a specific pin. It can accept a variable voltage and convert it into an integer—that is, a whole number—usually ranging from 0 to at least 1,000 (in decimal notation).

If you power a phototransistor with 5VDC as I suggested in the experiments here, its output should be compatible with the input of a 5V microcontroller. That sounds nice and simple: just link the phototransistor with the microcontroller. (Actually, you may wish to insert a resistor of 5K or 10K between them, to protect the input of the microcontroller. Because the input of the chip has such a high impedance, the resistor won't drop the voltage at the input pin significantly.)

Your next step would be to expose your phototransistor to a light that is midway between "dim" and "bright," to establish a transition point. Above it, you want the microcontroller to do something. Below it, you want the microcontroller to stop doing something.

While your transitional light is on, you need to find out what number the ADC in the microcontroller is generating. The easiest way to handle this is to hook up the microcontroller with some kind of digital display, and write a little program that will show the number on the display.

Now you can write a new program that contains a conditional statement, telling the microcontroller to start doing something if the light is above the dividing line, and stop doing it if the light is below the dividing line.

So far, this doesn't sound like too much work—although if you change your mind about the transition point, you'll have to rewrite your program and reinstall it in the microcontroller all over again. Clearly, this is more of a hassle than just tweaking a trimmer potentiometer.

But you want hysteresis, don't you? In that case, you have to define two levels of light, to establish the upper and lower levels of the gray zone in which the microcontroller should ignore small variations. Basically the program will tell the chip, "If the light value rises above this upper level, start doing something; if the light value falls below this lower level, stop doing it; and if the light value is between the upper and lower levels, continue as before."

The real problem, once again, occurs if you change your mind and want to make modifications. Suppose you use your microcontroller in a phototransistor-based device to switch an outside light on at sunset, and off at dawn. You need some hysteresis so that the light won't flicker on and off in response to small random variations caused by differing amounts of cloud cover in twilight conditions. How can you figure this out on a workbench? Well, you can't. You'll have to assemble your hardware, move the device to the location where you want to use it, and see how it responds. To adjust the hysteresis, you'll need to use a laptop to install a new version of the program, setting new upper and lower light values.

To me, this doesn't sound like a lot of fun.

Microcontrollers are indispensable in many situations, but sometimes a simple analog circuit built around a chip that costs less than $1 will be a more practical option.

Make Even More: A Laser-Based Security System

You now have all the electronics knowledge you need to create a perimeter defense system using the circuit that you just built, with a cheap laser pen and some mirrors that will reflect the beam around the edges of the protected zone. Instead of using an LED attached to the output from the comparator, you'll attach the output to the base of a 2N2222 transistor that will route power to a coil in a latching relay. The relay will then sound an alarm—or notify you of the intrusion in a more discreet way.

If you want a more elaborate system, you could use multiple lasers and phototransistors so that if one is triggered, you'll know roughly where the intrusion occurred. Remember, there are four comparators inside the LM339 chip, and they can each function independently.

To make the system work well, you'll need to build the phototransistor into an enclosed box with just a small hole where the laser beam can get in. This way, the phototransistor will be protected from ambient light, and the system may work during the day. Even so, you will still need to set the sensitivity of your phototransistor and the extent of the hysteresis. The only way to do this is by trial and error.

How many other applications can you think of, for a phototransistor? If you use your imagination, I'm sure you can come up with many ideas. My favorite is the chronophotonic lamp switcher, which you will find in the very next experiment.

Experiment 7: It's Chronophotonic!

7

This experiment will make use of the information from previous experiments relating to transistors, phototransistors, 555 timers, and comparators. Yes, there was a good reason for laying all that groundwork: you will now be able to utilize that knowledge to build a gadget that has a practical application. Moreover, there will be a bonus: the fun of cracking open a digital alarm clock, figuring out how it works, and repurposing it.

A slightly different and much shorter version of the project was published in *Make* magazine, where I had to omit a lot of explanations for reasons of space. This new version contains several improvements, is much more detailed, should be easier to understand, and will work with a wider variety of clocks.

The objective here is simple enough: to create a device that will switch a lamp on and off in your home when you're not there. Of course you can buy a variety of cheap gadgets to make it look as if you are at home, but for me, they don't do the job right. Where I live, the sun sets two hours later in the summer than in the winter, and if I use a timer, I have to reset it manually several times a year to allow for this.

Really, the lamp should turn itself on by sensing the dimming of the light that occurs at sunset, and a phototransistor connected with a comparator can be adapted to perform this task. You can buy a gadget to do this, but it turns the lamp off after a fixed interval. This seems inappropriate to me. Most people go to sleep on a fairly consistent schedule. They don't turn a lamp off later just because the sun sets later. Therefore, to be realistic, the lamp should turn itself off at the same time every night.

So, here's the specification for my kind of lamp controller: it should use a light sensor to switch the lamp on, and a timer to switch it off. Does such a device exist? Apparently not. Thus I created the chronophotonic lamp controller, because I had no other choice.

Warning: Avoid Dangerous Voltage

This circuit is capable of controlling a lamp of up to 60W powered by house current. If that's what you want to do, I can't stop you from doing it, but I think it's a much better idea to use a 12V LED lamp or 12V halogen lamp. House current at 110VAC or 120VAC really is dangerous. If you're a young reader, please ask for parental advice before messing with it. No matter how old you are, you can always make mistakes, and your life may last longer if you make mistakes with low voltages than with house current.

If you really want to switch house current, a good compromise is to buy something designed for this purpose, such as the PowerSwitch Tail from MakerShed. This requires a 3VDC to 12VDC input that passes through an internal opto-isolator, keeping you (and your breadboarded circuit) safe from higher voltages. You can use the 6V bus voltage in the Chronophotonic Lamp Switcher circuit as the power to be switched by the relay in the circuit, connecting directly to the PowerSwitch Tail. Of course, you will pay extra for this protection.

The Circuit Basics

Figure 7-1 shows a schematic that will get you acquainted with some of the components. The top half is very similar to the comparator circuit from Experiment 6 (in Figure 6-3), with some of the wires rerouted. The principal differences are that the LED attached to the LM339 output has gone, the pullup resistor to the left of the output of the LM339 has been increased from 470Ω to 10K, and the 500K potentiometer providing adjustable positive feedback between the output and the input has been replaced with a fixed 220K resistor, because this will set an appropriate amount of hysteresis here.

After you make these alterations, you'll be ready to add the rest of the components in the circuit. Figure 7-2 shows how a breadboarded version can look.

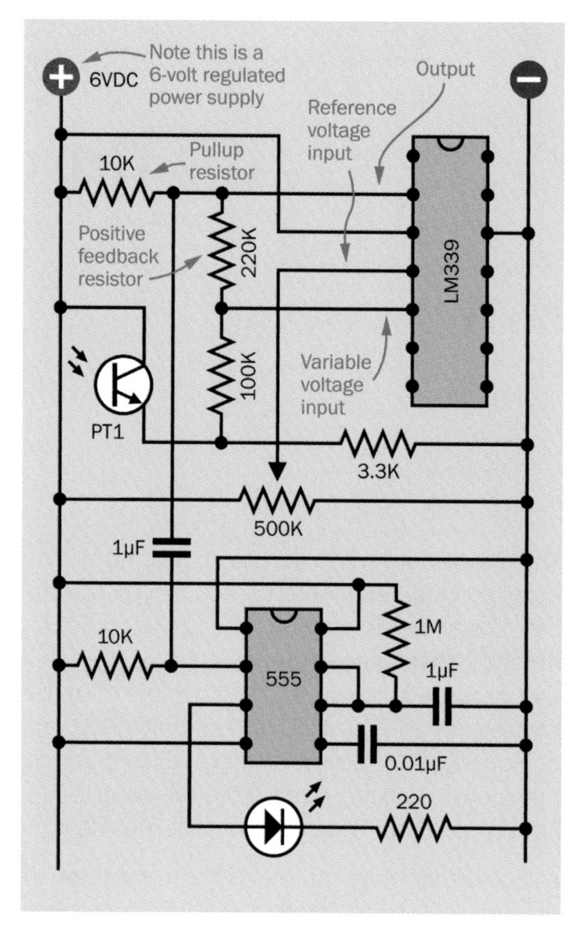

Figure 7-1. *The phototransistor and comparator from Experiment 6 now trigger a 555 timer that emits a one-second pulse. The circuit is now powered by 6VDC.*

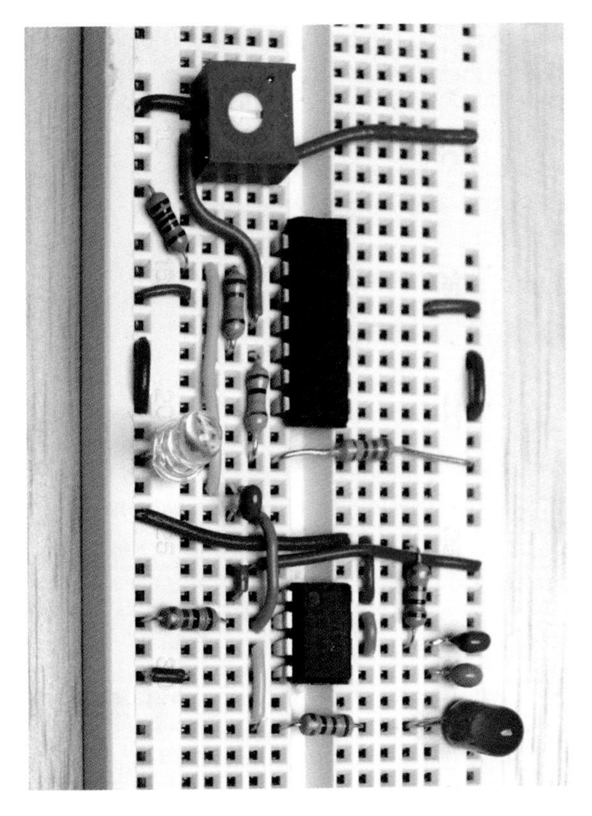

Figure 7-2. *The first section of the chronophotonic lamp switcher.*

- *This is now a 6VDC circuit.* Because I will be adding a transistor-driven relay that has a common ground between its coils, transistors driving the relay must be used in common-collector mode, which entails a significant voltage drop. The 6V supply will compensate for this.

You need to remove the LM7805 voltage regulator from your power supply and substitute an LM7806. This is a simple substitution, because their pinouts are identical. I have not bothered to include the power supply in the schematic, because it's so simple.

The output from pin 2 of the LM339 now runs down the left side of the schematic, through a 1μF capacitor, to the trigger pin of the 555 timer. The timer has its own 10K pullup resistor to keep it normally positive so that its output is normally low. Remember:

- When a 555 timer is wired to run in monostable mode, its output stays low so long as the voltage on its trigger pin remains high.
- When the trigger pin is pulled low, the output goes high, for a duration determined by the capacitor and resistor attached to the timer.

The idea is that the phototransistor causes a change in the output from the LM339 when the light gets dim, and the LM339 triggers the timer, which will emit a pulse lasting for about one second. This in turn will activate a latching relay (not shown yet) that will turn on a lamp. For the time being, I have shown an LED attached to the output of the timer, just so that you can see that it's working.

Apply the power and wait for the timer to reset. Shine a bright light on the phototransistor, then slowly move the light away (or shadow it with your hand) to simulate the dimming that occurs at sunset. You should see the LED pulse for 1 second. Now try adjusting the sensitivity of the phototransistor with the trimmer potentiometer, and repeat. Make sure this circuit works reliably before you continue.

Step Two

The next step in this project is shown in Figure 7-3. The output from the timer now goes through a 1K resistor to the base of a transistor, and the transistor will switch one coil of a 3VDC latching relay. The other coil of the relay is activated by a pushbutton, such as a tactile switch. The 47Ω resistor is necessary to protect the relay from the full voltage of the power supply. This pushbutton will be eliminated in the final version of the circuit, but is useful for demonstration purposes. Likewise, LEDs have been added to the relay outputs, to show their status.

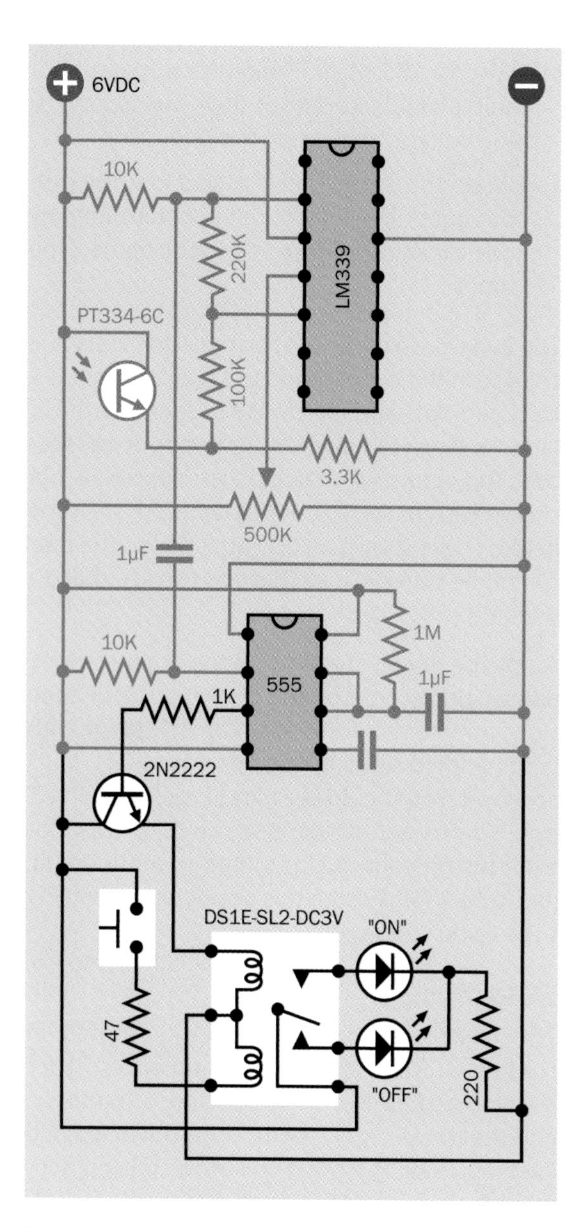

Figure 7-3. *The previous schematic has been extended by adding a relay. Previous wiring is gray.*

You may remember from *Make: Electronics* that a latching relay "sticks" in either of its two states without consuming any power. It just needs a brief pulse to flip it to and fro. Consequently, it's ideal for a circuit that has to switch something on (in this case, a lamp) for long periods, while minimizing power consumption.

You may wonder why we need a transistor between the relay and the output from the timer. Isn't a bipolar 555 timer sufficient to drive a small relay directly? Well, theoretically, yes; but when a relatively low power-supply voltage is being used, the relay can make the timer behave erratically. This is not described in the timer's datasheet, but I've seen it happen.

Circuit Testing

Follow these steps to check out your circuit:

- Press the button, and the lower LED beside the relay should light up. Eventually this setting of the relay will turn an external lamp off, when we connect it.
- Release the button, and gradually dim the light that falls on the phototransistor, simulating the setting of the sun.
- Eventually the relay should click into its other position, causing the upper LED beside the relay to light up. Eventually you will replace this LED with a lamp that comes on at sunset.
- Press the button again. The button will eventually be replaced by a clock that turns the lamp off at a predetermined time.
- Allow the light on the phototransistor to increase gradually, simulating the beginning of a new day. Nothing should happen.
- Dim the light again, and the cycle should repeat.

Relay Details

The relay I have chosen is a Panasonic DS1E-SL2-DC3V that has coils designed for 3VDC, because the output from the transistor will be around 4V, which will not switch a 5V relay reliably. According to the datasheet for the 3VDC relay, it can tolerate a coil voltage as high as 4.8VDC, so it is appropriate for this job.

The relay is shown in Figure 7-3 with the pins positioned as you will use them (seen from above). Check Figure 7-4 if you are uncertain about their function. The numbers beside the pins are the numbers you will find molded into epoxy underneath the relay. If you're wondering why the pins are not just numbered 1 through 6, it's because Panasonic wanted a numbering scheme that would be consistent throughout all of its relays, some of which have as many as 12 pins.

If you substitute a different relay, check its datasheet for the pin locations and functions, as they are not standardized among different manufacturers. You must use a 3VDC dual-coil latching relay, and it should be rated to switch 2A.

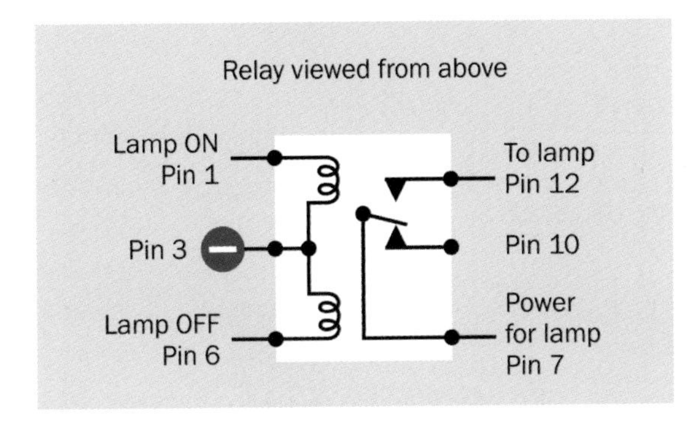

Figure 7-4. *Pinouts of the Panasonic DS1E-SL2-DC3V relay, seen from above. If a substitute relay is used, its pin layout will almost certainly be different.*

Note that the coils inside a relay may not be bidirectional. For the Panasonic relay, you have to apply negative ground in the position shown. If you put the positive supply there instead, it won't work.

The Coupling Capacitor

A key concept in this circuit is the 1µF capacitor between the output from the comparator and the trigger pin of the 555 timer. Remember that a capacitor in this configuration blocks DC, but when the voltage applied to it changes, the capacitor passes the pulse through.

Here's how it works:

- Bright light on the phototransistor creates a high input to the comparator.
- The high input eventually causes the comparator to have a high output, keeping a positive charge on that side of the capacitor.
- The 555 timer has a high input because a 10K pullup resistor maintains it.
- The relay is in its "off" position.
- Nothing happens.

Now when the light on the phototransistor starts to dim:

- The voltage from the phototransistor drops below the reference voltage on the comparator.
- The comparator output flips low.
- The coupling capacitor passes this fluctuation through to the timer, momentarily overwhelming the 10K pullup resistor.
- The timer reacts by emitting a high output pulse that triggers the relay. The relay moves to its "on" position (which can turn on a lamp).
- After that, the coupling capacitor resumes its function of blocking DC.

Make sure that the circuit works. So far, it is only being triggered by photons (which you can think of as being particles of light) and the tactile switch. The next step is to add the "chrono" part of "chronophotonic."

Cracking a Clock

If you wanted to build your own programmable timer, you would buy a timer on a chip, and a numeric display, and some pushbuttons to set the timer—but that sounds expensive and complicated to me. Alternatively you can use a microcontroller with an external clock crystal, but you would still need a numeric display, and the setup is still more complicated than I prefer.

At my local Walmart, Target, or Walgreen's, I can buy a battery-powered digital alarm clock for about $5 that has its display and buttons built in.

Can it be used somehow with the chronophotonic lamp controller circuit? I think it can.

Just make sure that you choose a clock that runs off *two* 1.5V batteries. Be careful about this: Some alarm clocks only use a single 1.5V battery, which will not work in this circuit. Travel clocks often are powered by one 1.5V battery. Read the box carefully!

Warning: No AC-Powered Clocks!

Please *don't* try to adapt a clock that plugs into a wall outlet. Internally the clock probably transforms 110VAC to a safe voltage, but there's a significant risk of making a connection with the higher voltage by error.

Looking Inside

So long as you have a 3V battery-powered clock, the brand and model shouldn't matter, because any digital clock must switch power to a beeper internally, and this switching operation can be tapped for the needs of our circuit.

Your first step is to open the plastic case of your clock. The black clock in Figure 7-5 has four screws underneath (circled), three of them deeply recessed. The white clock in Figure 7-6 has only one screw, hidden inside the battery compartment. The picture shows it being removed with the kind of miniature Phillips screwdriver that you will probably need. They are sold in sets for a couple of dollars at your local hardware store.

Figure 7-5. *All four screws (circled) must be removed to open this clock.*

Figure 7-6. *Only one screw secures the case of this clock, but it is hidden inside the battery compartment.*

Clock Voltage

Once you open the case, your first priority is to check the power. Insert the batteries and then look at the underside of the battery compartment. Three clocks are shown in in Figure 7-7,

Figure 7-8, and Figure 7-9. In each photograph, the tabs labelled A and B deliver +3V and 0V, respectively. Use your meter to check the tabs in your clock.

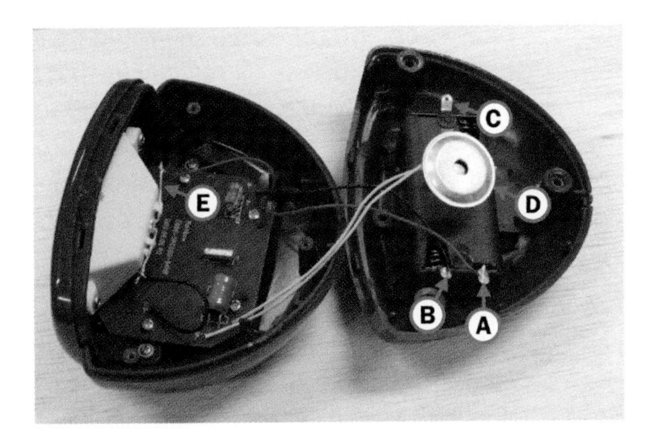

Figure 7-7. *3V battery power is delivered through tabs A and B. Tab C has no connection. D identifies the beeper. E connects with an LED that lights the display when the alarm goes off.*

Figure 7-8. *3V battery power is delivered through tabs A and B. Tab C provides 1.5V for the clock chip. D identifies the beeper.*

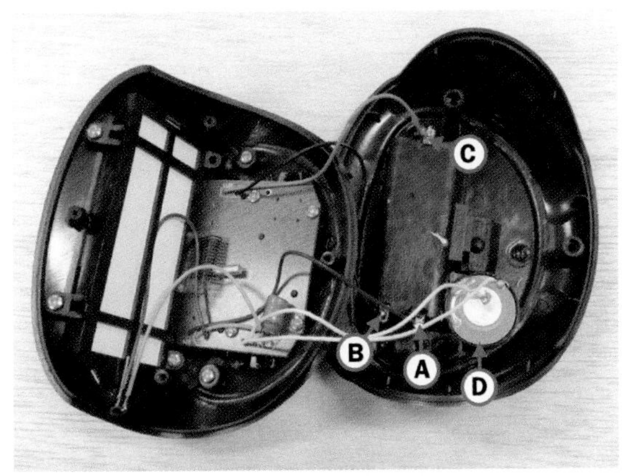

Figure 7-9. *3V battery power is delivered through tabs A and B. Tab C provides 1.5V for the clock chip. D identifies the beeper.*

In the three photographs, the tab labelled C can supply 1.5VDC by tapping into the connection between the batteries. Some clocks don't make use of this feature, while others use it to run chips that are designed for low voltages. It is of no interest to us, because we need the 3VDC that the clocks use to activate their alarm beepers.

The beeper in each clock is labelled D. The red clock in Figure 7-7 also has a wire labelled E, which lights an LED.

Now you need to find out what the clock actually does when it starts beeping. With the batteries in the clock, hold your black meter probe against tab B, the negative side of the power supply. You may find this easier if you use a patch cord with an alligator clip on each end. One alligator grips the tab, while the other grips the black meter probe. This allows you to use both hands for the rest of the procedure. A setup to measure beeper behavior is shown in Figure 7-10.

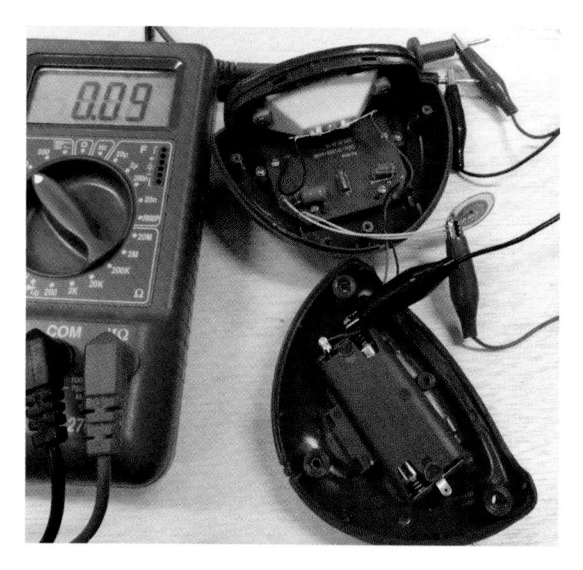

Figure 7-10. *Measuring voltage on the beeper inside a clock, using patch cords with alligator clips to allow hands-free operation. The beeper is the thin, circular object with a red alligator clip gripping one of its solder joints.*

Touch the red probe to tab A, just to check that you have at least 3VDC. Now move the red probe to one of the solder bumps on the back of the beeper, and most likely you will find the same 3V voltage as you measured from the batteries. Try the other solder bump, and it should be the same, too. Because there is full positive voltage on both sides of the beeper, there is no potential across it. This explains why it isn't beeping!

Set the alarm for one minute ahead of the current time, and make sure the alarm switch is on. Your black meter probe must still be securely attached to the negative side of the battery power supply. As soon as the alarm starts beeping, touch your red probe to each solder bump on the beeper again. This time I am betting that one side of the beeper will show an unstable, lower, fluctuating voltage, while the other side stays high. I'm going to refer to the unstable side of the beeper as the "low side."

Reset the meter to measure AC volts, and test the low side again, while the beeper keeps beeping. I think you will see an AC voltage lower than 3V but probably higher than 1V. It will be fluctuating within a narrower range than the DC reading.

How It Beeps

What is happening here? Well, something has to turn the beeper on and off, and that would be a transistor inside the clock. In all the clocks that I have investigated, the transistor is connected to the low side of the beeper (just like an open-collector output in a comparator) and sinks current through it to make it beep. This concept is suggested in Figure 7-11.

You won't be able to see an actual transistor, because it will be embedded in the main chip that controls all the clock functions. It's very likely to be a CMOS transistor rather than the bipolar type illustrated in Figure 7-11, but the principle is still the same. I'll refer to it as the "beeper transistor."

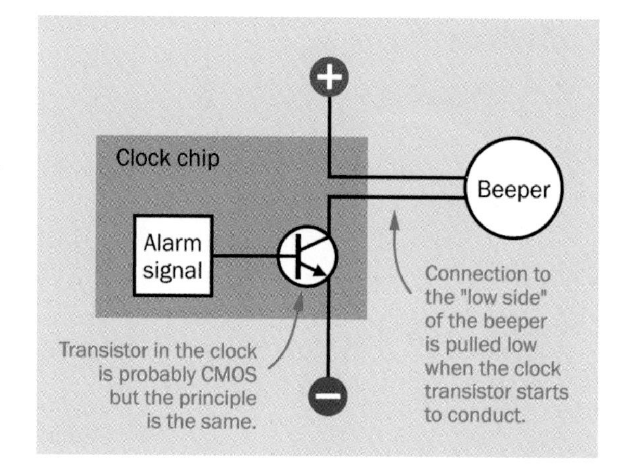

Figure 7-11. *Typical configuration to sound a beeper inside the alarm clock. In reality, a CMOS transistor may be used, but the principle remains the same.*

When the alarm is not going off, the beeper transistor blocks current. The power from the battery has nowhere to go, which is why you could measure full voltage with your meter. You were measuring the voltage either on the high side of the beeper, or on the low side, passing through the beeper.

When the alarm does go off, the transistor sinks power through the beeper, and also sinks power from your meter probe, enabling you to measure a lower voltage on the low side of the beeper. But the voltage didn't just diminish, it also fluctuated; why was that?

You can buy beepers that create their own audible frequency when they receive plain-and-simple DC power. But they are more expensive than beepers that are passive, like loudspeakers. A cheap clock will have a cheap beeper in it, and the chip in the clock will have to do the work of creating an audio frequency. This is a form of alternating current, probably between 1KHz and 2KHz, which is why you obtained a more meaningful reading with your meter set to AC volts.

I'm betting that the voltage fluctuates between a high near 3V and a low near 0V. You didn't see this on your meter, because it cannot react fast enough to show it.

Using the Beeps

How can we use the fluctuating beeper signal? Well, we have four comparators inside the LM339, and we're only using one of them so far, to work with the phototransistor. I'll call it Comparator A. We can use another, which I will call Comparator B, to work with the clock. In response to a signal from the clock, Comparator B will trigger another 555 timer, which will energize the second relay coil, and switch off the lamp.

The only remaining question is the difficult one: How do we connect the clock to Comparator B? The clock uses 3VDC, while the comparator circuit uses 6VDC, and we have to protect the clock from the higher voltage. The way to do this is by using the convenient feature of the comparator that I mentioned previously: the voltage that the comparator controls can be completely different from the voltage that activates the comparator.

Take a look at Figure 7-12 and the breadboard photograph in Figure 7-13. The three white labels at the top of the schematic indicate connections coming via wires that you will attach to the clock's positive power, negative ground, and the low side of the beeper. The signal from the beeper passes through a coupling capacitor to the noninverting input of Comparator B on pin 11 of the LM339. This voltage will be 3VDC or less, and activates the comparator.

Pin 13 is the output pin from Comparator B. It uses 6VDC (passed through a 10K resistor) to trigger a second 555 timer, which has been inserted directly above the relay, and now controls the second relay coil through a bipolar transistor.

Incidentally, when you're wiring this circuit, notice that on the LM339, the pin for the noninverting input that you used previously is *not* directly opposite the pin for the noninverting input on the righthand side. Check the pinouts of the LM339 in Figure 6-12 to make sure you don't get the inputs mixed up. Remember, the "plus" input is the noninverting input.

To make this work, the clock and the breadboard must share a negative ground. All voltages have to be relative to the same ground. But the 3V positive voltage from the clock must be kept separate from components on the breadboard, except for the inputs of the LM339. As previously noted, the voltage associated with the current *passing through* the comparator can be separate from the voltage that is *powering* the comparator.

When you attach a wire to the beeper in the clock, make sure you use its low side—the solder bump where you detected a fluctuating voltage when the beeper made noise.

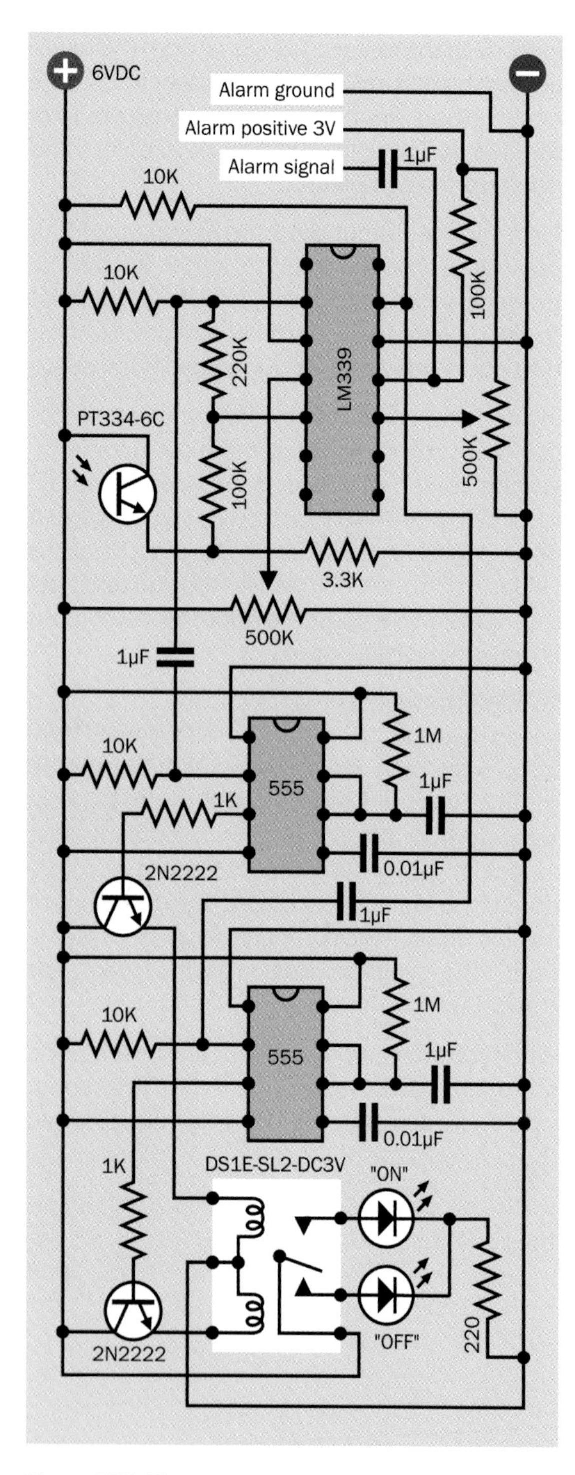

Figure 7-12. *The complete schematic for the chronophotonic lamp switcher.*

Figure 7-13. *Breadboarded version of the final chronophotonic lamp switcher, omitting the alarm clock and the power supply that are necessary. The three colored wires disappearing off the edge of the photograph at top-right will be connected with the clock.*

Adding a wire to that solder bump may damage the beeper with excessive heat—or, equally problematic, you may end up unsoldering the existing wire while you are trying to attach your own wire. Therefore I used wire strippers to open a segment of the existing wire, and I tapped into that. This is shown in Figure 7-14.

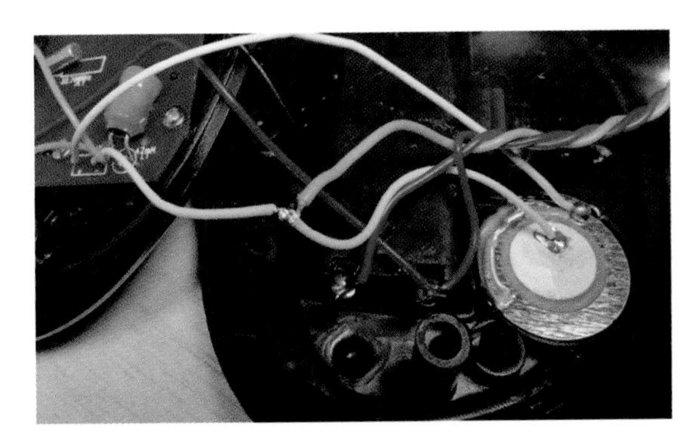

Figure 7-14. *The yellow wire is attached to the existing white wire connecting with the low side of the beeper inside the alarm clock. The blue and red wires have been attached to the battery carrier.*

Eventually you may want to disconnect the beeper, because its noise is irrelevant to the performance of the lamp controller. But for the time being, the noise is useful to notify you of clock activity while you're getting everything to work properly.

Hooking Up the Clock

Here are the precise steps to upgrade the circuit. Remove batteries from the clock while you follow these steps, until you reach step 6:

1. Connect the negative side of the battery compartment in the clock to the negative bus on the breadboard.
2. Connect positive power from the clock's battery compartment to one side of a 500K trimmer potentiometer, which will provide the reference voltage for Comparator B. Connect the other side of the trimmer to the negative bus on the breadboard. Connect the center pin of the trimmer to pin 10 of the LM339—the inverting input, which will receive the reference voltage. Set the trimmer to the midpoint of its range. These connections are shown on the righthand side of the schematic.
3. Connect a wire from the low side of the beeper to a 1µF capacitor on the breadboard. (This is another example of a coupling capacitor.) Connect the other side of the capacitor with pin 11 of the LM339, which is the noninverting input. The capacitor will pass pulses from the clock to the comparator, while blocking DC voltage.
4. Add two pullup resistors to pins 11 and 13. Note that one of them is 100K, and its power source is the 3V from the clock, not the 6V on the breadboard. This is important.
5. Power up your breadboard, and check all the voltages carefully, especially on the wires that lead to the clock. You don't want to burn out your 3V clock with 6V from your breadboard!
6. Insert batteries in your clock and check that there is 3V on the wire carrying power from the clock to the breadboard. Check that the negative ground from the clock is connected with negative ground on the breadboard.
7. Set the alarm for one minute ahead, and wait until it sounds. Your red meter probe should now show a fluctuating output from pin 13 of Comparator B.

This all sounds complicated, but once you get it working, it will be reliable.

The next step is to add your second 555 timer. This is wired to the righthand side of the LM339 in exactly the same way that the first 555 timer was wired to the lefthand side.

How It Ought to Work

While the alarm clock is not sounding its alarm, the positive power from the clock batteries passes through the 100K pullup resistor and keeps the noninverting input of Comparator B at about 3VDC. The impedance of the LM339 is so high, it only draws a few microamps. Then when the alarm goes off, the beeper transistor inside the clock will start oscillating at an audio frequency, sending bursts of pulses through to the noninverting input of the comparator. The comparator will see that during the short intervals between each pulse and the next, the voltage drops below the 1.5V reference that you set with the trimmer on the righthand side of the circuit. Consequently the comparator will trigger the 555 timer, which will activate the relay, which will turn off the lamp.

To a comparator, an audio frequency is quite slow. As soon as the voltage dips below 1.5V just for a fraction of a second, the comparator will pull down the voltage on its output, triggering the 555 timer. The timer, like the comparator, has no problem responding to a rapid input. It will send a 1-second pulse to reset the relay.

As the clock alarm keeps on beeping, it will cause the comparator to retrigger the timer, and the timer will continue to send a high output to the relay—but that doesn't matter. The relay has already moved to its "lamp off" position, and a continuing high input will just tell it to do what it has already done. After a minute or so, the clock will get tired of beeping, and will stop. The circuit will be stable for the rest of the night.

What happens next? Dawn light will wake up the phototransistor, and Comparator A will respond by changing its output from low to high. This will send a positive signal to the first 555 timer, which the timer will ignore, because it already has a steady positive input from its pullup resistor.

During the day, nothing happens. Then sunset arrives, causing a low output from the phototransistor to Comparator A. The comparator's open-collector output now sinks current, which is interpreted as a low pulse to the first 555 timer, momentarily overwhelming its 10K pullup resistor. The timer is triggered, and it sends a pulse to the relay, which turns the lamp on.

Now the lamp will stay on until the alarm clock turns it off. Then the cycle will repeat.

You may be wondering, at this point—is this really going to work? Well, my version has worked (with three different clocks), and I think yours will, too. It doesn't matter what kind of clock you use, so long as it is battery-powered and digital (not some ancient clock with hands that move). Any digital alarm clock must contain a beeper. The voltage on the beeper must change when the alarm goes off, and if you tap into that voltage change, the clock won't know the difference (so long as you connect it with a very high-impedance device that draws hardly any current —such as a comparator).

Perhaps there is a clock out there in which the beeper voltage goes from low to high, and perhaps it will be a DC voltage instead of the rapid cycles that I have talked about. But all digital alarm beepers sound intermittently, so there will be high and low pulses, and the first low pulse will trigger Comparator B.

Testing

To test the circuit, apply power, shade the phototransistor, then expose it to bright light, then darken it again. This should switch the relay into the "lamp on" position. Now set the alarm for one minute ahead, and when the alarm sounds, it should flip the relay to "lamp off" position. If the on or off cycle doesn't work, use your meter to check voltages at every point around the circuit. The key to success is being slow, calm, and persistent!

Once the circuit is working, you can remove the LEDs, which are not needed anymore.

For reliable operation, and to minimize power consumption, it's a good idea to stop the unused inputs of the LM339 chip from "chattering" because they have an undefined state when they are unconnected. Figure 7-15 shows how they can be terminated. One input should have a clearly defined high state, while the other has a clearly defined low state. It doesn't matter which is which.

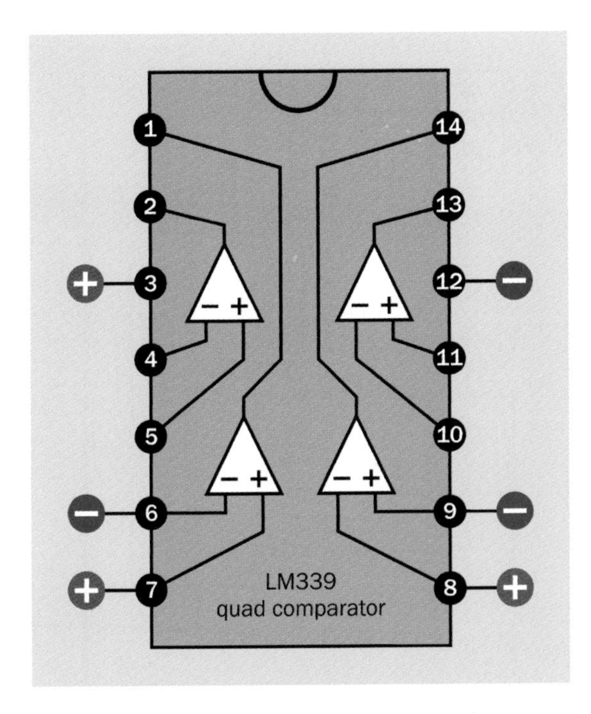

Figure 7-15. *How to deactivate the two unused comparators in the LM339.*

Connecting Relay to Lamp

Disconnect the wire that feeds 6VDC to the bottom-right terminal of the relay. Connect this terminal with one side of a power supply for your lamp, and run a wire to the lamp from the top-right terminal of the relay. The other side of your lamp is connected back to the other side of its power supply. Be very careful to keep the lamp's power supply separate from all other components and conductors on the board. Figure 7-16 shows the circuit.

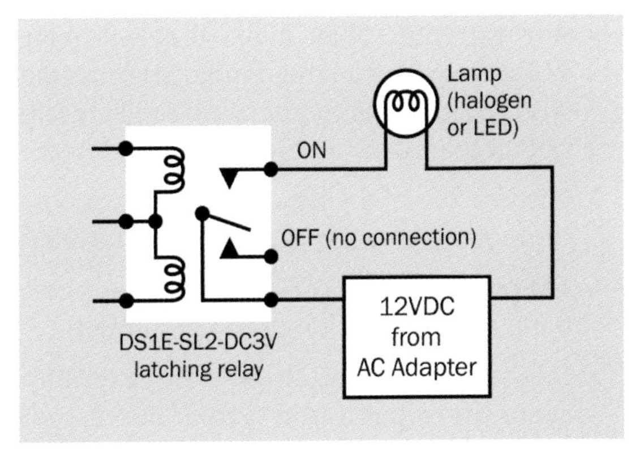

Figure 7-16. *After the circuit has been fully tested, the LED indicators can be removed from the relay, and a lamp can be connected as shown here.*

As noted previously, I strongly suggest that you use a 12V lamp. You should find various types of 12V LED lighting cheaply available, and 12VDC power supplies are easy to find because they have been manufactured in huge numbers for laptop computers. Check eBay for "12V AC adapter."

Once you get the chronophotonic lamp controller to behave properly, you need to decide where to put it. Ideally it should look out through a window facing north. The phototransistor should be protected from direct sunlight and also should not "see" the lamp that it switches.

Wait until the sun is setting, and adjust the left-hand trimmer potentiometer, which sets the reference voltage for the phototransistor. Turn up the trimmer until the lamp comes on, and then back it off just a fraction.

Warning: AC Precautions

If you insist on powering a lamp with AC house current, please take these precautions:

- Make a permanently soldered version of the circuit. Never supply house current to a breadboard, because it's too easy to push a wire into the wrong hole. You don't want components literally blowing up in your face. Also, wires can come loose too easily.

- Any exposed solder joints that will have 110VAC or more on them should be covered with liquid insulation or some similar compound that becomes an insulator as it sets.
- The live side of the power supply must pass through a 1A fuse before it reaches the relay.
- The circuit must be enclosed in a project box. If the box is metal, it must be grounded.
- Don't attempt to switch more than a 60W incandescent bulb, and avoid fluorescent bulbs. They contain a ballast that can draw an initial surge of current. This will be bad for the contacts in your relay.

Make Even More

The circuit has a reasonably low power consumption. My version draws about 11mA overall in standby mode, after removing the LEDs from it. The relay uses about 65mA when it is changing from "on" to "off" or back again, but that only occurs twice a day. Therefore, the lamp switcher can be battery powered—but only on a temporary basis. A 9V battery will last for about 24 hours.

You need an AC adapter to provide long-term power. At the same time, if you live in an area where outages are relatively common, you may want to keep a 9V battery in the circuit for emergency backup.

Figure 7-17 shows how it can be done. So long as there is at least 10V going to the 6V voltage regulator, the 9V battery has no load on it and should remain good for at least a couple of years. (Use an alkaline battery, not a rechargeable battery. Rechargeables don't keep their charge for very long.) The battery might not react well if the AC adapter tries to push current into it, so a diode is put in the way. If the AC power supply fails, the battery takes over, and a second diode prevents it from wasting energy by trying to pass current through the output end of the AC adapter.

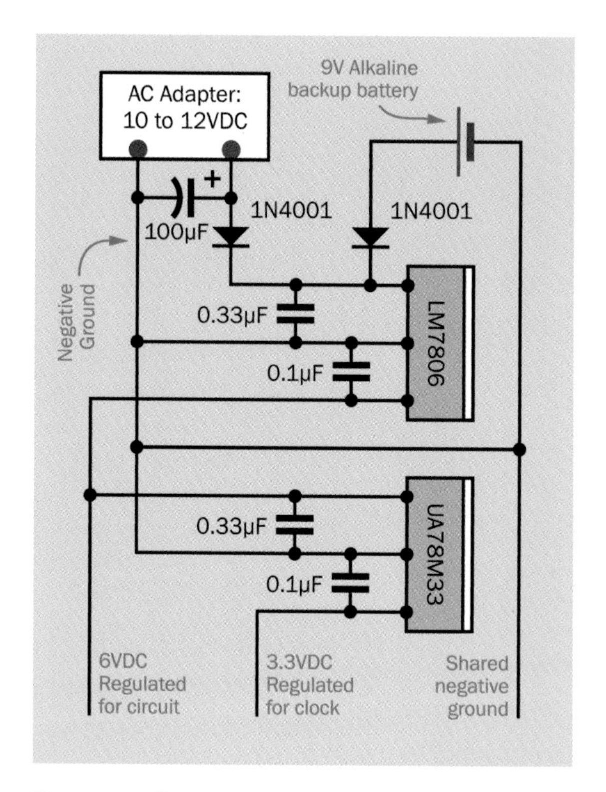

Figure 7-17. *Enhancements to the lamp switcher can include power from an AC adapter, 9V battery backup, and a 3.3V voltage regulator that powers the alarm clock, eliminating the need for its batteries.*

If you buy an AC adapter that provides 12VDC, you may also use it to power a 12V LED or halogen lamp. You should add a 100µF (minimum) capacitor across the output from the adapter, just in case it needs some smoothing.

You can get rid of the clock batteries by adding a 3.3V voltage regulator to your breadboard, as shown in Figure 7-17. Its 3.3V output will be acceptable to the clock, as fresh batteries provide almost this much voltage. The regulator will connect with the wires labelled "Alarm ground" and "Alarm positive 3V" in the schematic. The wires that run to the clock will remain in place, because now they will be sending power to the clock instead of taking power from it.

The input to the 3.3V regulator can come from your existing 6VDC power supply. The ground must be the same, but you must be very careful to keep the output separate from the 6V bus.

Also, you have to include the usual 0.1μF and 0.33μF capacitors to guarantee an accurate output from the regulator. See Figure 7-17 for details.

What's Next?

This was a fairly substantial project. Time now for something a bit more "lite" in nature: the interesting things that you can do with an electret microphone costing less than $1, in conjunction with an op-amp, which functions very similarly to a comparator—although with a different kind of feedback.

Experiment 8: Adventures in Audio

8

It's time to venture into the fascinating world of analog devices. In an analog circuit, voltages can be below zero, as well as above; they can fluctuate in mysterious and unpredictable ways; and the voltage you get at the output can be 100 times the voltage at the input—or more.

Our journey will begin with a microphone and an amplifier. Because of the fickle behavior of components in the analog world, you'll need ways to find out exactly what is happening in a circuit, and this will require a detour into methods of measurement (which is why I included an exercise in transistor measurements in Experiment 2).

After acquiring some necessary knowledge, in Experiments 13 and 14 you can ultimately build an entertaining gadget that fights noise by making more noise. I have to warn you, though, that like any quest into the unknown, this one will involve at least one wrong turn before we find our way through to a successful conclusion.

Amping Up

Near the center of the analog world we find one component: the op-amp. Its name is an abbreviation for "operational amplifier."

The op-amp existed before the comparator. In fact, the comparator evolved from the op-amp, but I showed you the comparator first because its simple high or low output provided an easier introduction.

Both components share the same schematic symbol, because they both work by comparing two inputs. Their purposes are different, though. Comparators are mostly used to get rid of annoying intermediate voltages by using positive feedback. Op-amps usually need to *preserve* every little intermediate fluctuation in the input, and for this purpose, as you'll see, they use negative feedback.

Introducing the Electret

A microphone provides a simple and convenient way to demonstrate the capabilities of an op-amp, so that's where I'm going to start. An *electret* type of microphone is available very cheaply (often, for less than $1), and its performance is sufficiently good that it is used in dozens of consumer-electronics devices, from cell phones to intercoms to gaming headsets.

Why is an electret called an electret? It contains a piece of ELECTRostatically charged film that behaves a bit like a magnET, being permanently charged. Sound waves change the capacitance between the film and another element adjacent to it. A tiny preamplifier built into the microphone senses these changes and generates an output. The output is still very small, which is why we need an op-amp to amplify it some more.

Some electrets have three terminals, but the two-terminal type is more common, and I'll be using it here. One of those terminals has to be connected to negative ground, but at first glance, there's no way to distinguish it from the other. Worse still, the manufacturer's datasheet probably won't tell you which is which. For reasons that remain unknown (to me, anyway), electret documentation is very uninformative compared with the datasheets for most other types of components.

Fortunately, the ground terminal can be identified with just a little detective work. When you look at the underside of an electret, you should find that there is a translucent insulating layer, beneath which you can see some metal fingers reaching out from one of the terminals to the cylindrical shell. This is the negative, or ground, terminal.

Take a look at the undersides of two electret microphones in Figure 8-1. One of them is manufactured with leads attached, while the other just has solder pads for surface-mount applications. In each case, you can see little green metal fingers extending from the righthand terminal, which is the ground terminal.

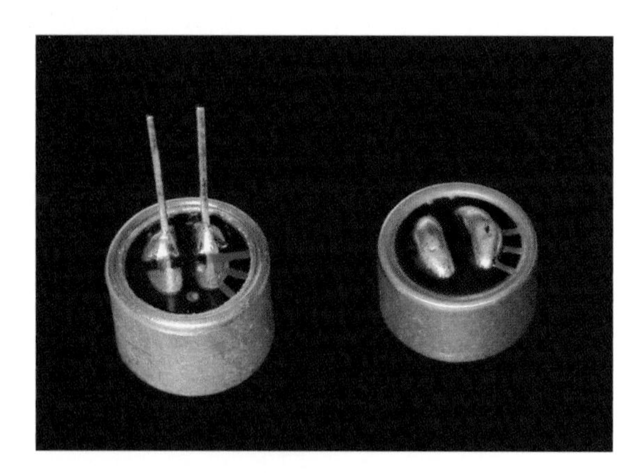

Figure 8-1. *The undersides of two typical electret microphones, one with leads, the other with solder pads for surface-mount. The ground terminal is on the right in each case, as indicated by the metal "fingers" visible through the green translucent insulating layer.*

There are a few electrets that don't resemble the ones in my photographs. They may have larger terminals, or an insulating layer that isn't green. You should still be able to see a silver-colored or gold-contact connection between one terminal and the shell of the microphone, just under the insulating layer.

If your electret doesn't have leads attached, you'll need to solder on your own, so that you can plug it into a breadboard. You can use a couple of pieces of 24-gauge wire, with appropriately colored insulation to remind you which terminal is which.

Figure 8-2 shows the desired result.

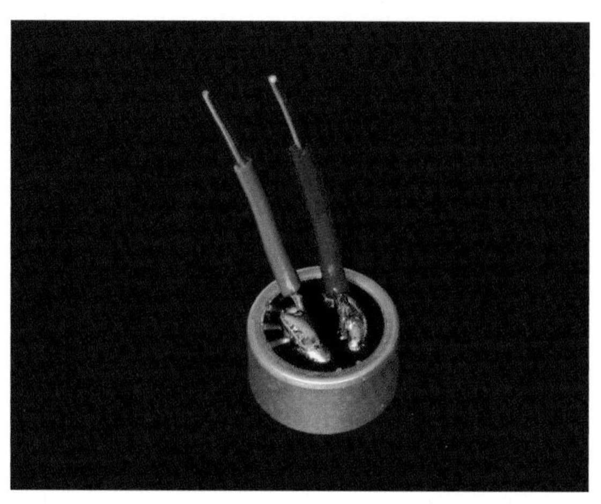

Figure 8-2. *An electret that lacks leads must have short pieces of wire soldered to it for breadboard work, as shown here. The wires should have appropriately colored insulation.*

Like any small component, an electret microphone can be damaged by heat, and if you've added your own leads, you may be wondering if the component has survived its ordeal. So let's find out.

Can You Hear Me?

In a circuit schematic, a microphone can be represented by any of the symbols shown in Figure 8-3. The symbols in the top row can be used for any microphone, while those in the bottom row are used specifically for electrets. The part inside the circle that looks like a capacitor represents the plates that are contained in the electret.

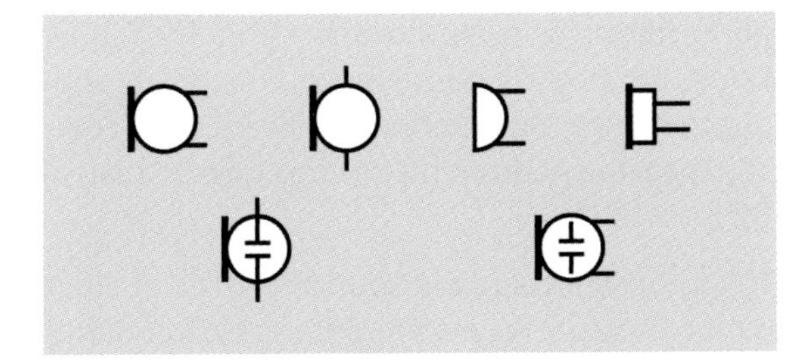

Figure 8-3. *Various schematic symbols are used to represent a microphone. Almost always, sound waves are imagined to enter from the left. The symbol at top-right can be confusing, as it may also represent an earphone or headphone if it faces the other way. The two symbols at the bottom specifically represent electret microphones, but many schematics that require an electret use a generic symbol to represent it.*

I'll use the symbol at bottom-left, as it's slightly more common than the symbol at bottom-right.

In Figure 8-4 you'll find a schematic for the simplest possible microphone test. You'll see that it bears an uncanny resemblance to the test circuit that was shown for a phototransistor in Figure 4-2. This is because both components contain an integrated transistor amplifier with an open-collector output. By the time you reach the end of this book, you'll find that almost every sensing device these days has that kind of output.

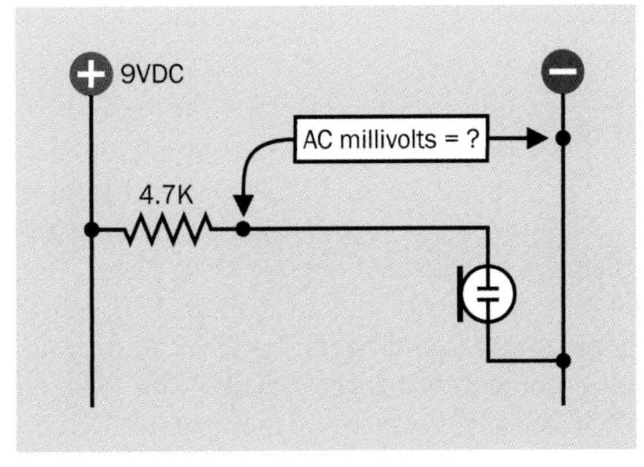

Figure 8-4. *The simplest possible circuit to verify the functionality of an electret microphone.*

Note that I am suggesting a 9VDC power supply. A 9V battery will do the job, and you don't need to add a regulator or smoothing capacitors. I've shown a 4.7K resistor, but you may substitute a resistor with a value as low as 1K. Here again the datasheets for microphones tend to be uninformative, and after you build the audio circuits that I'm going to describe, you can try various resistor values to find out which works best with your electret. By this I mean which value results in the best combination of sound volume and quality.

Install your electret with correct polarity, and set your meter to measure AC voltage. Yes, AC, not DC! Any DC voltage that you measure will not be meaningful.

If your meter does not have autoranging, be sure to select millivolts, not volts.

Apply the meter probes, and after the reading has stabilized, you should see a very small voltage—perhaps 0.1mV. Now make an "Aaah" sound into the microphone, and the voltage should jump up to between 10mV and 20mV. Your electret is listening and responding to you.

- Because microphones are relatively sensitive devices, it's not a good idea to test them by tapping them or blowing on them. You should test them with sound waves, the stimulus for which they are intended.

Background: Microphone Miscellany

The first practical, mass-produced microphone was developed for use in telephones. Patented by Thomas Edison in 1877, it consisted of carbon granules compressed between two plates. One of the plates vibrated in response to sound waves, and each tremor pushed the carbon granules momentarily closer together. This lowered their overall resistance and modulated a DC current passing through them.

Carbon microphones were primitive devices with a very limited frequency response, but they were cheap and rugged, and were still being used in telephones as late as the 1950s (even later, in some countries).

The condenser microphone was an innovation that varied the capacitance of two electrically charged plates in response to sound waves. It functioned similarly to an electret, but required a constant polarizing voltage. "Condenser" was an early term for a capacitor.

Ribbon microphones, such as the early Shure series used by rock artists of the 1950s (including Elvis Presley and James Brown), contained a metallic ribbon that vibrated in response to sound. This design was displaced by moving coil microphones, which function like a loudspeaker or headphone in reverse. A diaphragm causes a coil to vibrate in a magnetic field, inducing current in the coil.

The big challenge in microphone technology has always been to create a mechanical design that responds equally to a wide range of sound frequencies. When the electret microphone was developed at Bell Labs in the 1960s, its performance was limited by available materials. Developments in the 1990s greatly enhanced the component to the point where it now performs almost as well as the old high-end moving coil microphones, but at a fraction of the price.

Ups and Downs of Sound

In Figure 8-4, the electret responds to an external signal by sinking current through an external resistor. As I just mentioned, this is an open-collector system very similar to that used in a phototransistor, but quite apart from the higher values for the pullup resistor and the power supply, there is a much more significant difference. You measure AC from the microphone instead of DC.

This is because audible sound consists of alternating pressure waves, which range in frequency from around 20Hz to 15KHz (although some people claim to be able to hear up to 20KHz). By comparison, a phototransistor responds to light waves, which have such a high frequency, they can be thought of as being a steady source of energy. This is why the phototransistor appeared to create a DC voltage.

The frequency of sound is much lower, and because it induces nerve impulses by vibrating a diaphragm in the ear, we have to preserve the fluctuations to maintain their audibility.

In Figure 8-5, the upper section of the figure shows a person making a sharp sound that travels as a high-pressure wave, shown in white. Because the vocal cords fluctuate to and fro to make the sound, the wave of relatively high pressure is followed by a wave of relatively low pressure, shown in black in the figure.

Because I am talking about "relative" pressure, you may be wondering, "Relative to what?" The answer is, relative to ambient air pressure—the pressure that is all around us. This is shown in gray in the figure.

The lower half of the figure shows the ideal electrical output that should result from the sound input. The voltage varies as a precise imitation of the varying pressure of the sound waves, fluctuating above *and below* a reference level of 0 volts that corresponds with ambient air pressure. This means that our op-amp must accept voltages that are both more-positive and more-negative than the reference level—and indeed, most op-amps are designed to do this.

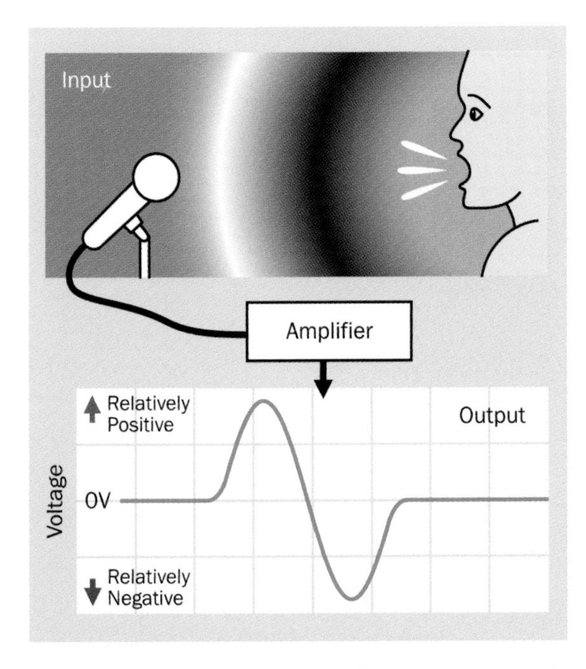

Figure 8-5. *A good amplifier will create an output in which the variation in voltage matches the variation in pressure created by the sound input.*

For this purpose, they often require what is known as a *split power supply*. A typical supply could provide +12VDC, 0V, and −12VDC. For the experiment in this section of the book, we'll need +4.5VDC, 0V, and −4.5VDC. This is an annoying requirement, because most other types of components do not require a split supply, and I'm sure you don't want to buy an entirely separate power supply just to satisfy the needs of an op-amp. Fortunately, there is a workaround that I will use in our op-amp experiments.

The workaround is simple enough in theory, because the high voltages and low voltages are only "relative" to the zero voltage, just as high pressure and low pressure are relative to ambient pressure. So, for example, instead of +4.5VDC, 0V, and −4.5VDC, we could use +9VDC, +4.5VDC, and 0V. Because the differences between high, middle, and low voltages are the same, the components in the circuit won't notice any difference.

But if we only have a 9V battery, how do we create that intermediate +4.5VDC voltage? Figure 8-6 shows the answer. The top section of this figure is what we want, while the middle section shows how we can simulate it with a simple voltage divider, using two resistors of equal value.

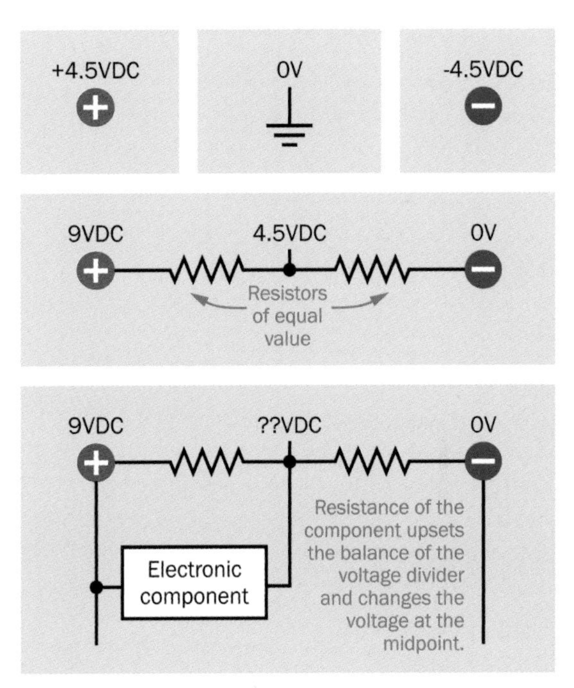

Figure 8-6. *Ideally an op-amp should have a split power supply, with a neutral center reference value that is depicted by a ground symbol in schematics, as shown in the top section of this figure. The split supply can be emulated with a voltage divider, as in the middle section. But the center value will be affected by any component sinking power into it (or drawing current from it), as shown in the bottom section.*

Unfortunately there is a snag to this. If you attach a component between the 9VDC bus and the 4.5VDC midpoint of the voltage divider, the resistance of the component is now in parallel with the left half of the voltage divider. This is shown in the bottom section of Figure 8-6. Now we don't know exactly what the voltage at the central point is, because the component has altered the resistance between the positive power supply and the midpoint of the voltage divider. This will tend to increase the midpoint voltage above 4.5VDC.

The best we can do to deal with this problem is use relatively low resistor values, while any component attached to the midpoint of the voltage divider should have an effective internal resistance that is as high as possible. It will still affect the midpoint voltage to some extent, but the effect will be minimized.

I'll be coming back to this issue in our next experiment.

Experiment 9: From Millivolts to Volts

9

In the previous experiment, you determined that your electret was working. Now we can begin to make it do something useful.

Putting a Cap on It

The first step is to install a coupling capacitor, as shown in Figure 9-1. If you remember the basics, you already know that it will block DC voltage while allowing a pulse to pass through. In fact, depending on the size of the capacitor, it will be transparent to many small voltage fluctuations —such as those in an AC audio signal.

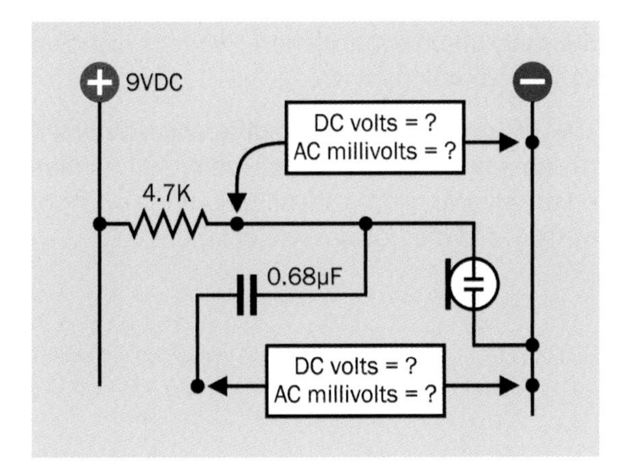

Figure 9-1. *This simple test with a multimeter demonstrates that the capacitor blocks DC voltage while passing an AC audio signal.*

First try measuring DC volts (relative to negative ground) above and below the capacitor. You should find that at the upper measurement point, you have almost the full 9VDC from your battery. At the lower point, you only have a tiny fraction of a volt, because the capacitor is blocking the DC voltage. (If the capacitor is doing its job well, you should not be able to measure any voltage at all.)

Now if you reset your meter to measure AC millivolts, and do your "Aaah" test again with the microphone, you should get almost the same reading from above and below the capacitor. The lessons are simple but important:

- The capacitor removes the 9VDC from the signal.
- The capacitor allows the AC signal from the microphone to pass through.

We do not want our op-amp to amplify the DC voltage. But we do want it to amplify the AC signal from the microphone. Therefore, the capacitor is just what we need to couple the microphone with an amplifier.

If you're wondering why the capacitor value happens to be 0.68μF, that's a more difficult question. A larger-value capacitor generally should work better in this kind of application, but bigger capacitors tend to cost more than smaller capacitors, and a smaller capacitor should filter out some high frequencies, which may be desirable. You can try various capacitor values, repeat the "Aaah" test, and see if you measure any differences.

Introducing the Op-Amp

Time, now, to amplify the microphone signal. A bipolar transistor is not the right tool for this job, because it will amplify current rather than voltage. We need a voltage amplifier. You already saw in Experiment 6 that a comparator will amplify voltage, because its power supply does not have to be the same as the voltage applied to its open-collector output. An op-amp can function in a similar way. It can convert the plus-or-minus 20 millivolts from your microphone into an output of plus-or-minus 2 to 3 volts.

I'm going to use the LM741 as our op-amp. It's one of the oldest chips around but is still manufactured and used in large quantities because it's cheap, easily obtainable, and does the job. The pinouts are shown in Figure 9-2, and you'll see that unlike the LM339, which contained four comparators, the LM741 contains just one op-amp. As I noted before, the symbol for an op-amp is the same as the symbol for a comparator, because they both work by comparing two inputs. If you are looking at a schematic and wondering whether the triangular symbol is identifying an op-amp or a comparator, the part number and the accompanying text should make that clear.

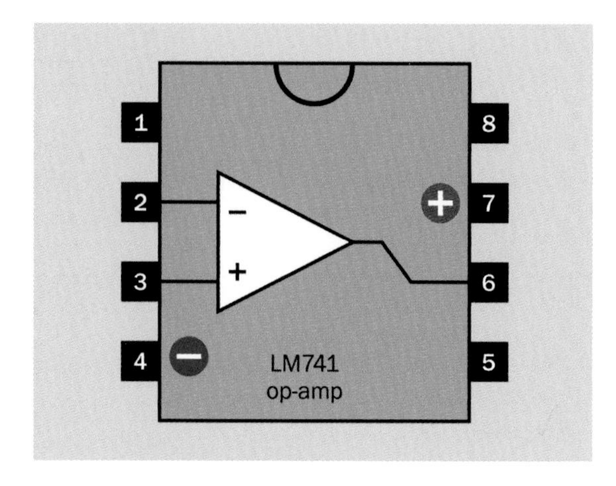

Figure 9-2. *The primary pin functions of an LM741 op-amp. Pins 1 and 5 are included for calibration, but are not often used. Pin 8 has no internal connection.*

What's the Difference?

Here's the plan. I'm going to apply a 4.5VDC voltage to the inverting input of the op-amp (the one with the minus sign) by using a voltage divider. This will be my reference voltage. I'm going to apply 4.5VDC separately to the noninverting input (the one with a plus sign) using another voltage divider, and will add the signal from the microphone (through the coupling capacitor) to the noninverting input. This will make the voltage on the noninverting input fluctuate above and below the 4.5VDC level, as illustrated in the top half of Figure 9-3, where the input signal is the wavy green line and the 4.5 voltage is shown as a horizontal black line.

The op-amp will amplify the differences between the noninverting input and the reference voltage on the inverting input, ideally creating an output as shown in the bottom half of Figure 9-3.

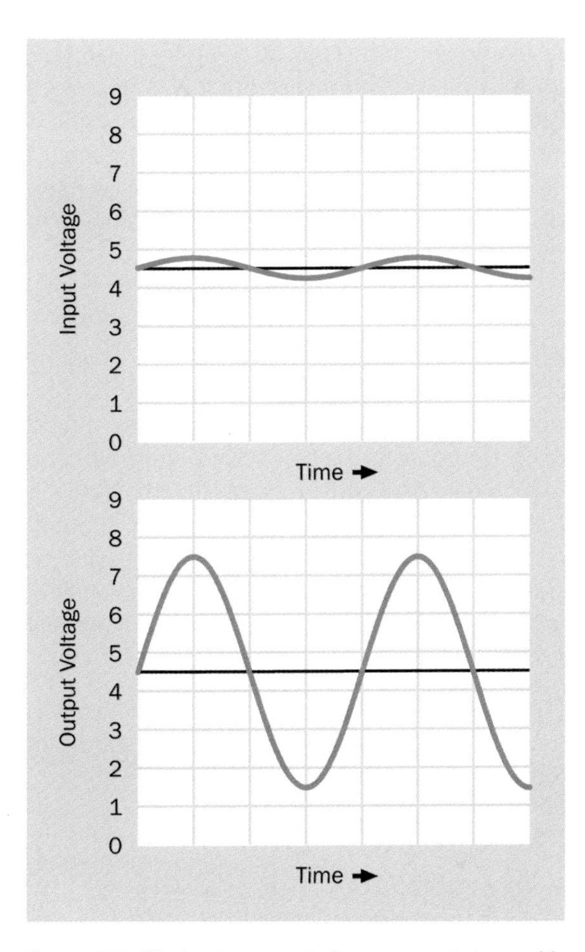

Figure 9-3. *The basic concept of an op-amp is to amplify the differences between an input signal and a reference voltage (4.5VDC in this example). The input is shown here in green, the output in orange, and the reference voltage is a horizontal black line. The variations shown by the green line have been exaggerated to be visible.*

For this to work, I need to apply the same basic reference voltage to both of the inputs of the op-amp—although one will have the alternating voltage from the microphone added to it. This means I need two separate supplies of the same voltage. I can use two separate voltage dividers, but to make sure they deliver the same voltage, the two resistors in each divider must be accurately matched with each other.

Figure 9-4 shows how the circuit will look. It can be built as an add-on to the simple electret test circuit that you created previously.

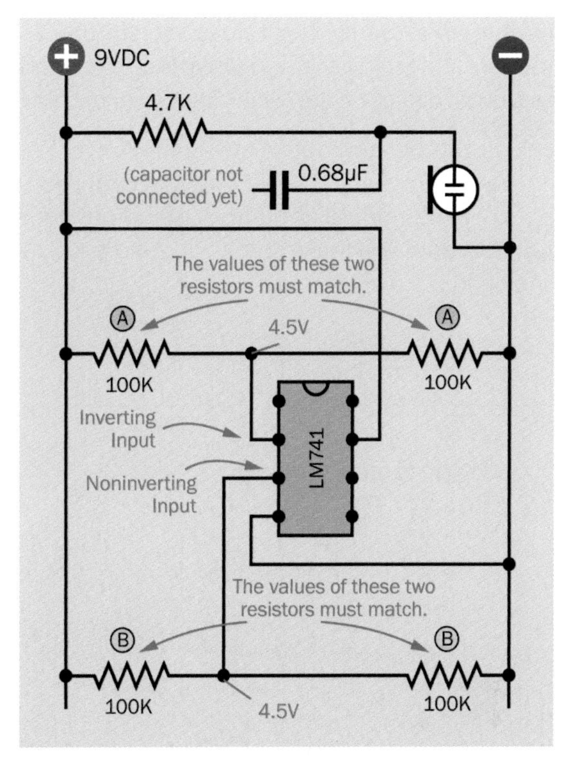

Figure 9-4. *Setting up two voltage dividers to test the op-amp. If your battery does not provide exactly 9VDC, the voltage at the midpoint of the A divider will not be exactly 4.5VDC. It should be half of the actual battery voltage. (This circuit is intended to clarify the way the op-amp functions. It is not typical of real-world applications.)*

The microphone isn't connected yet. I'll get to that in the next step. First we have to deal with the problem of matching the resistors.

A Perfect Pair

Because of manufacturing imperfections, the actual values of resistors will vary. If they have a 5% tolerance, 100K resistors can have actual values as low as 95K and as high as 105K. Even if they are manufactured to a 1% tolerance, they can range between 99K and 101K.

To deal with this, you will have to use your meter to find a "matched pair" of resistors. Here's how you do it.

Measure the values of, say, ten 100K resistors. Select two that have identical values, at least within the limits of accuracy of your meter. It doesn't matter what the resistor values are, so long as they are almost the same as each other. Use them as the "A" pair in Figure 9-4.

Select another two that have identical values, and use them as the "B" pair. To avoid confusion, try to be methodical about this, laying out your resistors until you find a couple of good pairs, as shown in Figure 9-5.

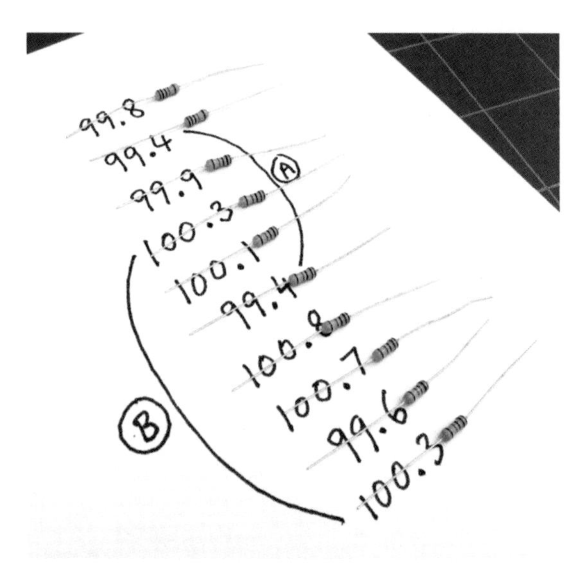

Figure 9-5. *Finding two matching pairs of 100K resistors.*

Notice that each "A" resistor does not have to be the same as each "B" resistor. The "A" resistors just have to have the same resistance as each other, and the same is true of the "B" resistors.

- When measuring resistance, don't press the probes of your meter against the leads of a resistor while holding them in your fingers. The resistance of your skin will give you an erroneous value. Lay each resistor on an insulator such as a dry piece of plastic, paper, cardboard, or wood before you press the probes against its leads. Alternatively, consider setting up the resistor-testing mini-breadboard that I described in the introduction to this book. See Figure S-16.

Are you wondering if you will have to go through this time-consuming process of selecting resistors every time you use an op-amp? No, definitely not! Matching the resistors is necessary only in this test circuit, because I will be using it later to make some accurate measurements of op-amp performance. Also, having two voltage dividers makes it easy to see what's going on in the circuit.

Measuring the Output

After you create the circuit in Figure 9-4, test the voltages on the inverting and noninverting inputs, relative to negative ground. They should both be equal to half of your battery voltage (which may be slightly more or less than 9VDC). If there is a tiny difference, don't worry about it. If there is a significant difference (such as one input being 4.7VDC and the other being 4.4VDC), you didn't select your resistor pairs carefully enough.

Now you're ready to complete the circuit by wiring the microphone into it. This is shown in Figure 9-6.

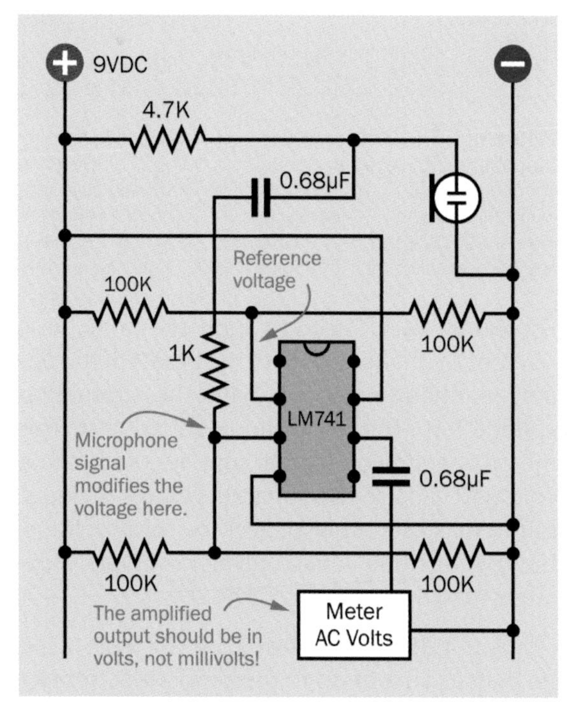

Figure 9-6. *Complete circuit for verifying op-amp performance with an electret microphone.*

A photograph of the breadboarded version appears in Figure 9-7.

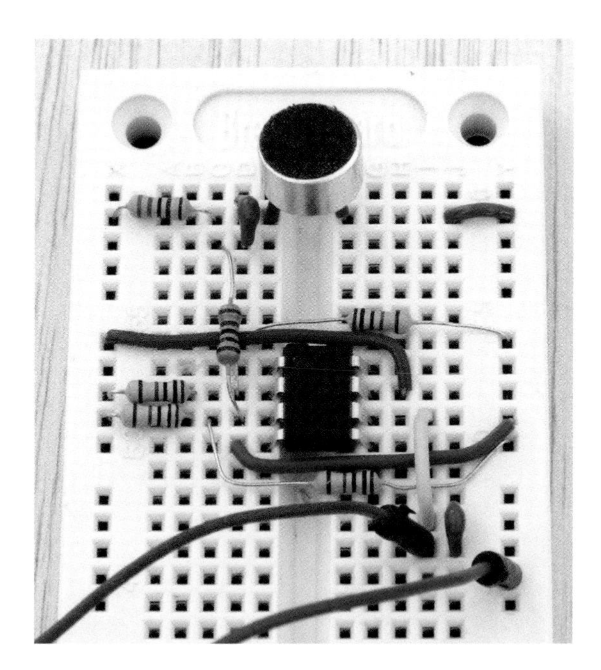

Figure 9-7. *A circuit to assess the amplified output from an electret microphone. The red and blue jumper wires below the circuit are connected with a meter set to measure AC volts. The power supply for the buses of the breadboard is a 9V battery (not shown here).*

The microphone section has been linked with the noninverting input of the op-amp through the coupling capacitor and a 1K resistor. The output from the op-amp passes through another coupling capacitor, and you measure it with your meter set to AC volts—not millivolts, anymore, because the voltage will be amplified.

- Note that unlike the LM339 comparator, which has an open-collector output that requires a pullup resistor, the LM741 has a "real" output capable of delivering a small amount of current. No pullup resistor is needed.

If you attached your meter between the output pin of the op-amp and the center point of the "A" voltage comparator, you'd be measuring deviations above and below the reference voltage. But after the op-amp output has passed through another coupling capacitor, the DC component of the signal is blocked, and you can now measure the signal with reference to 0V ground.

Make an "Aaah" sound into the microphone, and sustain it to give your meter time to respond and stabilize. You should find that an input of around 20mV from the microphone creates an output of greater than 2V. The op-amp is increasing the voltage by a factor of more than 100:1. This is known as its *gain*.

Now, how can we use this amplified output? We can use it in many ways, beginning with the very next experiment.

Experiment 10: From Sound to Light

10

You can now create a noise-activated LED. Figure 10-1 shows the same circuit as before, with just five components added. A photograph is included in Figure 10-2.

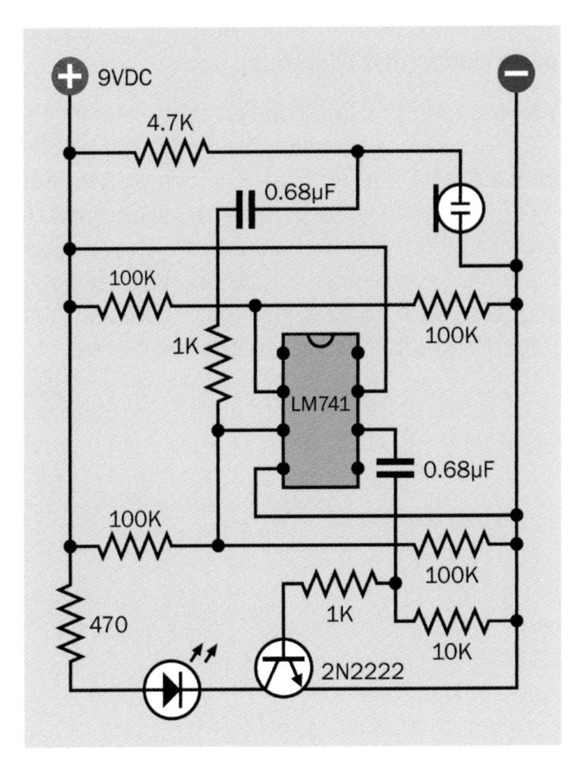

Figure 10-1. *This op-amp test circuit will flash the LED whenever the electret microphone picks up sound of moderate intensity.*

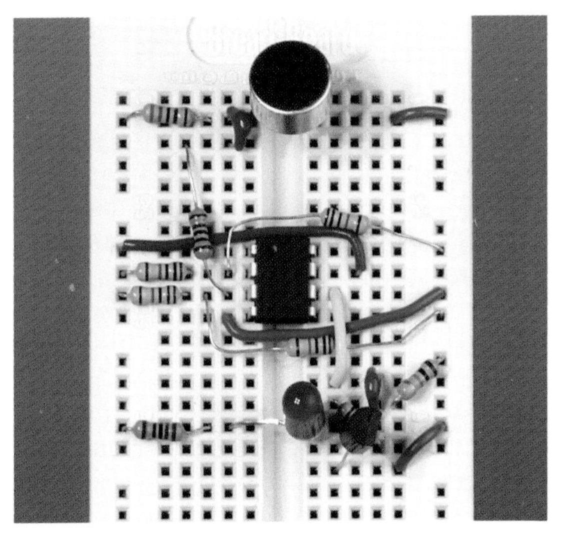

Figure 10-2. *The circuit with a noise-activated LED, breadboarded (9V battery is not shown).*

An LED-Transistor Combination

After the output from the LM741 passes through the coupling capacitor, it reaches the midpoint between a 1K resistor and a 10K resistor. Consequently, voltage fluctuations created by the op-amp are now relative to ground. They pass through the 1K resistor to the base of that plain-and-simple workhorse, a 2N2222 transistor. The transistor responds by amplifying current through an LED.

The circuit now flashes the LED in sync with your voice when you talk into the microphone. For some reason I find this magical. Maybe I'm just easy to please! If you don't share my sense of wonder about such simple things, rest assured we have only just started on a mission toward a much more ambitious circuit.

If the circuit doesn't perform exactly as I just described, here is a list of possible issues:

The LED won't light up
: This is almost certainly the result of a wiring error. Check everything slowly and carefully, and apply your meter to each stage of the circuit, remembering to test for AC as well as DC.

The LED stays on all the time
: This is unlikely to happen, but variations in components can have unexpected results. The particular 2N2222 transistor that you use, and the particular LED, may affect the behavior of the circuit slightly. If your LED suffers from the always-on problem, most likely the voltage at the base of the transistor is just high enough, even without a signal from the op-amp, to enable some current to pass from the collector to the emitter. Substitute a higher-value resistor instead of the 1K resistor at the base, and this problem should be resolved.

The LED flashes rhythmically
: This kind of oscillation can be a problem in op-amp circuits. When the LED is bright, it draws a little more current, which can pull down the voltage from a 9V battery. This affects the voltages in the voltage dividers. A lower voltage difference causes the LED to dim, at which point it draws less current from the battery—and the cycle repeats. This is most likely to occur when you make a relatively low noise into the microphone. It's less likely to occur if you power the circuit with an AC adapter that has a more stable DC output than a battery.

Before we move on, there's one other thing for you to try. Remove the LED and substitute a very small loudspeaker. Hold the loudspeaker up to your ear. Now when you speak into the microphone, you should just be able to hear a faint, really horrible scratchy version of your voice—although it is likely to suffer from the same kind of oscillation that I described above.

You may need a loudspeaker with a relatively high impedance to make this work. I got results with a 2" 63Ω speaker. If you try to make it louder by reducing the value of the 470Ω resistor, you'll probably find that the sound from the speaker becomes even more distorted. Why does it sound so bad? To fix it, we're going to need the negative feedback that I mentioned earlier.

Experiment 11: The Need for Negativity

11

Now that you've seen that an op-amp can amplify, I want to address two questions:

1. How can we determine how much it is amplifying?
2. How can the output become a more accurate copy of the input, so that if we listen to it through a loudspeaker, the noise won't sound scratchy?

In this experiment I'm going to guide you through the process of answering the first question. I'll deal with the second question in Experiment 12.

Messing with Measurement

In an ideal world, measuring the amplifying power of an op-amp would be a no-brainer. You'd have a signal generator to create a steady sinewave input. You'd also have an oscilloscope, which would display the sine wave on its screen. You would check its amplitude visually against a scale on the screen, then check the output. To calculate the amplification factor, you would divide the output amplitude by the input amplitude. Simple!

(The "amplitude" of a signal basically describes how big it is, although this is not an entirely simple matter when dealing with a complex AC waveform. It can be the maximum voltage of each pulse, or the average voltage, or the root-mean-square voltage. I'll leave you to search for that last one, if you're interested.)

Unfortunately, most people do not possess a signal generator or an oscilloscope. Can you measure the performance of your op-amp using nothing more than a multimeter? I think you can, although it will not be so easy, because the meter won't make accurate measurements of a signal produced by a microphone when you say "Aaah" into it.

The answer to this problem is to forget about AC for a moment. If we remove the capacitors that were installed to block DC, the op-amp can amplify a steady DC voltage difference, and you can use your meter to measure that.

Or can you? Here's another problem: merely touching your meter probe to one of the inputs of the op-amp can cause a tiny change in voltage, and this will be amplified to become a bigger change, along with the signal. As I mentioned in Experiment 2, measuring affects measurement (see Chapter 2). The process of trying to discover the voltage will change the voltage.

Fortunately there are ways around this, and I think the process of solving the problems will be interesting and instructive in itself. So let's get started.

DC Amplification

At the risk of causing inconvenience, I'm going to ask you to remove and set aside some of the components that you installed previously. The new circuit that I'd like you to build is shown in Figure 11-1. Please make sure that there are no leftover components from the old circuit that will cause confusion.

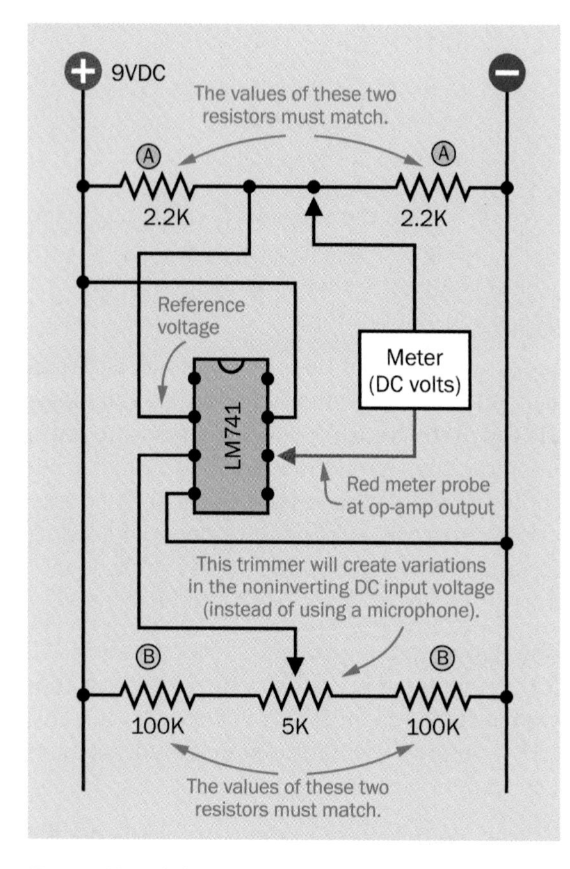

Figure 11-1. *A basic circuit for measuring the performance of an op-amp.*

Notice that you must now use two 2.2K resistors in your "A" voltage divider. This is because we're going to be sinking a little current into this divider, so I want the "A" resistors to have a relatively low value. (In the previous experiment, you could use 100K resistors because the impedance of the op-amp inputs is extremely high, and was applied equally to both voltage dividers.)

You'll need to go through the same procedure for matching two 2.2K resistors as you used for matching two 100K resistors in the previous experiment. I promise that this will be the last time you'll need to deal with this little task.

The microphone and the coupling capacitors have been removed, because we're only interested in DC signal amplification in this experiment. But without a microphone, how are we going to create a difference between the two inputs of the op-amp, to give it something to amplify?

A 5K trimmer is the answer. Notice that it is inserted between the two "B" resistors. By adjusting the trimmer, you change the balance of the voltage divider and vary the voltage to the noninverting input. When you turn the trimmer to and fro, you can think of it as being like a very, very slow microphone input—slow enough that your meter can keep up with it.

- Reduce the lengths of resistor wires to a minimum in this circuit so that they don't pick up stray electromagnetic fields that will be amplified by the op-amp. Keep all your resistors snug against the breadboard. Remember, an op-amp will amplify electronic noise as well as any signal that you put into it.

After you have set up the circuit, you touch the red probe on your meter to the op-amp output, while the black probe touches the connection between the "A" resistors. Remember to reset your meter to measure DC volts!

The Ins and Outs of Amplification

You are now measuring the difference between the op-amp output and 4.5VDC. Therefore, if the signal from the op-amp is relatively low, you should see a negative relative voltage. Fortunately, almost all digital meters can display a negative voltage as easily as a positive voltage, but you have to be watchful for the minus sign on your meter display.

You can begin using this circuit just by playing with it, varying the value of the trimmer slowly while observing the voltage on your meter. I'm betting that the voltage will jump from negative to positive, suddenly, halfway through the trimmer's range. Why does this happen?

The upper graph in Figure 11-2 is derived from my own actual readings in this circuit. The meter measured the output in volts, but I have converted them to millivolts so that the units in the input graph are the same as the units in the output graph.

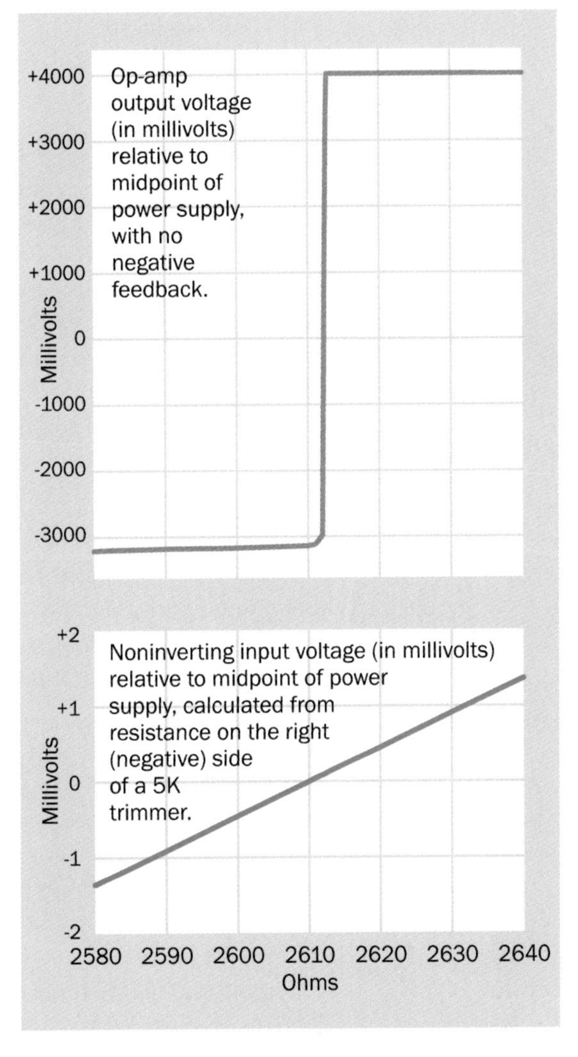

Figure 11-2. *The lower graph shows a calculated linear progression of voltage on the noninverting input of an op-amp. The upper graph shows the output, relative to a voltage midway between the two sides of the op-amp's power supply.*

As you can see, the output jumps from one extreme to the other. This is because the op-amp is applying an extreme amount of amplification. As the trimmer nears the midpoint of its resistance, the tiniest change in voltage creates a huge reaction.

If you replaced the 5K trimmer with perhaps a 5-ohm trimmer (if you could find one!) you might be able to flatten the gradient of the output a bit, but still the shape of it probably wouldn't match the shape of the input very well, and the extreme amplification of the circuit would tend to amplify a lot of electrical noise. This didn't matter when all you wanted was to see an LED light up, but it's not useful for the faithful reproduction of an audio signal. We need the output to be exactly the same shape as the input—in other words, they should have a linear relationship.

The way we achieve this is with *negative feedback*. While positive feedback was useful to clean up the output of a comparator, we need negative feedback to exert some discipline on the output from an op-amp.

Take a look at the graph in Figure 11-3, where a linear input causes an almost precisely linear output. This is exactly what we want and is surprisingly easy to achieve. To see it for yourself, make the following revisions to your circuit:

- Remove the jumper that connects the two "A" resistors with the op-amp's inverting input.
- Add two more resistors, identified as "F" and "G" in Figure 11-4. The photograph in Figure 11-5 shows how this can be breadboarded.

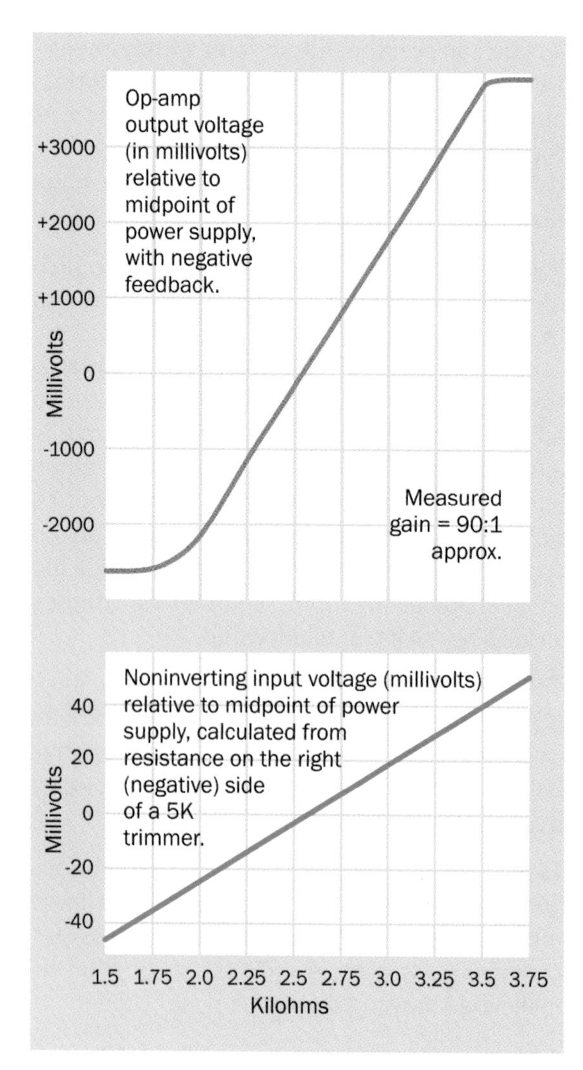

Figure 11-3. *The central area of the green input curve is now reproduced correctly in the output curve, as a result of negative feedback.*

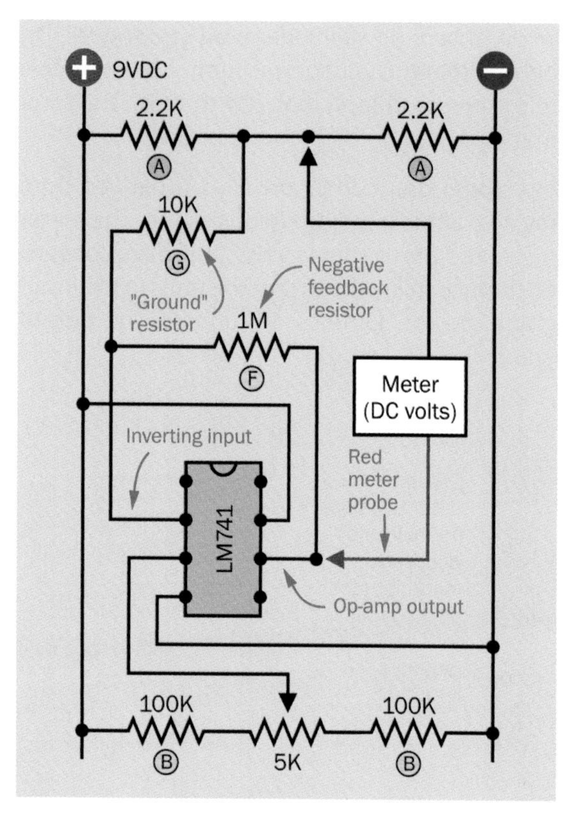

Figure 11-4. *The previous schematic has been modified to introduce negative feedback, by adding two new resistors.*

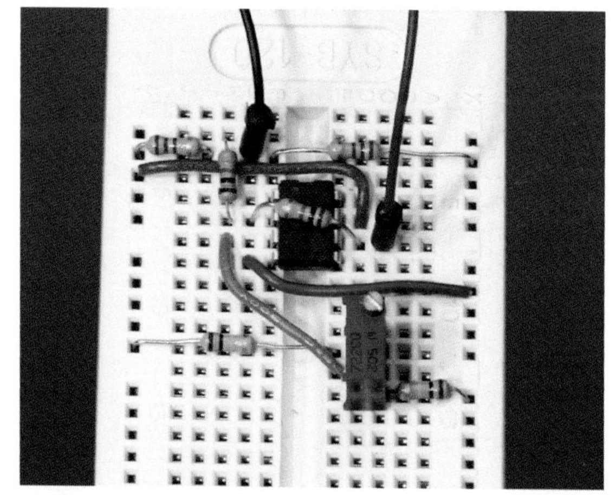

Figure 11-5. *The previous schematic to adjust and measure negative feedback, breadboarded. The red and black wires are connected with a meter that measures voltage.*

Resistor F is a 1M negative-feedback resistor, which taps into the output from the op-amp and feeds it back to the inverting input. Here's the concept:

- When you were using a comparator, you achieved *positive* feedback by routing some of its output back to the *noninverting* input.
- Now you are using an op-amp, you are achieving *negative* feedback by routing the output back to the *inverting* input.

Resistor G is a 10K "grounding resistor." It doesn't actually go to negative ground, just to the midpoint between the "A" resistors. This is because the op-amp is still using that midpoint as the reference voltage.

Take care to make the changes in the circuit. Now when you turn the trimmer, you should find that the meter shows the output voltage increasing very smoothly. It doesn't jump from one extreme to the other anymore. In Figure 11-3 there are some tiny deviations in the orange line (which may or may not be visible in the version reproduced here), but I think these are just caused by imperfections in the measuring process. On a breadboard, every socket has some small resistance, and merely wiggling the components will change the measurements that you make, at least to a small degree.

So how does negative feedback work?

Electronic Ritalin

An op-amp is capable of applying a huge amplification ratio—as much as 100,000:1. But negative feedback reduces the ratio like this:

- If the noninverting input is a little more positive than the inverting input, the op-amp output goes up.
- Some of the output feeds back to the inverting input. This reduces the difference between the two inputs.
- The reduced difference between the inputs causes a reduction in the op-amp output.

In other words, when the op-amp overreacts, negative feedback calms it down.

What if the noninverting input has a fraction less voltage than the inverting input? In that case, the output swings negative—and some of that feeds back to the inverting input, bringing it down so that, once again, the difference between the two inputs is reduced.

There is an additional factor, here, which is the grounding resistor, labelled G in Figure 11-4. This diverts some of the negative feedback to the midpoint of the "A" voltage divider. In other words:

- The negative feedback stops the output of the op-amp from getting out of hand.
- The grounding resistor stops the negative feedback from getting out of hand.

Gain

The term "gain" is usually used to mean the same thing as "amplification ratio," as it's less of a mouthful.

It's possible to prove mathematically that the gain of the amplifier can be derived from an utterly simple formula, using the values of the "F" resistor and the "G" resistor:

```
Gain = 1 + ( F / G )
```

In case this isn't entirely clear, take a look at Figure 11-6, where I have redrawn just the pieces of the the feedback circuit that are relevant. You now see that resistors F and G are really just another voltage divider.

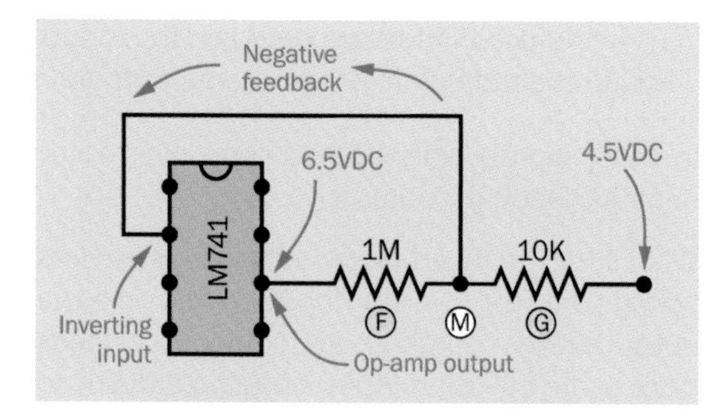

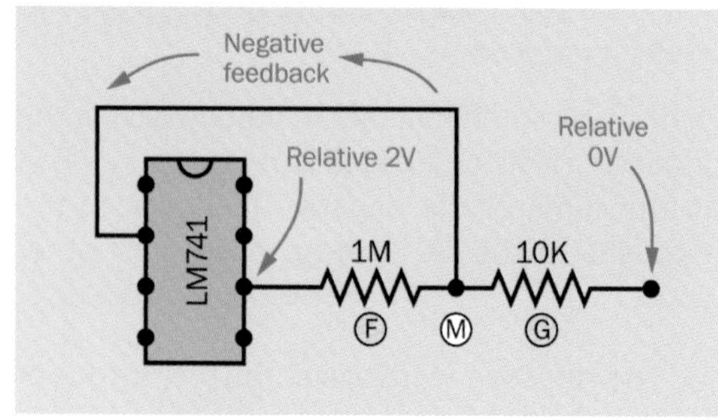

Figure 11-6. *A section of the op-amp test circuit has been redrawn to clarify the function of the two resistors that control negative feedback.*

In the top half of the figure, I've shown the actual voltages as you might measure them. In the bottom half of the figure, I've shown these voltages converted so that they will be easier to use in the calculation that follows. If the output from the op-amp is 6.5VDC and the other end of the "G" resistor is 4.5VDC, this is the same, relatively speaking, as if the output is 2VDC and the other end of the "G" resistor is 0VDC.

Do you remember the formula for calculating the voltage at the midpoint of a voltage divider? Turn back to "What About the Voltage?" on page 12 and you'll find it, although I'm using F and G, here, instead of R1 and R2, to identify the resistors. If V_M is the voltage at the midpoint:

```
VM = V * (G / ( F + G ) )
```

I am using letter V to mean the voltage from the op-amp output, at the left end of the pair of resistors. I can't call it V_{CC} (as I did in the previous version of the formula) because we only use V_{CC} to mean the full power-supply voltage. Here, V just means the voltage at the left end, relative to the voltage at the right end.

If I plug in the actual values from Figure 11-6 (expressing them in kilohms), I get this:

```
VM = 2 * (10 / ( 1,000 + 10) )
```

This is about 0.02 volts.

But now suppose I change the feedback resistor from 1M to 100K. The formula looks like this:

```
VM = (2 * (10 / 100 + 10) )
```

This is about 0.2 volts.

In other words, when I reduced the value of the feedback resistor (keeping the "ground" resistor the same), the negative feedback went up by a factor of about 10, and the gain of the amplifier went down by a similar amount.

Now imagine that instead of reducing the value of the "F" negative feedback resistor, I increase it. In fact, suppose it becomes almost infinitely high. In that case, the negative feedback voltage becomes close to 0. That was the situation when the feedback network wasn't installed at all, and the only connection between the output and the input was empty air. This is why the op-amp behaved in such an extreme manner: it had no negative feedback at all.

Here's the general rule:

- When the feedback resistor decreases in value relative to the ground resistor, the negative feedback increases, and the gain of the op-amp decreases.
- When the feedback resistor increases in value relative to the ground resistor, the negative feedback decreases, and the gain of the op-amp increases.

Background: Negative Origins

Negative feedback sounds like a simple idea, but it was extremely radical when it was developed at Bell Labs in the 1930s. In fact, initially the patent office was reluctant to issue a patent that seemed to have no application. Amplifiers are supposed to amplify, so why would you want a system that makes an amplifier amplify less? In fact, as you have seen, there's a very good reason. It is a simple way to control the output and force it to match the shape of the input.

The concept of negative feedback was developed for amplifiers before the actual op-amp existed. In fact, the op-amp wasn't even named until 1947, at which point it was used in analog computers to perform mathematical operations (which was how it came to be named as an "operational amplifier").

So long as vacuum tubes were used, the multiple components required in an op-amp took up a lot of space and generated a lot of heat. Op-amps were not fully developed and did not become widely used until the 1960s, when the advent of the integrated circuit chip made them cheap and practical.

Pushing the Limits

Looking back at Figure 11-3, the orange line levels off at each end. This seems to suggest that its smooth linear performance breaks down, and indeed it does, but for a very good reason. The output voltage range of the op-amp cannot be greater than the voltage being delivered between the high end and the low end of the power supply. In fact the maximum and minimum output voltages will always be slightly less than the range between the supply voltages, because the op-amp has to siphon off a little power for itself to perform its magic. Consequently, if you increase the input beyond a certain point, you don't get a higher output voltage. When this happens with an audio signal, you hear distortion.

In Figure 11-7 the upper graph shows an input signal, in green, with the output signal, in orange, superimposed. The height of the orange arrow, divided by the height of the green arrow, tells you how much gain we're getting. This particular op-amp seems to have a lot of negative feedback, so it only has a gain of around 6:1. What happens if we reduce the negative feedback? Now the input signal is not subjected to so much control, so it gets a bit bigger. The op-amp tries to amplify it, but the output reaches the maximum voltage before the signal reaches its peak. Consequently, the peak is chopped off.

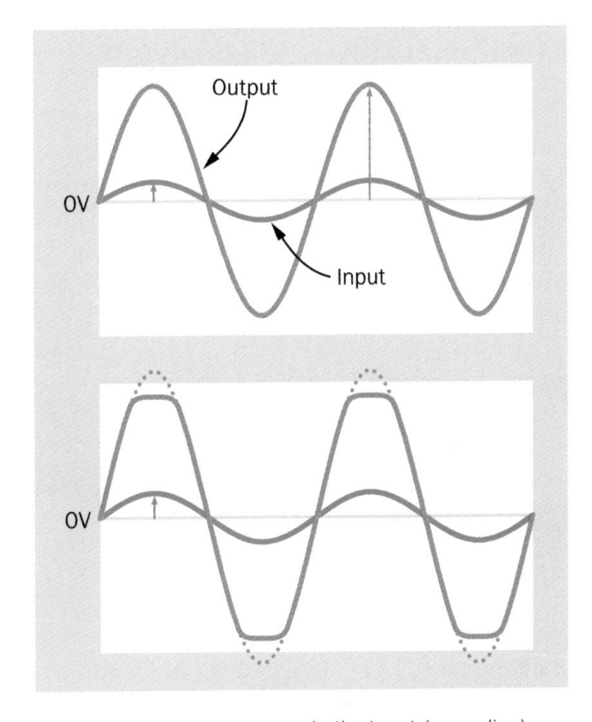

Figure 11-7. *In the upper graph, the input (green line) creates an amplified output (orange line) that just reaches the limit of the amplifier. In the lower graph, the amplitude of the input has increased, but because the output cannot exceed the limit allowed by the power supply, the amplifier clips the output. Dotted lines show where an accurately amplified reproduction of the input would be if there was enough voltage to create it.*

In this situation, the op-amp starts to behave like a transistor that saturates. It still amplifies the weaker parts of the signal, but when it gets to the peaks, it gives up. We say that the op-amp is being *overdriven*, and the result is known as *clipping*, because the peaks of the signal are clipped off. (I mentioned this briefly in *Make: Electronics*.) If it's an audio signal, it acquires a harsh, jagged sound, like a guitar being played through a fuzz box. In fact a fuzz box works by overdriving an amplifier.

No Pain, No Gain!

Now I want to get back to the topic of measuring the gain of an op-amp. Remember:

- The gain is the ratio of the output voltage to the input voltage.

Figure 11-8 shows some input voltages and output voltages that I measured myself, for various settings of the trimmer potentiometer in the circuit in Figure 11-4. These were the numbers that I used to draw the graphs in Figure 11-3. Then I used the graphs to calculate the gain of my op-amp.

Resistance on negative (right) side of trimmer in ohms	Op-amp output relative to "A" voltage divider in millivolts	Noninverting input relative to "A" voltage divider in millivolts
1500	-2630	-47.3
1750	-2590	-36.3
2000	-2150	-25.2
2250	-1140	-14.3
2500	-160	-3.3
2750	+810	+7.7
3000	+1790	+18.7
3250	+2800	+29.7
3500	+3810	+40.7
3750	+3900	+51.7

Figure 11-8. *These measured values were used to draw the previous graph showing op-amp performance with negative feedback.*

You can do the same thing with your op-amp. I'll take you through the process step by step. There will be four phases:

Phase 1
: You'll make measurements to build your version of the first two columns of the table in Figure 11-8.

Phase 2
: You'll calculate your version of the third column of the table. Because you can't actually measure the small input voltage to the op-amp without disturbing it, calculating the values is the alternative. Nothing more difficult than basic arithmetic will be necessary.

Phase 3
: Using the numbers in your table, you will draw two graphs, like the ones that I drew.

Phase 4
: You'll obtain the gain of the op-amp by comparing the slopes of the graphs.

The whole process will take about fifteen minutes. Are you ready? Let's find out what your op-amp is really doing.

Phase 1: Output Voltages

Touching the probes of your meter to the output side of the op-amp won't disturb it significantly, so you can measure the output voltage directly. Remember, though, you have to determine the change in resistance of the trimmer that corresponds with each change in the output. Here's how:

Step 1
: Pull the trimmer out of the circuit and measure the resistance between its wiper (the center terminal) and the terminal that connects with the righthand 100K "B" resistor, on the negative side. (See Figure 11-4.) Adjust the trimmer until you measure 1.5K (1,500 ohms).

  Don't touch the metal ends of the leads while you make the measurement. This is easier if you use a couple of alligator patch cords. The alligator clip at one end of a patch cord holds a meter probe, and the alligator clip at the other end of that patch cord attaches to a terminal of the trimmer.

Step 2
: Write down the resistance of the trimmer that you just measured.

Step 3

Put the trimmer back into the breadboard, keeping it the same way around as before.

Step 4

Measure the voltage of the op-amp output by resetting your meter to DC volts and positioning the probes, as shown in Figure 11-4. Make sure you have the red probe and the black probe in the positions shown. Take care to observe if there is a minus sign before the digits.

Step 5

Write down the output value, but change it from volts to millivolts by multiplying by 1,000—that is, moving the decimal point three spaces to the right. You're going to compare the output voltage with the input voltage later, so they must both be in the same units. For instance, if your measurement was –3.5V, you would write down –3500 millivolts.

Step 6

Pull out the trimmer and adjust it while measuring the resistance of its right (negative) side until it has increased by 250 ohms. Go back to step 2, above, and repeat.

Each time you change the trimmer resistance, you must increase it by exactly 250 ohms. Continue making measurements until the trimmer resistance reaches 3,750 ohms (3.75K). It's important that your trimmer values range from 1.5K to 3.75K, so that you can compare your table directly with mine.

Phase 2: Input Voltages

You can calculate the input voltage from the resistances and the supply voltage. This would all seem very easy if I could demonstrate it for you with a meter on a lab bench. I can't, so the next-best way is to show you in a diagram, which I have drawn in Figure 11-9. This is a piece of the circuit from Figure 11-4 that contains the "B" voltage divider.

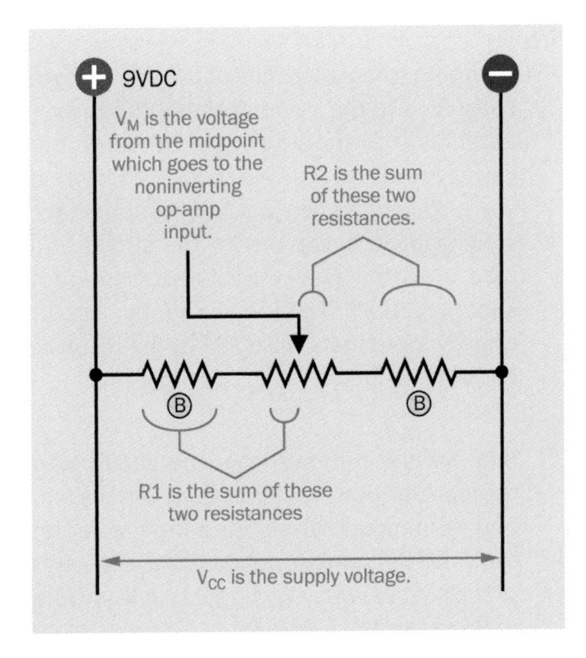

Figure 11-9. *The voltage to the noninverting input of the op-amp can be calculated if you know the values shown above.*

What we want to know is V_M, the voltage at the midpoint. This is the same as the voltage that is applied to the noninverting input of the op-amp. We're going to get it from the three values in the diagram identified as V_{CC}, R1, and R2, because R1 and R2 form a voltage divider. Notice that R1 is the value of the lefthand resistor *plus* the resistance of the lefthand segment of the trimmer potentiometer, while R2 is the value of the righthand resistor *plus* the resistance of the righthand segment of the trimmer potentiometer.

Here is the procedure:

Step 1

Measure V_{CC}. Set your meter to measure volts, and measure the voltage of the power supply between the positive bus of the circuit and the negative bus. If your 9V battery is reasonably fresh, you should get a value of at least 9.2V. Whatever it is, make a note of it. I'm calling it V_{CC}, because that's the abbreviation most commonly used for a power-supply voltage.

Step 2

I'm going to call the value of the lefthand resistor RL and the value of the righthand resistor RR. They should both be the same, because you chose them to be the same when you installed them in the circuit. But to make sure, pull each one out of the circuit and measure it now. Note the values in ohms (not kilohms). They should be slightly less or slightly more than 100,000. Put the resistors back in the circuit.

Step 3

Remove the trimmer from the circuit, and measure its total resistance between the two end terminals, in ohms, ignoring the center terminal. Remember not to touch the leads or the probes with your fingers while you are making the measurement.

My 5K trimmer turned out to have a resistance of 5,220 ohms (according to my meter). Yours may be slightly more or slightly less. The value doesn't matter so long as you know what it is. I'm going to call this number RT (Resistance of Trimmer).

Step 4

To find R2, you add RR (the real exact value of your 100K resistor) to the resistance between the center terminal and the righthand terminal of the trimmer. This resistance varied while you were performing the previous part of the experiment because you kept readjusting it. But initially you used 1.5K, or 1,500 ohms, so we'll start with that, in which case:

```
R2 = RR + 1500
```

Step 5

To find the resistance of the lefthand segment of the trimmer, we just subtract the resistance of the righthand segment from RT. In the illustration above, the righthand segment had a resistance of 1,500. So:

```
R1 = RL + RT - 1500
```

Step 6

Now we can apply the same old formula to find the voltage at the center of a voltage divider:

$$V_M = V_{CC} * (R2 / (R1 + R2))$$

The variables have different meanings from before because I am applying them to the "B" voltage divider, where V_M is the noninverting input voltage to the op-amp. Remember, V_{CC} is the power-supply voltage, which you measured in Step 1. You established R2 in step 4 and R1 in Step 5. So, you just plug these values into the formula. I can't do it for you because I don't know the exact resistances of your 100K resistor and your 5K trimmer. But you measured them yourself, so, you know what they are.

Step 7

V_M is what the noninverting input voltage must have been when your trimmer was set to 1.5K. But wait—the op-amp was not amplifying that voltage. It was amplifying the *difference* between that voltage and the reference voltage. Well, what was the reference voltage (at the midpoint of the "A" divider)? It should have been exactly half of the power supply. So, to find the difference between the op-amp inputs (which I will call V_I), you have to finish up like this:

$$V_I = (V_{CC} / 2) - V_M$$

Just divide V_{CC} by 2, then subtract V_M, and that's the difference between the two inputs. It is properly called the *voltage differential*, and should be a negative number, so remember to include the minus sign. Write it in the third column of your table, on the first line. In my copy of the table, I entered the value −47.3mV. Is yours something like that?

All this calculation seems a bit tiresome—but, you only have to do it once more. Because you moved your trimmer in precisely equal steps of 250 ohms when you were making measurements in Phase 1, you can be fairly sure that the input voltage also increased in equal steps. In other words, the input voltage must have increased in a straight line. Therefore, you only need to calculate the lowest and the highest input voltages, after which you will draw a straight line between them.

So, now, go back to Step 4 and recalculate a new value for R2, like this:

```
R2 = RR + 3750
```

And in Step 5, calculate a new value for R1 like this:

```
R1 = RL + RT - 3750
```

Now use the new values for R1 and R2 in Step 6, find the new value for V_I in Step 7, and write it in the third column of your table, on the last line.

Phase 3: Graphing It

I asked you to graph the beta value of your transistor in Experiment 2 so that you would have a bit of practice before reaching this point in the book. Remember, you can get free graph paper online by searching for a term such as "print graph paper."

Mark the horizontal scale in thousands of ohms, and the vertical scale in thousands of millivolts, to make your version of the top half of the graph that I drew in Figure 11-3. Use the numbers from the second column of your table to draw your graph.

Now make your version of my graph in the bottom half of Figure 11-3, using numbers from the third column of your table.

Phase 4: The Gain

You need to compare the slopes of your two graphs. This only makes sense where the graph lines are reasonably straight, and you'll remember from Figure 11-3 that the output graph is curved at each end. So, I'll choose a center section. But this is important: choose the same range of resistance values in your input graph.

The segments that I took from my graphs are shown in Figure 11-10. You'll see that for each segment, the resistance range goes from 2.25K to 3.25K. You can use a different range, so long as it is the same for the output graph and the input graph.

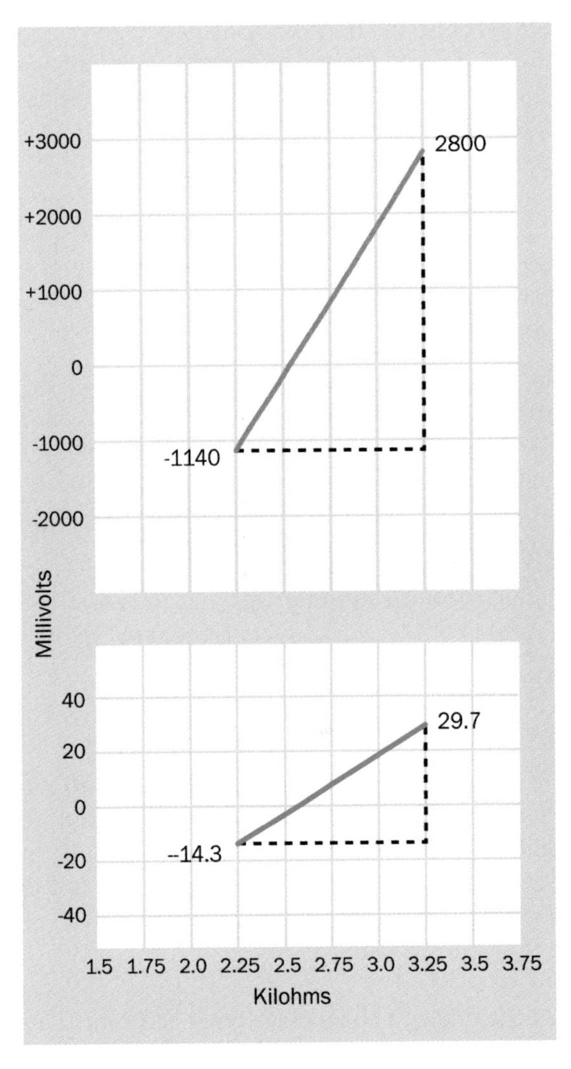

Figure 11-10. *Gradients of input and output voltages. See text for details.*

The slope (which I will call S) of a straight-line graph can be defined as the vertical increase (V) divided by the horizontal increase (H):

```
S = V / H
```

So, you can calculate the slope of the output graph (call it S1), and divide it by the slope of the input graph (S2), to get the gain of the op-amp:

```
Gain = S1 / S2
```

If you calculate the two slopes of your graph lines and divide one by the other, you'll get the answer. But because the horizontal increase is the same in each case, they cancel out, and the simple version of the formula looks like this, where V1 is the vertical increase in your output graph and V2 is the vertical increase in your input graph:

```
Gain = V1 / V2
```

So long as V1 and V2 are both measured in the same units, this will work. (You remember I asked you to write down all your values in millivolts?)

In Figure 11-10, my range of V1 goes from -1,140 to +2,800. Remember, the first number means "1,140 millivolts below the midpoint voltage," while the second number means "2,800 millivolts above the midpoint." Therefore the total increase was 2,800 + 1,140, which is 3,940mV.

Similarly for V2, the range was -14.3 to +29.7. That is a total of 14.3 + 29.7, which is 44.0mV.

Now, finally, I can calculate the gain!

```
Gain = 3940 / 44
```

The answer is about 89.6. I'll round that to 90, because my measurements were not accurate enough to justify a decimal place. What result did you get from your calculations? And more to the point—do you think it's right?

Is It Right?

At the end of a chain of arithmetic like this, I always wonder if I made a mistake. Well, I can compare the answer with what I know the gain should have been. Remember, I said that if F is the value of the negative-feedback resistor, and G is the value of the "ground" resistor (as shown in Figure 11-4):

```
Gain = 1 + ( G / F )
```

Because the "G" resistor is 1M (1,000,000 ohms) and the "F" resistor is 10K (10,000 ohms) the theoretically correct value for the gain is 101:1.

I think 90:1 is pretty close, considering the primitive methods that were used.

Too bad you didn't have the signal generator and the oscilloscope that I mentioned at the beginning. But even if you did, and you were working in a well-equipped electronics lab, you would still be making measurements, doing some simple math, and taking precautions to minimize inaccuracies. These kinds of procedures are inescapable in science and engineering. That's why I included them here.

Of course, it's more fun just to put things together and see them work. If that's what you would prefer—go ahead, I don't mind! You can skip the sections of this book where measurements are involved, and just have the pleasure of building the circuits and applying power.

The trouble is, you won't know *why* they work. You won't be able to measure their performance, or to design circuits of your own. If you're serious about electronics, some basic calculations and measurements will be necessary, especially when dealing with analog signals and amplification.

Splitting the Difference

If you're wondering where the main source of inaccuracy was in this experiment, which led me to a value of 90:1 instead of 101:1, my guess is that the "A" voltage divider was the culprit. The "G" resistor sank voltage into the midpoint of the "A" divider. This must have raised the value in the voltage divider slightly, so that it was not exactly half of the supply voltage. I don't know for sure, because trying to measure it would (once again) change it slightly.

The most obvious way to reduce the inaccuracy is to use a proper split power supply. And, indeed, if you look at an assortment of op-amp schematics, many of them will assume that you have such a supply.

For instance, the circuit in Figure 11-4, which was drawn in a layout suitable for breadboarding, could be redrawn in a more conventional format as shown in Figure 11-11. The "B" voltage divider is still necessary to apply a variable voltage to the noninverting input, but the "A" voltage divider has disappeared. In its place you see a ground symbol (at bottom-center of the figure). This symbol means that you connect this part of the circuit with the neutral ground on your split supply, to create the reference voltage.

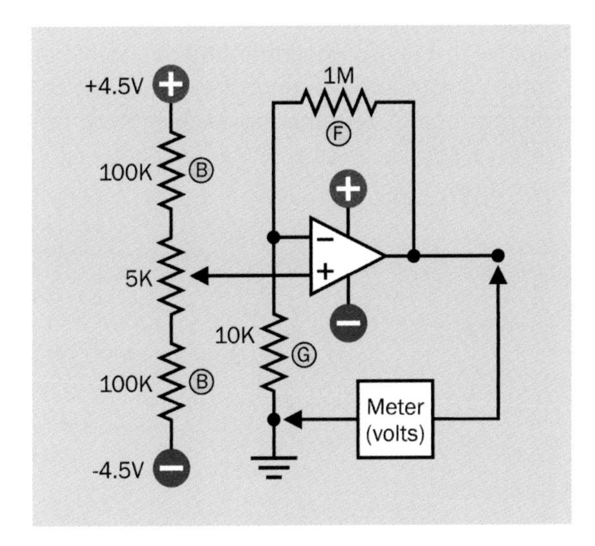

Figure 11-11. *The previous schematic used a layout appropriate for a breadboard, and assumed that only a 9V battery was available as the power supply. That schematic is redrawn here in conventional format, assuming availability of a split power supply. Note the ground symbol, which implies a potential of 0V.*

This can be confusing, because in many circuits where a regular, single power supply (not a split supply) is used, a ground symbol identifies negative ground. The general rule is, the ground symbol always means 0 volts.

This is about as far as I'm going to go into the theory of op-amps. Really I have just scratched the surface, but this is primarily a hands-on book. For instance, I don't have space to prove why the gain is found by 1 + (F/G). (You can look that up in any electronics book that has a section on op-amps.) If you have followed me step by step so far, you'll have a big advantage in that you have actually seen how an op-amp works, which I hope will make the explanations in other books easier to understand.

The Basics

At this point I'm going to do what most books do at the beginning: show you the two most common, simplified, basic op-amp circuits. The reason I didn't go through this before is that my approach is always to do some hands-on experiments first. The basic op-amp circuits aren't much use to you until you know about the extra components that must be added, to get them to do something.

The two basic circuits are shown in Figure 11-12 and Figure 11-13. In Figure 11-12 the signal goes straight into the noninverting input. This is the configuration that we have used so far. The feedback resistor, which I called F, is usually referred to as R2. I called it F because R1 and R2 already have been used for two other purposes in this book, creating confusion. But you should know how the rest of the world refers to feedback and "ground" resistances.

The gain of the op-amp is now:

```
Gain = 1 + ( R2 / R1 )
```

Which is the same as the formula Gain = 1 + (F / G) that I used earlier.

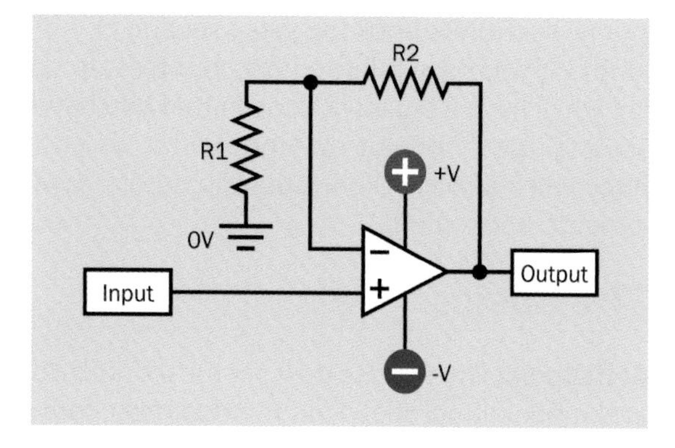

Figure 11-12. *The simplest possible representation of an op-amp circuit where the signal is applied to the noninverting input.*

In Figure 11-13 you will find an alternate configuration, which I haven't mentioned before. The signal goes through resistor R1 to the inverting input, which it shares with the feedback resistor, R2. The noninverting input connects with neutral ground to provide the reference voltage. As before, when the value of R2 is high relative to R1, you get less negative feedback and more amplification, but the output is turned upside down, literally—because the signal was applied to the inverting input. A higher voltage on the inverting input creates a lower output, and vice versa. Consequently, the formula for finding the gain is preceded with a minus sign:

```
Gain = -(R2 / R1)
```

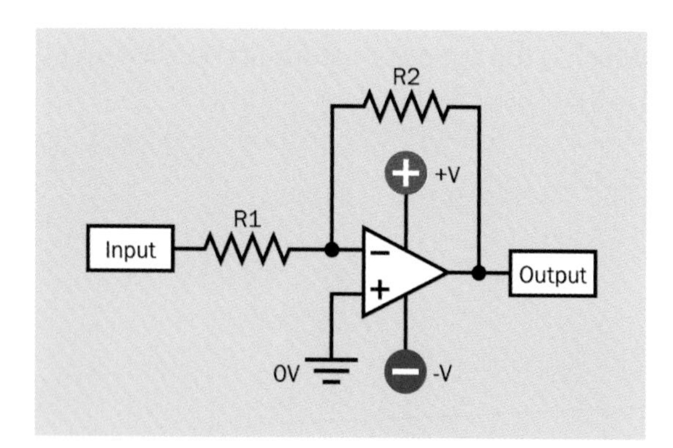

Figure 11-13. *The simplest possible representation of an op-amp circuit where the signal is applied to the inverting input.*

In either of these schematics, if R2 is omitted, the negative feedback resistance becomes almost infinite, and consequently the gain becomes almost infinite, too. This was the mode in which you used the op-amp in the first experiment. Without negative feedback, the op-amp will overreact to the tiniest variations between its input terminals.

Basic with No Split

The basic circuits assume that you have a split power supply. Some books will suggest that you can create this by using a pair of 9V batteries. The concept is illustrated in Figure 11-14. You connect the positive terminal of one battery with the negative terminal of the other, and this point becomes your theoretical neutral ground. The remaining "spare" terminal on each battery becomes a source of plus 9V and negative 9V, respectively.

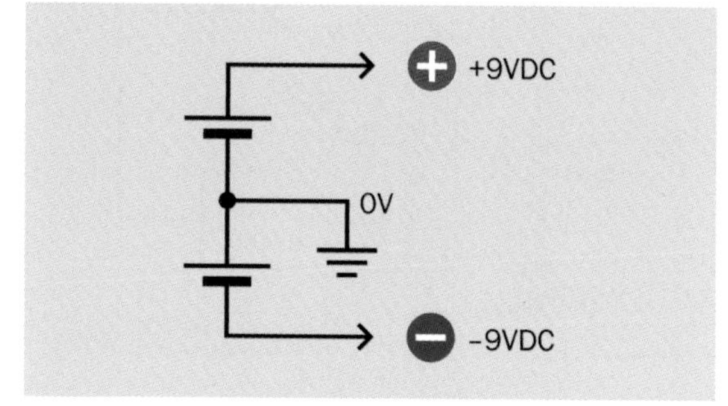

Figure 11-14. *Two 9V batteries can be used to create a split power supply, although this arrangement has some disadvantages (described in the text).*

Well, this sounds simple, so why didn't I suggest it at the beginning? Several reasons:

1. I felt that using voltage dividers to create an intermediate voltage was a useful concept to learn.
2. For all I know, you may prefer to use an AC adapter, and I didn't think it was reasonable to ask you to use two AC adapters.

3. Batteries cannot be matched to each other, and the 0V midpoint voltage will not be exactly 0V.
4. Creating a 0V reference requires adding another bus to your breadboard, which becomes confusing and is difficult to represent in a breadboard-style schematic.
5. I don't think you really get better performance by using two batteries in this way. Batteries always suffer from voltages that vary depending how old they are and how much current you draw from them.

In any case, the general need for a split power supply has been reduced by some well-known workarounds, which remove the hassle of choosing an accurately matched "A" pair and "B" pair of resistors.

Figure 11-15 shows a real-life version of Figure 11-12 without a split power supply, using component values that are suitable for an audio signal. Two 68K resistors form a voltage divider to provide a reference voltage, but this is the only place where you need a divider, and the resistors do not have to be accurately matched. The input can be relative to any voltage, because the 1µF input coupling capacitor isolates the circuit from the DC component of the input. Similarly, a 10µF output coupling capacitor separates the output from subsequent components. The only remaining problem is the 10K ground resistor, which should connect with neutral ground in a split power supply, but the 10µF capacitor isolates it so that neutral ground is not necessary.

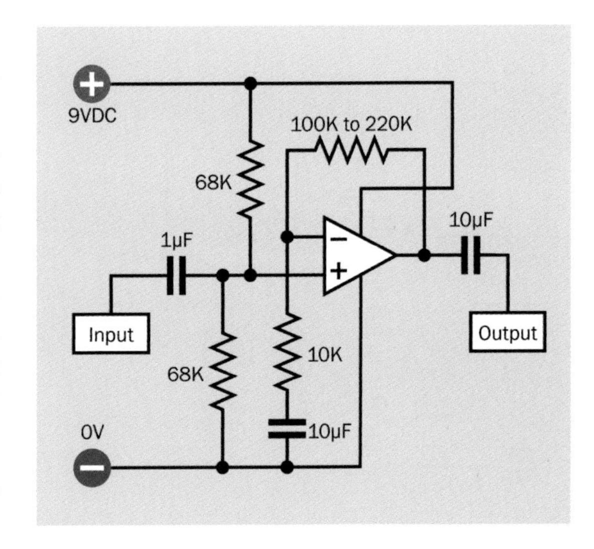

Figure 11-15. *An op-amp configured for audio amplification with a single-voltage power supply and the signal connected with the noninverting input.*

By using a feedback resistor ranging in value from 100K to 220K, in combination with a 10K ground resistor, you obtain a gain ranging from 11:1 to 23:1. The lower amplification value is suitable when someone is speaking close to the microphone, while the higher value would be appropriate for monitoring background noises in a room.

Figure 11-16 shows a circuit that makes similar compromises, this time for amplification from the inverting input, using a single-voltage power supply. Compare this with the theoretical version in Figure 11-13.

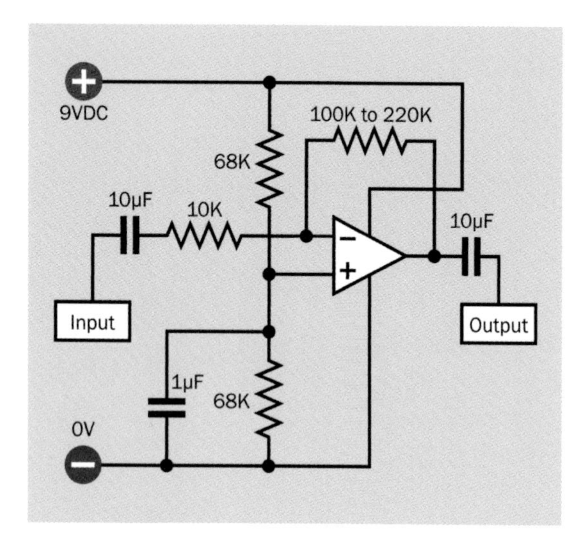

Figure 11-16. *An op-amp configured for audio amplification with a single-voltage power supply and the signal connected with the inverting input.*

Quick Facts About Op-Amps

It's time to sum up, so that you can come back to this section and refresh your memory about some of the tricky concepts:

- An op-amp is a similar to a comparator. They both have the same schematic symbol. They both have a pair of inputs, and they both compare the voltages on the inputs to produce an output.
- One of the inputs is a noninverting input, while the other is an inverting input.
- Op-amps and comparators both use feedback to modify one of the inputs. However, a comparator uses positive feedback to create a clean high or low output, while an op-amp uses negative feedback so that its output is an accurate copy of its input.
- Negative feedback is always applied to the inverting input.
- Positive feedback enables a comparator to ignore small variations in its input. This is known as hysteresis. In an op-amp, hysteresis is very undesirable, because the op-amp must amplify (not ignore) every little detail in its input.
- The primary application for op-amps is to amplify alternating signals, such as audio signals. Comparators are mostly used with DC inputs.
- The input signal for an op-amp can be applied to the noninverting input, but may alternatively be applied to the inverting input. The op-amp amplifies the difference, either way. However, if you apply the signal to the inverting input, the output will be inverted (which is why it is named the "inverting" input).
- The voltage on the input that you are not using for the signal has to be controlled in some way, to serve as a reference voltage.
- The output from a comparator, usually employing a pullup resistor, can be compatible with digital chips. The output from an op-amp is usually not suitable for a digital input, because it is an analog signal that contains many small variations.
- A comparator usually needs a pullup resistor to produce an output. An op-amp usually generates its own output.

Now that I've gone through all the op-amp theory, it's time for some practice: building a functional audio amplifier.

Experiment 12: A Functional Amplifier

12

As you saw in Experiment 10, an LM741 is not appropriate to drive a loudspeaker, even through a 2N2222 transistor. Really the LM741 is intended to be a bare-bones *preamplifier*, which increases the voltage from a very small input signal but cannot deliver significant power. A preamplifier is often referred to as a "preamp."

Fully-featured preamplifiers used to be available in the consumer-electronics world as standalone devices, to amplify inputs from a tape deck, a microphone, or a phonograph cartridge. They included some adjustments for volume, bass, and treble, and then passed the signal along to a power amplifier, which was designed to drive loudspeakers. Today a preamplifier and a power amplifier are usually combined in one unit, as in a typical stereo receiver or home entertainment system. For our purposes, though, the LM741 is the equivalent of a preamp, and we now need a power amplifier.

This can be another bare-bones device, the LM386 chip. It has the same dual-input format as an op-amp but is designed to drive a small loudspeaker with about 300mW of power. That sounds puny compared with the wattage that you can expect from a modern music system, but in fact, 300mW is enough for many practical purposes.

Introducing the 386 Chip

The pinouts of the LM386 are shown in Figure 12-1. It's similar to the LM741, but not exactly the same, so be careful when you attach the power supply and the output.

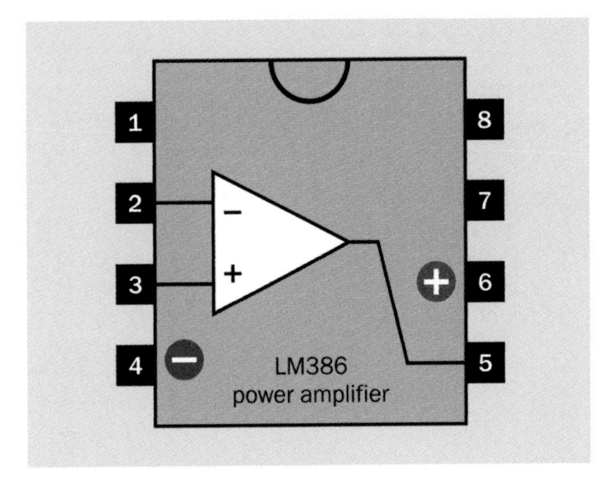

Figure 12-1. *The interior workings of an LM386 amplifier chip. Pins 1 and 8 are reserved for an external capacitor that can increase the gain of the chip from its default gain of 20:1 all the way up to 200:1.*

Pins 1, 7, and 8 are shown with no connections. They can be used in conjunction with external components if you wish to increase the gain of the amplifier from 20:1 (its default value) all the way up to 200:1. This is liable to create a lot of distortion, but if you want to try it, add a 10µF capacitor between pins 1 and 8, and put a 0.1µF capacitor between pin 7 and ground. This configuration has the advantage of diminishing the tendency of the LM386 to go into oscillations or create noise. If 200:1 is too high an amplification ratio, put a resistor of 1K or higher in series with the 10µF capacitor between pins 1 and 8.

The Amplification Circuit

If you look now at the schematic in Figure 12-2, several aspects of it should appear familiar. You've seen the simple microphone section before. Its output modifies the middle voltage created by the two 68K resistors functioning as a voltage divider. The voltage fluctuations are then connected with the noninverting input of the LM741.

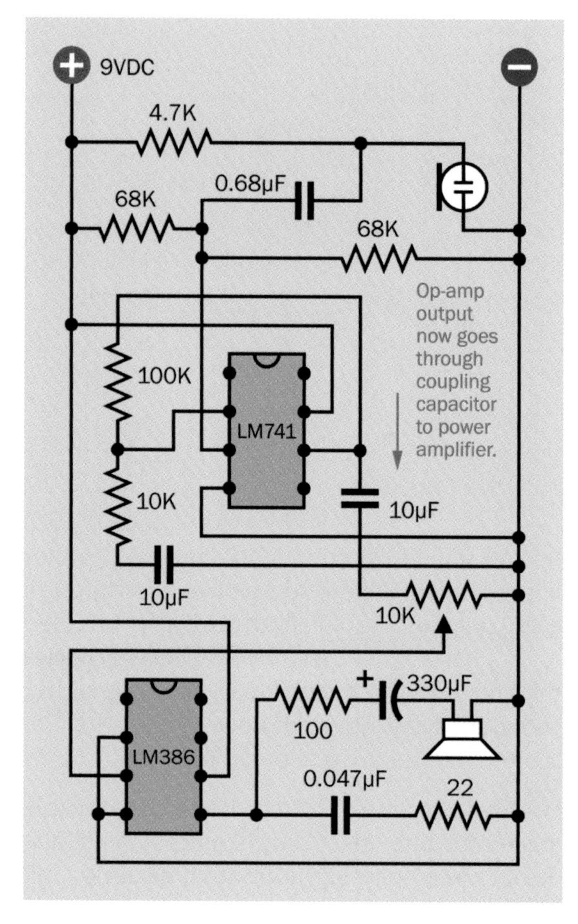

Figure 12-2. *A complete audio amplifier with microphone and loudspeaker. The LM741 forms the preamplifier stage, while the LM386 is the power amplifier.*

Feedback from pin 6 of the LM741 circles around through the 100K feedback resistor to the inverting input, and is modified by the 10K "ground" resistor. The 10µF capacitor between this resistor and the negative bus allows the inverting input of the LM741 to "float" independent of the power supply so that a voltage divider is not needed in this input. (If these terms don't entirely make sense to you, I have to ask you to go back and reread the previous technical stuff beginning in Experiment 11—assuming you want to know what they mean.)

The output from the LM741 passes through a 10µF coupling capacitor to a 10K trimmer potentiometer, which functions as a volume control for the LM386. The wiper from this potentiometer is connected with pin 3 of the LM386, which is its noninverting input. Pin 2, its inverting input, is connected with negative ground. The LM386 amplifies the variations between its two inputs, just like the LM741, except that its output is just powerful enough to drive a small loudspeaker. Note the substantial 330µF capacitor and the 100Ω resistor in series with the loudspeaker.

This circuit will demonstrate the tradeoff between gain and distortion. If you substitute a lower value for the 100Ω resistor, your voice will sound louder, but more distorted. You can experiment with the value of this resistor, and with different loudspeakers, as well as the trimmer that adjusts the gain.

A breadboarded version of Figure 12-2 is shown in Figure 12-3.

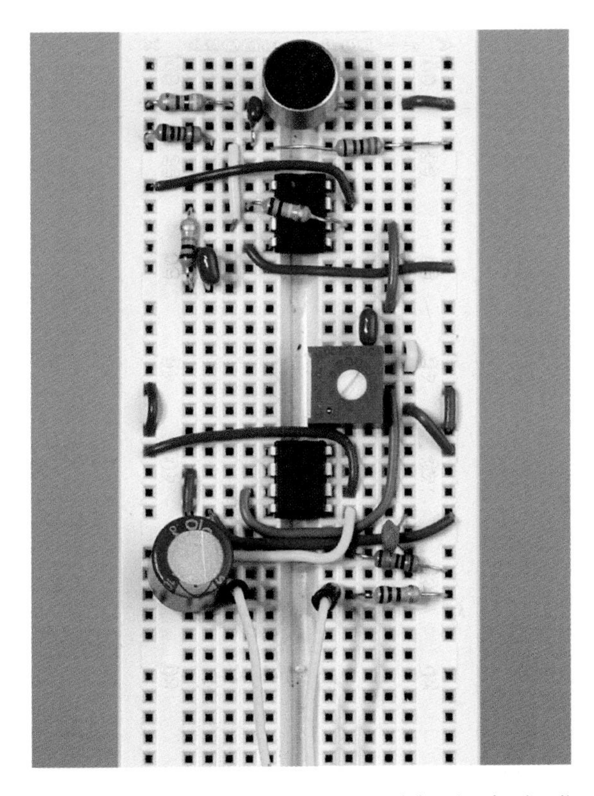

Figure 12-3. *A breadboarded version of the simple circuit using an LM386 amplifier. Yellow wires at the bottom of the picture are connected with a loudspeaker, not shown. The circuit can run for a limited time from a 9V battery.*

Troubleshooting the Amplifier

Here are a few ways to troubleshoot the amplifier:

Noise

If you hear whistles, buzzing, or rhythmic oscillations when you power up the circuit, this can be the result of various factors:

- Bad wiring. If you use jumper wires with plugs at each end, the tangle that they create is almost guaranteed to pick up electrical noise. This circuit should be wired with jumpers cut to size and placed flat against the breadboard. Components should be spaced as closely as possible.
- Acoustic feedback. Try moving the speaker farther away from the microphone.
- Add the components to pins 1, 7, and 8 of the LM386, as I described previously. Grounding pin 7 of the LM386 through a 0.1µF capacitor can suppress noise if you do this in conjunction with the capacitor and optional series resistor linking pins 1 and 8.
- Increase the 0.047µF capacitor to 0.1µF.
- If you are using a benchtop power supply instead of a battery, this may add some humming or buzzing through the speaker. Add the highest value capacitor you can find between the positive and negative buses of the circuit board. Although I have a fairly high-qualify power supply myself, I still found that a 4,700µF capacitor reduced some background hum.

Distortion

Because this is such a basic little circuit, some distortion is inevitable. You can try these options:

- Substitute a 3.3K resistor for the 4.7K resistor in series with the electret microphone.

- Insert a 10K resistor in series between the wiper of the 10K trimmer potentiometer and the input of the LM386 on pin 3.

Insufficient sound

You won't get much volume from this little circuit, but here are a couple of options to try:

- The combination of resistors with the LM741 creates a gain of 1 + (100/10) = 11:1 in the preamp stage. Try substituting a 150K resistor for the 100K resistor, to increase the gain to 16:1. You can also reduce the value of the 10K "grounding" resistor.
- You can increase the perceived loudness by placing the loudspeaker in a small closed box or tube. I find that 6" length of 2" diameter PVC water pipe works well as an enclosure for a 2" speaker.
- I got good results by connecting the output from the LM386, and a ground wire, to the input plug of a computer speaker system (which contains its own little amplifier). Try this at your own risk! The 330µF capacitor should protect your speaker system, but if you make a wiring error, the results will be unpredictable.

Experiment 13: No Loud Speaking!

13

The culmination in this series of audio projects is a device that goes back to the concept of using sound to switch something on and off. I showed how this could be done in Experiment 10, but I'm going to take it much further, now. This project was inspired by a story about one of the pioneers of analog integrated circuits.

Background: The Widlar Story

A legendary engineer named Bob Widlar did a lot of the early development work on op-amps during the first semiconductor boom period of the 1960s. He was an important figure at early Silicon Valley startups such as Fairchild and National Semiconductor, where he was memorable not just for his breakthrough designs but for his bad behavior. He had a lifelong love affair with alcohol, and was described by coworkers as paranoid, reclusive, and impossible to deal with—although they dealt with him anyway, because he was such a brilliant engineer. In those days, an unstable, abrasive personality would be tolerated in Silicon Valley, because electronics was still a field for mavericks, and human resources departments had little control over the hiring process.

Widlar's intolerance for defective parts and malfunctioning prototypes was so intense, he was in the habit of destroying them with a sledge hammer. This came to be known as "widlarizing" them. He was also very intolerant of noise, and installed a device that emitted a piercing whistle if a visitor in his office raised his voice and shouted at him. A Fairchild engineer told me that the gadget was known in the company as "The Hassler."

I'm going to call it the Noise Protest Device, and I'll show you how to build your own version of it. Since Widlar achieved his reputation by designing op-amps, it seems appropriate that this project is built around an op-amp.

Step by Step

I'll go through the process of designing and building this circuit in steps, to give you an idea of the procedure that you might follow if you were designing it yourself.

Maybe the idea of circuit design seems intimidating. Like—where do you start? But so long as a circuit can be broken down into sections, and you can make them communicate reliably with each other, and you can test them one at a time, the design process doesn't have to be too difficult.

Of course, the initial attempt at designing a prototype may not be totally successful. But that's what you should expect from a prototype.

The first step is to think about what the circuit has to do, and make a list of components that may be able to do it. For the Noise Protest Device, my list looks like this:

1. A device to detect noise and turn it into an electrical signal. That would be an electret microphone.
2. A preamplifier for the signal. That's an LM741 op-amp.
3. A current amplifier. We can use the same 2N2222 transistor as before.
4. When voltage or current exceeds an adjustable limit, it has to trigger something. Not sure what this will be, yet.
5. The triggered device has to make a noise in protest. I'll call this the Protest Output. I could use just a 555 timer running in astable mode at an audible frequency.

Sensing

The key to making this work is the fourth step, above. How can this be done?

Well, let's think how a 555 timer works when it is running in astable mode. It starts itself as soon as it receives power, and it stops when the power is interrupted. But this isn't the whole story. It also has a reset pin. When the reset pin receives a positive input, the timer is enabled. When the reset pin is pulled low, the timer is interrupted.

Maybe I can convert the output from the op-amp to drive the reset pin. This sounds plausible:

- When the electret microphone doesn't pick up any noise, the output from the op-amp will be low, and a low input to the reset pin will prevent the timer from creating its Protest Output.
- When the electret microphone hears someone shouting, the output from the op-amp will go high, and a high input to the reset pin will enable the timer and its Protest Output.

The only problem is, the output from the op-amp is AC. How about if I put it through a coupling capacitor, biased to negative ground, so the output just varies relative to ground, as in Experiment 10? Then I can add a smoothing capacitor to remove the ripples from the signal—somewhat, at least.

If you read *Make: Electronics*, a smoothing capacitor is typically placed between the signal and ground, to smooth out the peaks and dips. The result might be clean enough to work the timer's reset pin.

Time to check the 555 datasheet. It says that the reset pin stops the timer when the potential is 1VDC or lower. Otherwise, it allows the timer to run.

So, if I can process the output from the op-amp to be greater than 1VDC when someone is shouting, but less than 1VDC when no one is shouting, the timer should react appropriately.

Will It Really Work?

At this point, I could create a simulation in software such as SPICE, showing the interaction of components. But since I'm using quirky analog signals, really I'll have to put the components together to find out whether they will do what I want them to do.

The first part of the Noise Protest Device circuit is shown in Figure 13-1. It's very similar to the upper half of the circuit from Experiment 12, shown in Figure 12-2. The primary difference is that the feedback resistor has been changed to a 1M potentiometer so that I can vary the sensitivity. A 10K resistor has been added below the output coupling capacitor, between the output and ground, to bias the capacitor.

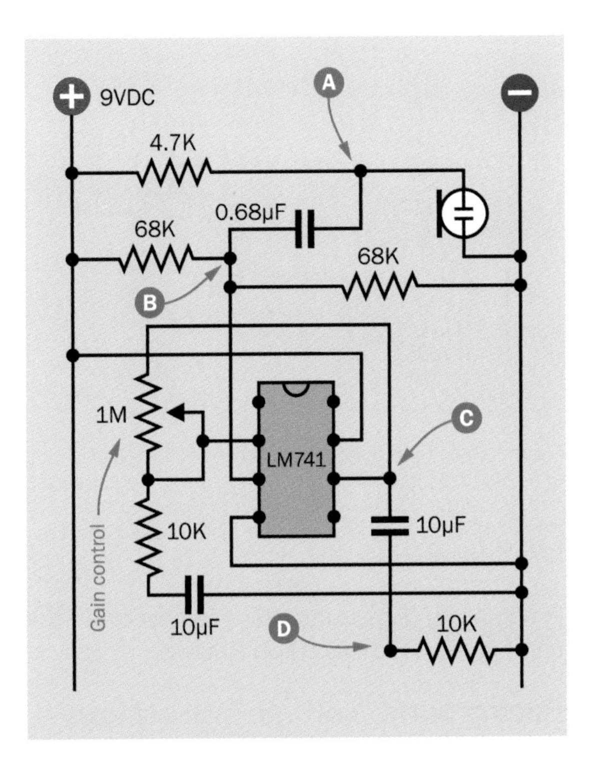

Figure 13-1. *The first step in constructing a noise-against-noise circuit.*

The next step is to check what actually happens in the circuit. Here again an oscilloscope would be useful, but since you probably don't have one, I'm not going to use one either.

Assemble this circuit and make sure that the 1M potentiometer is providing maximum resistance between the op-amp output and the noninverting input, to minimize the negative feedback and maximize the gain.

Now do the "Aaah" test, or whistle, while you check the voltages at points A, B, C, and D in the circuit, with your red probe touching the places indicated by the arrows and your black probe touching negative ground. You should get voltages like those listed in Figure 13-2. (You may wonder why the table lists a point E, even though there is no point E in Figure 13-1. There will be a point E when the circuit is extended downward.)

Locations in Circuit	No Input Noise		With Input Noise	
	AC	DC	AC	DC
A	0mV	7.5V	30mV	7.5V
B	0mV	4.5V	30mV	4.5V
C	0V	4.5V	2V	4.5V
D	0V	0V	2.5V	0V
E	0V	0V	0V	3.0V

Figure 13-2. *Voltage readings at points in the circuit, showing how a very small alternating signal is amplified and converted to a usable DC output.*

If your readings aren't exactly the same as mine, this may be for several reasons. You may not be making the "Aaah" sound as loudly as I did (or, you may be making it louder). Your microphone may be more or less sensitive than mine, or your meter may take longer to stabilize, or it may measure AC differently. Either way, small deviations are not important.

Initially, the microphone generates around 30mVAC at point A, when you speak loudly into it. This value remains the same at the other side of the capacitor, at point B, because it's an AC value. But the DC voltage has been reset to 4.5V by the voltage-divider resistors. Continuing on down, the output directly from the LM741 at point C is amplified to approximately 2.5V AC when the microphone is picking up noise. There's still a DC component of 4.5V at point C, but the output coupling capacitor blocks the DC, so at point D, we now just have an AC reading, which changes from 0V to 2.5V when you make noise. So far, so good. A swing from 0VAC to 2.5VAC looks promising.

Now check the next version of the schematic, shown in Figure 13-3. I have added a transistor, which provides more power, and a 100µF smoothing capacitor. Why 100µF? Experience suggests that this will be about right, for audio frequencies. But I'll have to try it to make sure and change it if necessary.

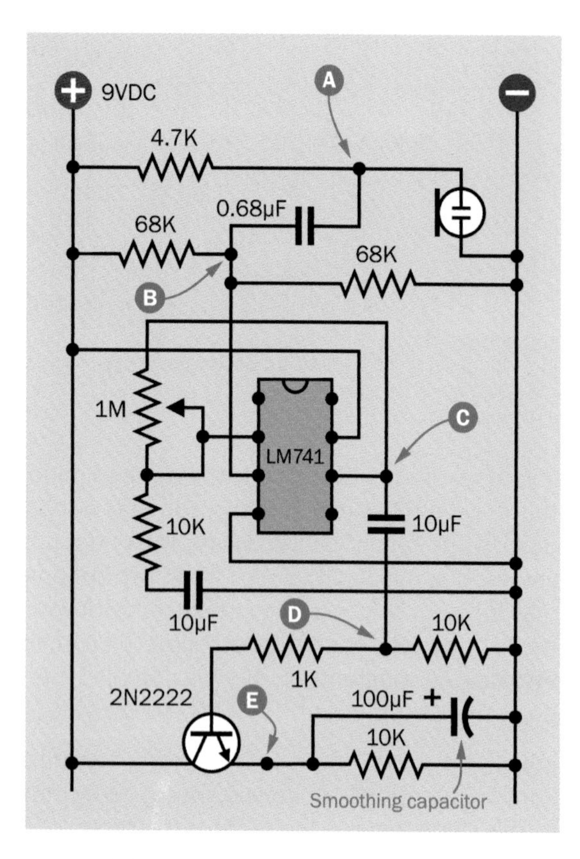

Figure 13-3. *The second step in building the Noise Protest Device.*

In the circuit in Figure 13-3, the transistor allows current to reach point E with very little resistance when it is "on." But when the transistor is nonconductive, its effective resistance becomes high relative to the 10K resistor, so the voltage at point E goes low.

This is an *emitter follower* configuration, because the voltage at the emitter follows the voltage at the base (with a small deduction imposed by the transistor). Of course, the transistor also amplifies current.

If you swap the transistor and the 10K resistor, the effect is reversed. This is suggested in Figure 13-4, which shows that a transistor can either pass along a voltage transition or invert it, depending on the configuration.

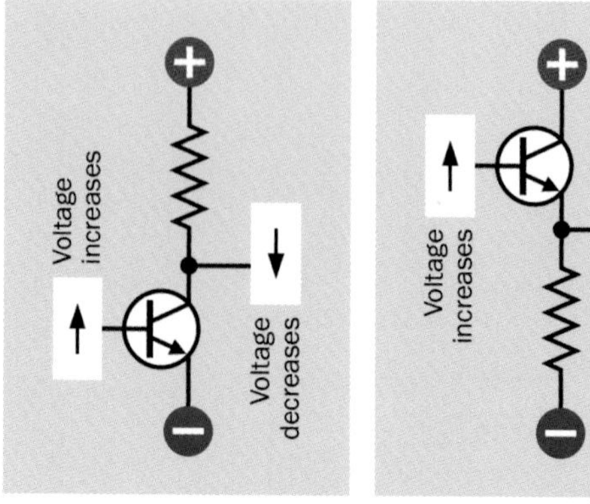

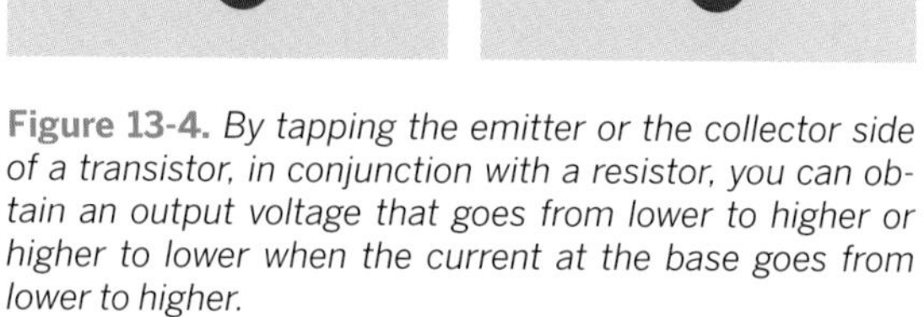

Figure 13-4. *By tapping the emitter or the collector side of a transistor, in conjunction with a resistor, you can obtain an output voltage that goes from lower to higher or higher to lower when the current at the base goes from lower to higher.*

I'm going to digress into this topic, because the technique has so many applications.

Background: Voltage Translation

If you want to see this for yourself, you can do it very easily using just one transistor and two resistors.

In Figure 13-5 you'll see some actual meter readings that I obtained when taking output from the collector side of a 2N2222 transistor. In each schematic, the base resistor is 1K, and it is connected first with negative ground, and then with the positive side of the power supply. Energizing the base changes the transistor from its "off" (nonconducting) mode to its "on" (conducting) mode. The transistor is shown gray when it is nonconductive.

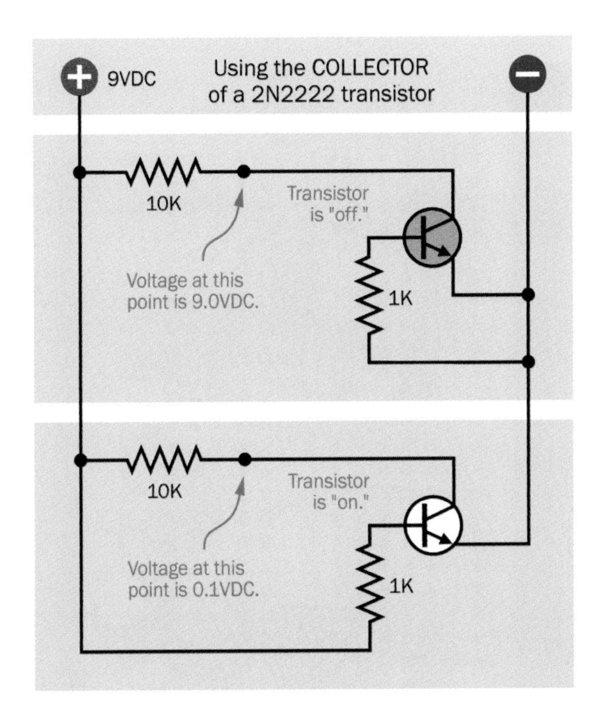

Figure 13-5. *Actual values measured with a transistor configured to deliver output from the collector side. All figures rounded to one decimal place.*

The numbers are all rounded to one decimal place, so the actual highest output with 9VDC power will be a fraction less than 9VDC:

- When you tap into the collector of a transistor in this way, you *invert* the input.

But bear in mind:

- The actual performance will depend on the device attached to the output from the circuit. The numbers were measured with a meter, which has a very high impedance. If you use a different device, the numbers will change. On the other hand, many devices, including op-amps, comparators, and a lot of digital chips, also have a very high input impedence.
- The numbers are obtained with the transistor in saturated mode. If the current through the gate is lower, the outputs will be different.
- Care must be taken to treat the transistor kindly without exceeding its maximum values. When the transistor is conducting, you should not sink excessive current through it. Check the datasheet to make sure!

Now turn to Figure 13-6. Here I have swapped the position of the 10K resistor relative to the transistor, and I tapped into the emitter side of the transistor. In this emitter-follower configuration, the transistor does not invert the voltage anymore. The voltage follows the polarity of the input, but the spread of the output voltages isn't quite as wide. Once again, they are rounded to one decimal place.

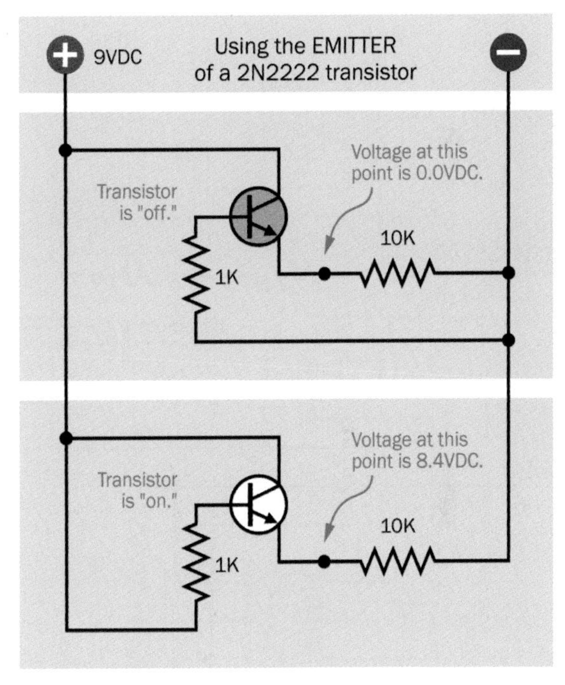

Figure 13-6. *Actual values measured with a transistor configured to deliver output from the emitter side.*

You can still adjust the voltages in each case by adding a resistor on the other side of the transistor. The resistors will form (you guessed it) another voltage divider.

Naturally, in every case, the transistor still functions as a current amplifier.

Noise Protest, Continued

The complete circuit for the Noise Protest Device is now shown in Figure 13-7, with a photograph of the breadboarded version in Figure 13-8.

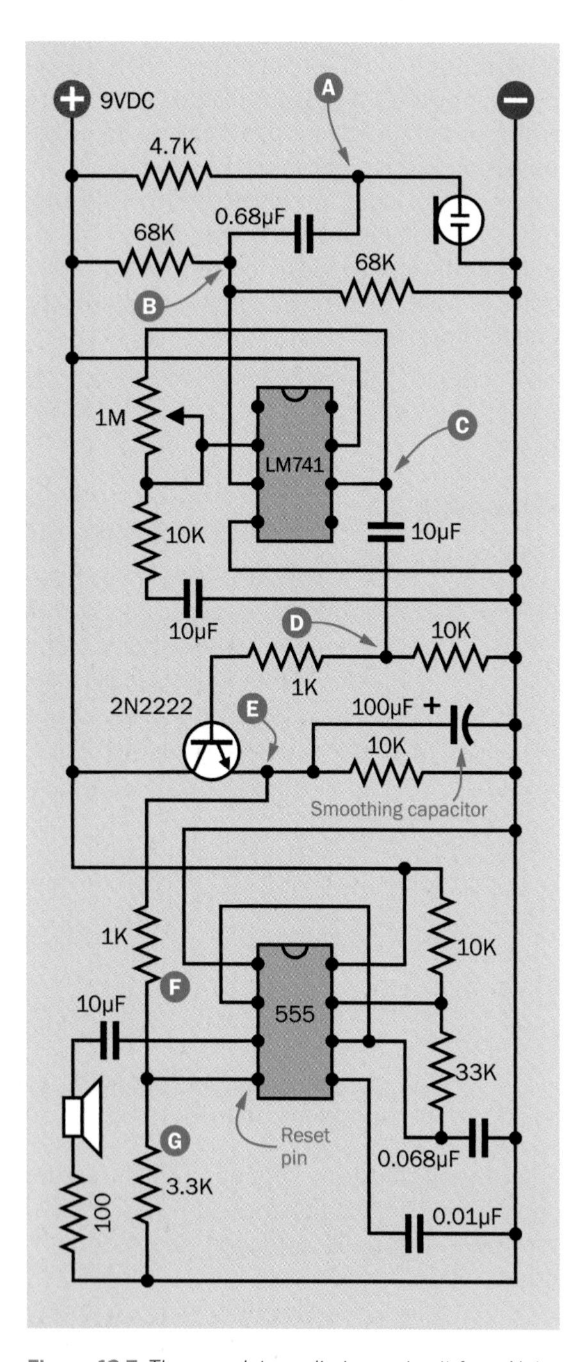

Figure 13-7. *The complete preliminary circuit for a Noise Protest Device.*

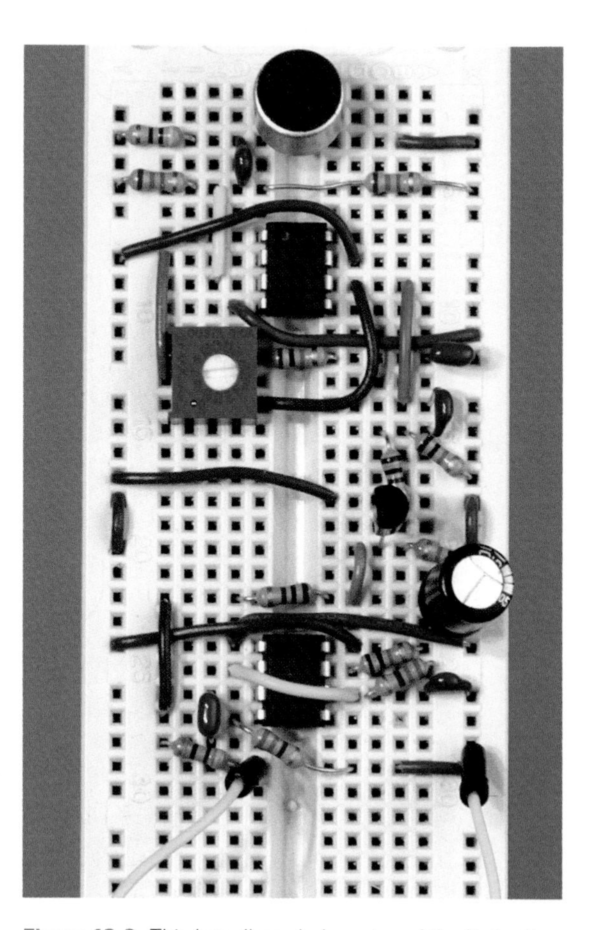

Figure 13-8. *This breadboarded version of the Noise Protest Device circuit is designed to run from a 9V battery. The yellow wires at the bottom of the photograph are connected with a loudspeaker (not shown).*

In case the basic concepts are still not entirely clear, a flow chart illustrating the logic of the thing is shown in Figure 13-9.

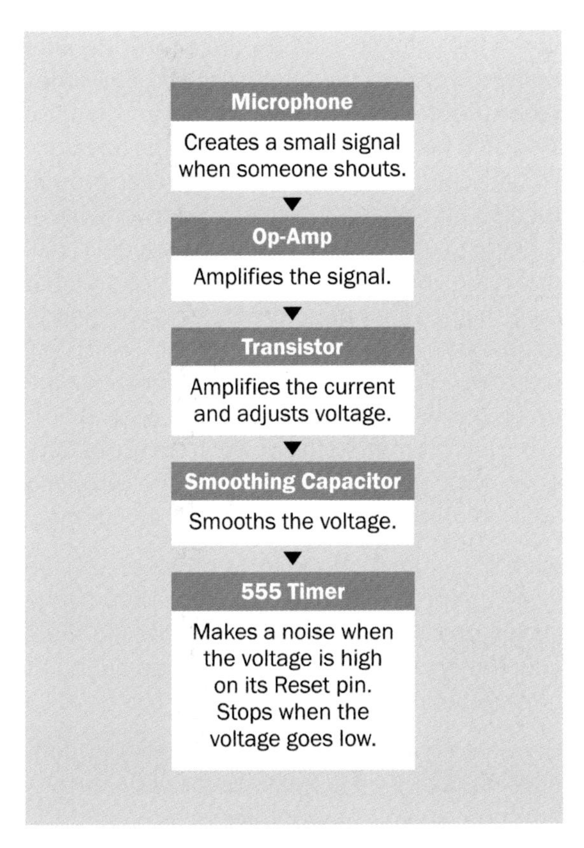

Figure 13-9. *This diagram illustrates how each section of the Noise Protest circuit communicates with the next section.*

You should go ahead and finish your own version of the circuit to find out if it works. Once again, remember to avoid using the kinds of jumper wires that have a plug at each end. They will create a tangle of loops of wire that will interact with each other electromagnetically, causing noise and erratic behavior in general. Op-amp circuits should always be built with wires that are as short as possible, and you should keep the components as close together as possible.

First check the voltages into and out of the op-amp. Compare them with the table in Figure 13-2. If they are comparable, the next step is to make sure that you have the 555 wired correctly. Unplug the connection to pin 4, and you should hear an annoying, high-pitched whistling tone from the loudspeaker. This is the 555's Protest Output. If you don't hear anything, you need to track down your wiring error(s) before continuing.

Now when you reconnect pin 4, the whistling tone should stop, although there may be a short delay while the 100µF capacitor is discharging. Adjust the 1M trimmer for maximum feedback resistance, and speak loudly into the microphone. There may be another short delay while the 100µF capacitor is charging, but then the timer should start whistling until you stop speaking loudly.

This is how the circuit is supposed to work. Indeed, my version works—but only just, and only with a good benchtop power supply. When I substitute a 9V battery, the behavior of the circuit becomes erratic.

This is very disappointing, but as I said at the beginning, the first version of a prototype doesn't necessarily work properly.

So, I have to figure out what is causing the trouble. By testing with my meter, the answer seems very obvious: the voltage range at point E was fine so long as point E wasn't connected with anything other than a meter, but as soon as I fed its output to the reset pin of the 555 timer, everything changed.

The 555 datasheet didn't tell me everything. I assumed the reset pin would have a high impedance, like the input of a logic chip, but—apparently not. Also, I think the 100µF smoothing capacitor is inadequate. It allows ripples or spikes in the current, which are sufficient to raise the voltage on the reset pin of the timer. This allows the timer to continue emitting a noise even when there is very little sound being detected by the microphone.

Either way, the output from the transistor turns out to be incompatible with the timer. What to do? In this kind of situation, there are two options:

- Fiddle around, trying to make it work.
- Try something completely different.

The first option always seems as if it should be quicker than rebuilding everything from scratch. Of course, often it isn't quicker, but I tried it anyway. I adjusted the voltage on the reset pin by adding yet another voltage divider. Its resistors are labelled "F" and "G" in the schematic. To determine the values of these resistors, I didn't use any calculations. I tried various values experimentally.

This helped, but I still don't regard the performance of the circuit as bulletproof. I found myself hearing clicking noises from the loudspeaker, or a rapid series of beeps. I also heard a scratchy version of the whistling tone, which suggested to me that it was being modulated by ripples in the signal from the transistor. I tried substituting a 330µF capacitor for the 100µF capacitor, but that just created oscillations. I also tried a 47µF capacitor. You can experiment with these values yourself to see if they make the circuit perform better, or not.

Power Problems

It's really annoying when a circuit doesn't work, but as always, you have to be methodical in your search for the answer.

I assume that you are powering your circuit with a 9V battery. Disconnect pin 4 of the 555 timer again so that the speaker starts making its annoying sound. Attach the black probe of your meter to the negative ground bus of your breadboard, and the red probe to point "B" where the two 68K resistors meet. Remember, this is where the circuit creates the reference voltage, which the op-amp compares with the microphone input. Set your meter to measure DC volts.

Check the voltage, and now disconnect the wire that provides positive power to the 555 timer. When the timer is deprived of power and its noise stops, I'm betting you will find that the reference voltage changes. The reason is that even though the 555 timer doesn't take much power while it is beeping (maybe 20mA), this is enough to pull down the voltage from the 9V battery, and that may change the reference voltage just enough to upset the output from the op-amp, which will reduce the voltage at point E, shutting down the timer. But when the timer stops whistling, it is using less current, so the power from the battery goes back up, and the timer starts whistling again—which is why you may get oscillations.

Even if you don't observe this problem, you are likely to encounter it (or scratchy noises in the circuit, or a beeping that won't stop at all) when your battery's voltage gradually drops during its normal lifetime.

Here are some possible fixes. I'll tell you right now, none of them appeals to me very much—but these are the easy options:

1. Always use a proper power supply instead of a little battery. My version of the circuit works quite reliably from a regulated benchtop power supply, and works with only a bit of hesitation from a RadioShack multivoltage AC adapter.
2. Use two 9V batteries, one for the top half of the circuit and the other for the bottom half. The positive side of one battery powers the op-amp, while the positive side of the other battery powers the timer. However, both batteries must share the same negative ground.
3. Use a higher-voltage power supply (12VDC or more), which then passes through a 9V voltage regulator. This should compensate for fluctuations in power consumption.
4. Increase the series resistance with the loudspeaker. But, wait a minute—this will reduce the volume from the speaker, and the whole

idea was to make a loud noise when someone speaks loudly!

You know, we shouldn't have to fuss around with a circuit like this to make it work properly. As I said before, it should be bulletproof. I decided that I had made a mistake, fiddling with the circuit to make it work. I should have tried something completely different.

Fail?

Does this mean that my circuit is a failure? No, I don't like that word, because it implies that something is worthless. In reality, almost every successful person has tried strategies that didn't work. People usually become successful because instead of giving up, they learn from their experience.

If something works outstandingly the first time, we learn very little from it. If we have problems, that's when the learning process really begins. So what have you learned, here?

- You have seen the instability that can occur in a circuit which contains amplification. Undesirable feedback and oscillations are common.
- You have seen that the power supply is not just a passive source of current. It is an active part of the circuit, and a battery has limitations which are not shared by an AC adapter.
- You have seen that you must verify the performance of a circuit using different options (such as different types of power supply), instead of just saying, "It works with my equipment, so if it doesn't work for you, that's your problem."
- You have seen that if one section of a circuit is only barely compatible with another section, this probably isn't good enough.

Just One More Little Thing

While I was tinkering with the circuit, something happened that I haven't even mentioned yet. I happened to leave the loudspeaker close to the microphone. Can you guess what happened next? Audio feedback, of course! When the loudspeaker made its Protest Output sound, the microphone picked this up. The microphone isn't smart enough to tell the difference between the Protest Output and someone shouting, so the circuit continued to be active. The noise went around and around and would never stop.

This is a conceptual problem, not a hardware problem. The concept of a device that responds to a shouting person by making a noise that is even louder was flawed from the start. The circuit ended up shouting at itself!

Did you foresee this, when I wrote out the specification? I didn't see it, because I had tunnel vision, which is a very common problem when a new device is being designed. I was focusing on the goal (in this case, making a noise in response to someone shouting) and forgot the larger picture.

Often you don't discover an obvious problem until a prototype is up and running. Then you feel embarrassed, because everyone will say, "That should have been obvious!"

This is another valuable learning process. No matter how experienced you are, you may fail to foresee an "obvious" problem. To take a random, classic example: the story goes that Steve Jobs was carrying one of the first iPhone prototypes around in his pocket for a couple of weeks, using it experimentally, less than two months before the product went into production. He found that the plastic screen on the phone became scuffed and scratched during that short period of time. Well, he should have expected that. It should have been obvious, right?

Perhaps the iPhone designers had assumed that plastic was the only way to go, because glass would break too easily. But when Jobs saw his scuffed screen, he wanted it to be changed to glass, even though the phone was almost in production, and the kind of thin, extremely strong glass that he needed was not even available in sufficient quantities. He had encountered an "obvious" problem that no one had really thought about, and instead of shrugging and accepting it, he initiated a huge redesign effort to fix it.

With this in mind, I am not going to just shrug and say, "Well, we can turn down the Protest Output so that the device won't retrigger itself, and that will just have to do." And I'm certainly not going to say, "The Noise Protest Device didn't work properly, so let's forget about it and move on." I'm going to do what you would do if you were developing a product that turned out to have a defect. I'm going to fix it.

Experiment 14: A Successful Protest

14

First I will restate the problem. Using the original concept of the Noise Protest Device that I developed in the previous experiment, if someone shouts loudly, the Protest Output will start and never stop. How can I deal with this?

One solution would be an audio filter on the microphone input so that the Noise Protest Device can't hear its own noise but still hears when someone shouts. This would be possible, but I'm not confident that it would work reliably.

Another solution would be simply to limit the duration of the Protest Output to, say, a couple of seconds. Then there's a pause while the Protest Output is suppressed. At the end of the pause, the Protest Output is now silent so it cannot retrigger the circuit, and if there's no more shouting, the device has done its job and can keep quiet. If someone is still shouting, however, the cycle will repeat.

This is more complicated than the original specification, but that's what usually happens after you build a prototype. Even when it works well, you find that it lacks some desirable features, and you have to build a second version. In every book I have ever written and every gadget that I have ever built, after the initial version has been delivered to the client, the client will want something more, or I will.

Timing Is Everything

How can I set a time limit on the Protest Output? By using a timer, of course! It will be in one-shot mode. And then how can I follow this with a pause? Well, how about if the first timer triggers a second timer, when the first timer reaches the end of its cycle? This is doable. In *Make: Electronics*, I showed how one timer can trigger another.

I'd better give my new pair of timers names to tell them apart. I'll call them the Noise Duration Timer and the Pause Duration Timer.

The Noise Duration Timer will be triggered by the op-amp. Wait a minute—that was where all my problems began, in the previous version. Yes, but the trigger pin on a 555 timer behaves quite differently from the reset pin. First, if the supply voltage is 9VDC, the trigger only requires a voltage below 3VDC, whereas the reset pin requires a voltage below 1VDC. And second, if there are any fluctuations in the voltage after the timer has been triggered, it ignores them until the end of its cycle.

You see why it's important to read datasheets to learn these details. Anyway, I believe that I can make the trigger pin compatible with the output from the LM741.

The output pin of the Noise Duration Timer will send power to an off-the-shelf noise maker, such as a beeper—or maybe a burglar alarm siren, which would really get people's attention. A beeper will cost maybe a couple of dollars, while a siren will be closer to $10. Either will work from a 9V supply.

When the Noise Duration Timer reaches the end of its period, its output will go low, which will shut down the external noise-making device. It will also pass through a coupling capacitor to the Pause Duration Timer, and the sudden drop in voltage will trigger it. Now while the Pause Duration Timer is going through its cycle, it must somehow inhibit the circuit from starting up again.

Maybe I can use the output from the Pause Duration Timer to pull down the reset pin of the Noise Duration Timer. What, we're back to using the reset pin again? Yes, but the output from the Pause Duration Timer will be DC, without any ripples in it, and it should be more robust than the output from an op-amp passing through a transistor. I think it will work.

At the end of the pause, the Pause Duration Timer releases the Noise Duration Timer, so that if there is still someone shouting, the process will repeat.

The final version of the circuit is in Figure 14-1. I have also updated the flow diagram from Figure 13-9 to show the new logic. The revised version is in Figure 14-2.

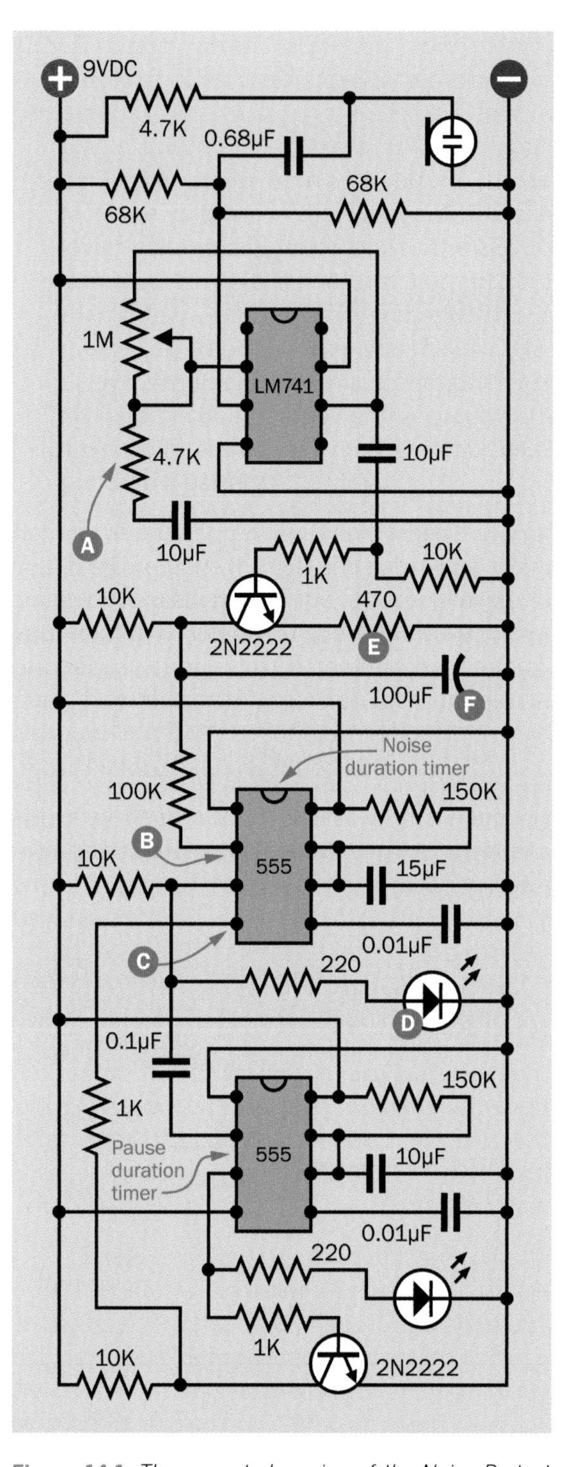

Figure 14-1. *The corrected version of the Noise Protest Device, to rectify flaws in the original version.*

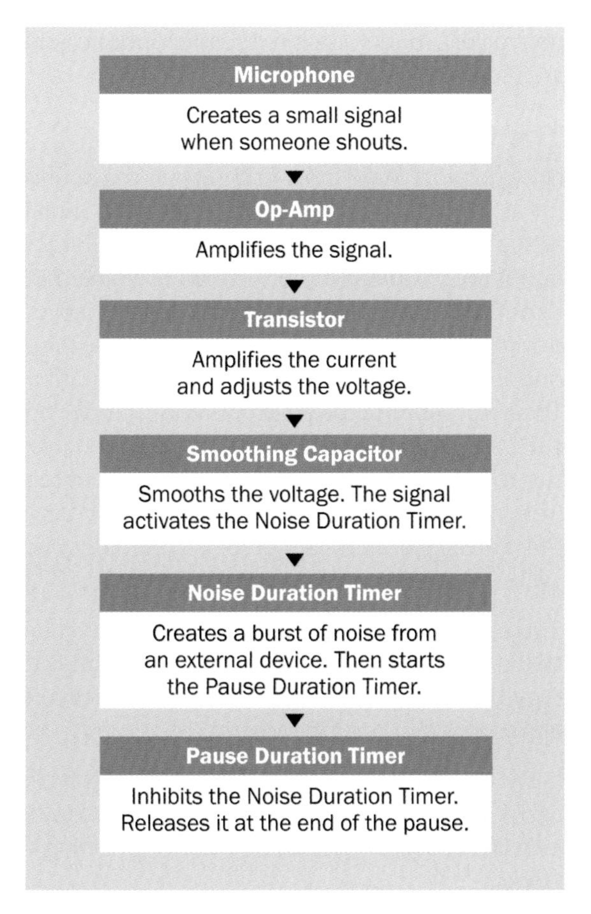

Figure 14-2. *A flowchart illustrating the logic of the final version of the Noise Protest Device.*

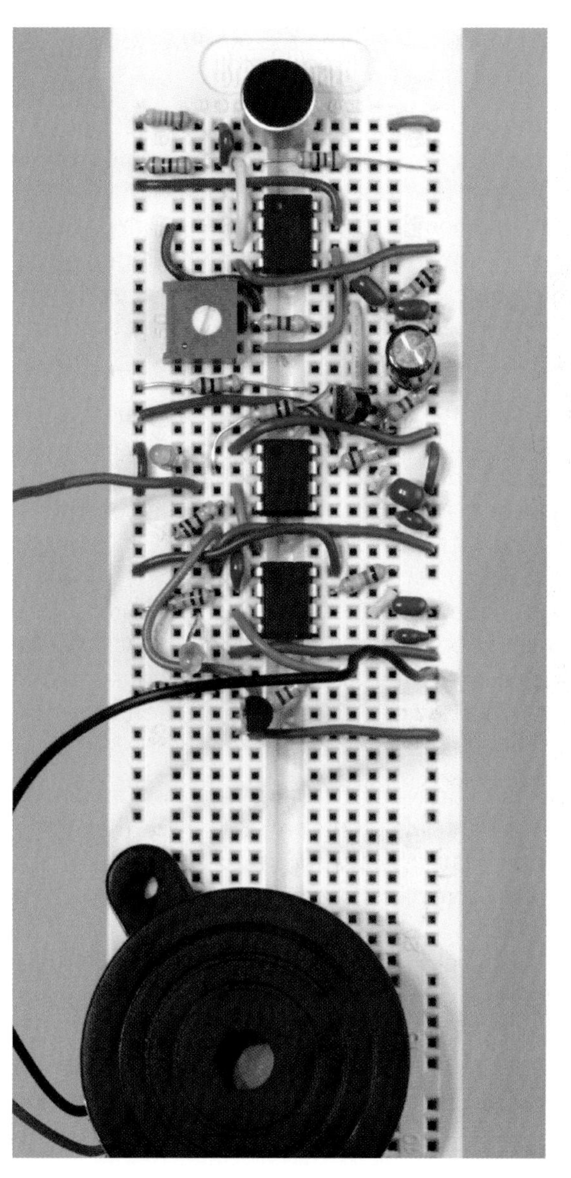

Figure 14-3. *The final Noise Protest Device. The large round object is a heavy-duty beeper. The circuit will run for a limited time from a 9V battery.*

Revision Summary

One important change that I had to make: the "grounding" resistor, which adjusts negative feedback to the LM741, has been changed from 10K to 4.7K. I've identified it in the new schematic with a green "A." The reason for this change is to increase the amplification of the op-amp to make it more sensitive. If it's too sensitive, you can just dial back the 1M trimmer.

The original 2N2222 transistor has been rewired, so that instead of inhibiting a 555 timer, it now triggers the Noise Duration Timer with a low signal. Remember, the voltage only has to go below 3VDC (one-third of the supply voltage). Apply your meter to the location marked "B" and you'll see it responding when you shout into the microphone.

The output from the Noise Duration Timer goes to an LED, just for demonstration purposes. I labelled this LED with a "D" in the schematic. If you want the circuit to perform usefully, you'll swap out the LED and substitute a beeper. If you want more noise, you may need a relay to trigger a more powerful external device. An optocoupler would be best, as it will completely isolate the device from the sensitive circuit containing the LM741. However, a very small electromechanical relay should work.

While the Noise Duration Timer is running, its output is high. At the end of its cycle, the output goes low. This transition passes through the 0.1µF coupling capacitor, momentarily overwhelms a 10K pullup resistor, and triggers the Pause Duration Timer. The output from this timer goes to a transistor at the bottom of the schematic, which converts it to a low output to inhibit the Noise Duration Timer by pulling its reset pin low. You can check at point "C" that this voltage is now in the correct range. The LED on the output from the Pause Duration Timer is just included so that you can see that it's working; it can be omitted from the final version of the circuit.

When you have finished wiring the circuit, apply power. The initial power surge may activate one timer or the other. You can ignore that.

To check that the timers are working, briefly ground pin 2 (the trigger pin) of each of them. This should make the LED light up in each case. You can also use your meter to verify the input voltage on the trigger pin of the first timer. When you make a loud noise into the microphone, you should see the voltage dip.

The Noise Test

Now do your "Aaah" test and sustain the sound for as long as you can. There may be an initial hesitation. Then you should see the first LED light up for approximately two seconds. Imagine that its output is activating the external noise-making device. Then that LED goes out, and the second one comes on, to tell you that the Pause Duration Timer is inhibiting the Noise Duration Timer. You can continue making as much noise as you like, but the Noise Duration Timer will be prevented from responding, and its LED will stay dark, until the Pause Duration Timer has completed its cycle.

If you hold the meter probe on point "B" you may still see some small oscillations in the voltage, but they don't matter anymore, because this circuit allows a bigger margin for error.

My copy of the Noise Protest Device works well, and I think your copy will work for you, too. However I have a few notes about the way in which it works.

The 100µF electrolytic capacitor (labelled "F" in the schematic) is necessary to smooth the AC signal that passes from the op-amp and through the transistor. However, this capacitor does take a second or so to charge. While it is charging, the Noise Duration Timer won't respond. This simply means that there is a short delay from the moment when someone starts shouting to the moment when the Protest Response begins. Similarly, when someone stops making noise, the capacitor takes a second to discharge, so you may get one additional Protest Response cycle.

Personally I like this behavior, because the circuit gives the shouting person a brief grace period in which to behave, but once the circuit decides that he's going to keep on shouting, it adds an extra cycle just to make sure that he's got the message.

If you prefer a more immediate response, you can substitute a 47µF electrolytic smoothing capacitor. This may cause the Noise Duration Timer to retrigger itself spontaneously, because the smaller smoothing capacitor is allowing more voltage spikes to get through. You can stop the retriggering by backing off the 1M potentiometer a bit. This should still allow a reasonably sensitive response.

Some marginal differences remain between the behavior of this circuit with a benchtop power supply and with a 9V battery. The battery takes longer to charge the 100µF capacitor, and the circuit may seem a little less sensitive. If the 1M trimmer doesn't provide you with enough range, you can always increase the sensitivity by reducing the value of the 4.7K resistor labelled "A" in the schematic.

I have tuned the circuit to be run from an AC adapter because it draws too much power to be used with a 9V battery for more than demo purposes.

I used the plastic-packaged version of the 2N2222 transistors. If you use the metal-can version, they have slightly more amplifying power, and you may have to adjust the 470Ω resistor labelled "E" in the schematic.

I didn't have any problems with oscillation, but if you do, try increasing the value of the 100µF capacitor labelled "F."

Make Even More

While I was working on this project, I started thinking of other things it could do. I have a friend who has two children who turn the family TV up loud. Instead of shouting to them to "Turn that thing down!" he can simply install a Noise Protest Device to do the job for him.

Alternatively you could use it as a car alarm, if you tape it securely to the inside of a window. Any sudden vibration should trigger the electret microphone.

If you have neighbors with a noisy barking dog, you could use the output from the Noise Protest Device to retaliate by triggering an ultrasonic transducer.

A friend of mine remarked that she could use the Noise Protest Device on herself, to remind her not to shout at her business partner when she gets frustrated because a work project isn't progressing quickly enough.

Personally, though, I like the original purpose of the Noise Protest Device. I like to think of electronics pioneer Bob Widlar installing something like this in his office, so that when he really irritated someone (which seems to have happened fairly often), and they came in to yell at him, all he had to do was sit back and wait for the decibels to reach the critical level, at which point his version of the Noise Protest Device would kick in.

Probably, that would have annoyed his visitor even more.

Can You Do It with a Microcontroller?

The analog-digital converter in a typical microcontroller expects a range of voltages that is large compared with the millivoltages from a microphone. So, I think you'd still want to pass the microphone output through an op-amp and connect the output from the op-amp to the microcontroller. In fact, you can buy an electret soldered to a small board with a surface-mount amplifier on it.

Alternatively, some microcontrollers have programmable gain built in. But you will still be dealing with an AC waveform, and you'd have to sample it very rapidly to determine its amplitude. Really it would be easier for the microcontroller to process a rectified or smoothed version of the signal. That would require using a transistor and capacitor, because the output from the op-amp doesn't have enough current for easy rectification.

So, you would have to use a lot of the same components that I put into the existing circuit.

After that, your task would be very simple, because it's easy to program a microcontroller to do something in response to an input. Creating a noise output, then pausing, and then waiting for another input would be easy. In fact, you could add more features.

For instance, you could write code that counts how many times someone shouts within a short period—and the more often they shout, the more frequenty the microcontroller will tell a noisemaking device to make noise. Alternatively, with some additional components, the microcontroller could make the Protest Output gradually louder each time it is triggered.

I'm sure you could invent similar options. The bottom line, though, is that regardless of whether you use a microcontroller, you still need to know how to use an op-amp.

What's Next?

Op-amps have many more uses, but many of the applications involve some fairly difficult concepts. I'll leave it to you to read up on this subject if it interests you. (One book that I like is *Make: Analog Synthesizers*.)

I'm going to move on, now, to digital chips. I prefer digital components in many ways. They talk to each other without any worries about incompatible voltages, and they don't overreact and amplify every tiny little ripple or glitch. Within reasonable limits, their inputs and outputs are either high or low. You can think of them as being on or off, or (in binary code) 1 or 0.

Bob Widlar had no interest in digital chips and the binary code that they use. Supposedly he once said, "Every idiot can count to 1." Some of us, though, aren't quite as smart as Bob was, and for us, digital circuits provide a welcome relief from the quixotic behavior of circuits in which everything fluctuates on an unpredictable basis.

Experiment 15: It's All So Logical!

15

In *Make: Electronics* I provided an introduction to digital logic, but I avoided the more challenging aspects, and I didn't deal with components such as multiplexers or shift registers. Logic chips of this kind are less widely used today than they used to be, but logic itself remains fundamental to all computing devices. So let's go deeper, now, into that world, to learn how it works—and have some fun.

Experiment 15: Telepathy Test

This first logic project seems ridiculously simple. All you need are four pushbuttons, a couple of chips, and an LED. As we delve deeper, though, you'll find that it's not quite so simple after all.

The ostensible purpose of this experiment is to test your capability for extrasensory perception, which is often known by its acronym, ESP. I'll call the circuit a Telepathy Tester.

Background: ESP

For many decades, researchers at the fringes of conventional science have looked for evidence of paranormal abilities in the human brain. J. B. Rhine was one of the pioneers, working at Duke University. His book *Extrasensory Perception* was published in 1934, and he continued to report serious research through the 1970s. A big criticism of his record has been that on several occasions, he discovered that his research assistants were cheating. This may have been because they actually believed in ESP and didn't want one unsuccessful day to bring down their average score. Bear this in mind, if you try this project: even people who have good intentions may try to fool you, if they are also fooling themselves.

The Setup

True "mind reading" doesn't happen often (if at all), and therefore we have to adopt a statistical approach, making dozens or even thousands of attempts, and comparing the success rate with the outcome that we would expect from chance alone.

The experiment that I'll describe is intended to be used on that basis. Two buttons are placed in front of one person, and two more buttons are placed in front of a second person, seated opposite. A screen prevents either person from seeing the other's hands. Figure 15-1 gives you an idea of what I have in mind, viewed from above. You may not feel inclined to build the whole thing, especially if you don't have a suitable collaborator. But you can improvise, and you should assemble the components of the circuit to see how it works.

Figure 15-1. *Viewed from above, Annabel and Boris get ready to test their capabilities in extrasensory perception.*

I'm going to name the participants Annabel and Boris, because I like distinctive names that are easy to remember. I'll abbreviate them as A and B in some of the diagrams.

You'll notice that the buttons in front of Annabel are identified as A0 and A1, while the buttons in front of Boris are B0 and B1. The goal is for them to press buttons directly opposite each other, by trying to sense each others' intentions telepathically. If Annabel presses A0 and Boris presses B0, or if they press A1 and B1, those are successes. Conversely, A0 and B1, or A1 and B0, are failures.

Because there are four possible combinations, and only two of them indicate a success, the odds are 50-50. A significant deviation from this result could be an indication of one person sensing the other person's intentions with psychic powers—or could indicate that someone is cheating. I'll deal with the problem of cheat detection a bit later.

Figure 15-2 shows how AND and OR logic gates can be used with the pushbuttons to confirm that a trial has been successful.

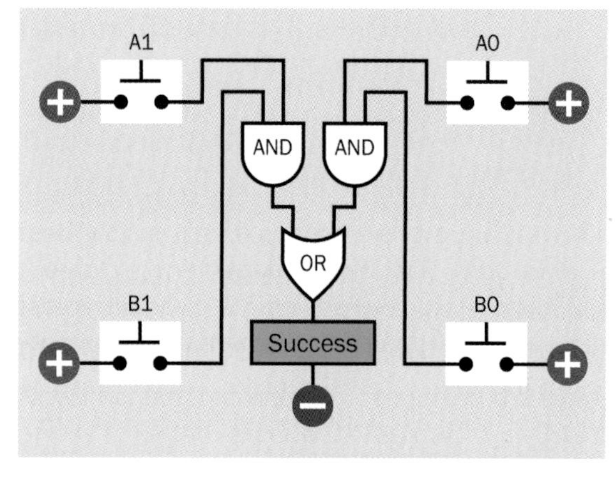

Figure 15-2. *These logic gates will activate a "success" indicator if A0 and B0 are both pressed, or if A1 and B1 are both pressed.*

If you don't remember the symbols and functions associated with logic gates, the six primary variants are shown in Figure 15-3 and Figure 15-4, with high and low inputs and outputs indicated by red and black, respectively.

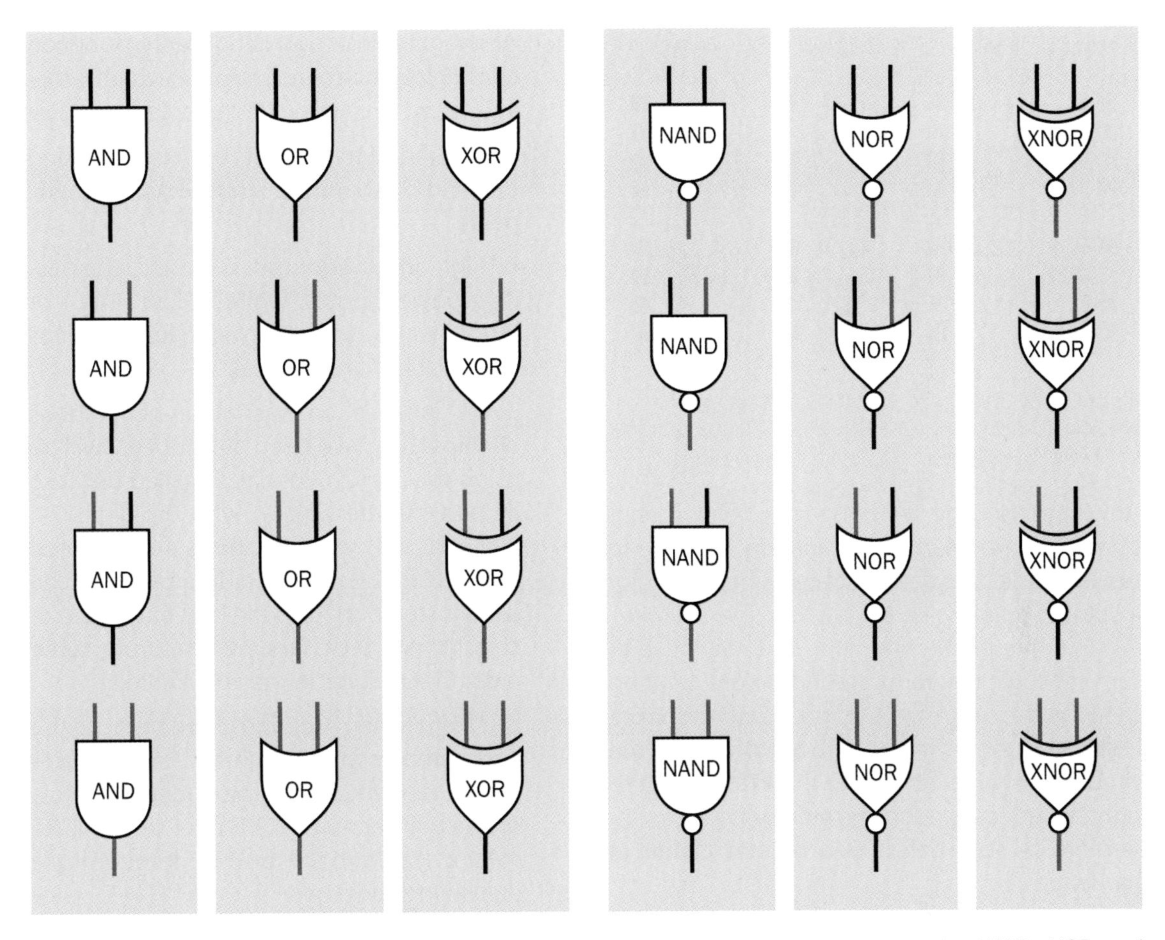

Figure 15-3. *Inputs are often shown above each logic gate, while the output of each gate is below it. If each of the two inputs can have a high or low state (represented by red or black in this figure), there are four possible combinations, each of which produces an output defined here.*

Figure 15-4. *Inputs and outputs for NAND, NOR, and XNOR gates. A red line indicates a high input or output, while a black line indicates a low input or output.*

For quick reference, Figure 15-5 shows the four possible input combinations in the column on the left, and the corresponding outputs for each gate, beside it.

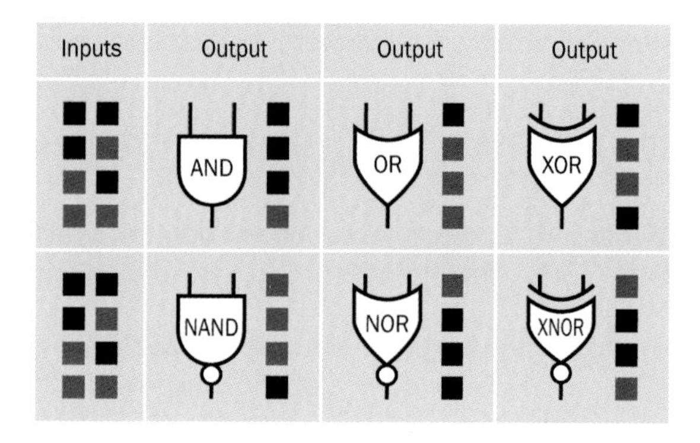

Figure 15-5. *This quick-reference chart shows the four possible input combinations and the resulting output that each gate will generate.*

Any time I use logic symbols in the development of a circuit, you can use simulation software to test the logic and demonstrate its behavior. For instance, *http://www.neuroproductions.be/logic-lab/* is a free online simulator. You may have to turn my logic diagrams around before creating them on the simulator, because I tend to show a logic flow running from top to bottom, while the simulator is more inclined to show it from left to right. In any case, a logic simulator is just an intermediate step before you build a circuit in hardware.

Quick Facts About Logic Chips

- Each logic gate is composed of multiple transistors etched into silicon. The original through-hole DIP packages for logic gates had 14 pins and are still being made today, although surface-mount versions have become widely used.
- Although computers are not built from logic chips anymore, gates still find some uses—as in "glue logic," which links different sections of a circuit board.
- A 14-pin chip may contain four two-input gates, three three-input gates, two four-input gates, or one eight-input gate. These configurations are known as quad, triple, dual, and single, describing how many gates are inside the chip.
- All the gates inside a multigate chip function completely independently of each other.
- The inputs of an unused gate in a chip should be grounded to prevent them from picking up and responding to electromagnetic fields.
- A "high" input or output is close to the positive power supply, while a "low" input or output is close to 0VDC. Negative-logic chips also exist but are unusual.
- Logic "families" are successive generations of chips that have been developed. I will be using the 74HC00 family, so called because all part numbers begin with the digits 74, while HC tells you that this is a high-speed CMOS chip. Two, three, or four digits can be used instead of 00 to identify each type of chip. Where necessary, I will also use some older CMOS chips in the 4000B family.
- Be careful when ordering chips, because the part numbers for through-hole and surface-mount variants can be almost identical. Most online vendors provide a filter that will restrict a search to DIP or PDIP through-hole dual inline packages.
- Logic chips are designed to be chained together so that the output of one connects directly to the input of the next, so long as they are both in the same family.
- A logic chip is said to "source" current in its high-output state, and "sink" current in its low-output state.
- Each logic chip in the HC family can source or sink up to 25mA of DC current, which is ample to power a generic LED. However, when you draw as much current as this from a chip, you pull down its output voltage. Use a meter to check the voltage when you are driving an LED, and be cautious of using that output also for an input to another logic chip. Increase the value of a series resistor with the LED if necessary.
- Where a pushbutton or SPST switch is connected with the input of a logic gate, you

must not allow the voltage on the input pin to "float" when the switch is open. Use a pulldown or pullup resistor to maintain a high or low voltage on the input pin. See Figure 15-6.

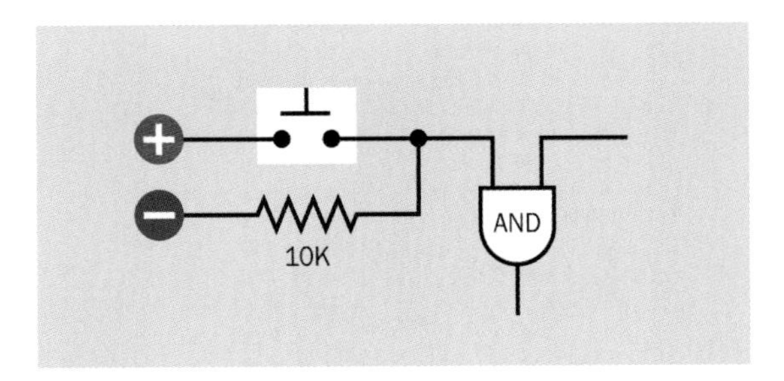

Figure 15-6. *When positive power is being connected with the input of a logic gate via a pushbutton or electromechanical switch, a pulldown resistor must be used to keep the input from "floating" when the connection is open. If the positive and negative symbols are swapped, the resistor becomes a pullup resistor.*

- A logic diagram is not the same as a component schematic. In a logic diagram, such as the one shown in Figure 15-2, power supplies to logic gates are not usually shown, and pullup or pulldown resistors are also omitted. In a component schematic, chips with pin connections are shown instead of logic gates, and all necessary power connections are included.

ESP Logic

The concept of Figure 15-2 is very simple. It can be summarized in this sentence:

If buttons A0 AND B0 are pressed, OR if buttons A1 AND B1 are pressed, the outcome is a success.

There's a clear comparison between the words AND and OR in that sentence, and the gates in the logic diagram.

I'm going to refer to the green rectangle in the diagram as an "indicator," but it could be simply an LED that lights up to tell Annabel and Boris that they guessed correctly (or read each others' minds, if you prefer to believe that this is what's happening).

So far, the logic is elementary, but please just take a moment to build the circuit. As always, the hands-on process is the best way to learn.

Building It

The inner workings of the quad two-input AND and OR chips are shown in Figure 15-7.

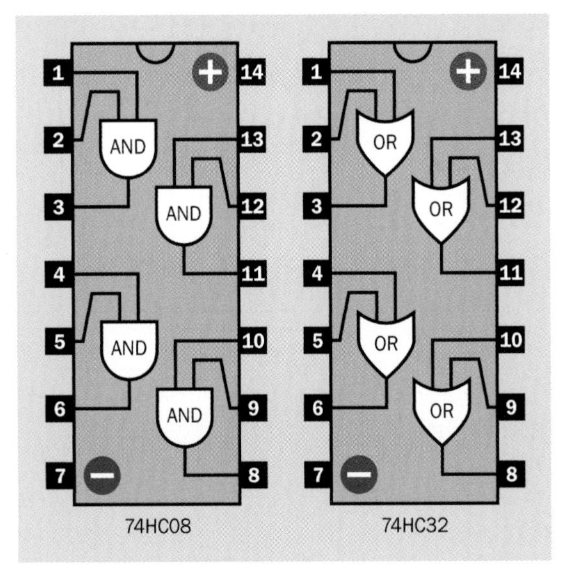

Figure 15-7. *Each 14-pin logic chip can contain four separate two-input AND or OR logic gates, as shown here. These types of chips are referred to as quad two-input.*

The schematic that corresponds with the logic diagram in Figure 15-2 is shown in Figure 15-8. I have squeezed some miniature representations of the gates into the chips to show what's going on. "&" denotes an AND gate, while "O" is an OR gate. (These are not standardized abbreviations.)

- The positive bus is now on the right-hand side of the schematic, as this is convenient to power the logic chips, which receive positive voltage on pin 14 in each case. Most of the circuits in the remainder of the book will have the positive bus on the right. Be careful not to power your chips with reversed polarity; they may never recover from the shock.

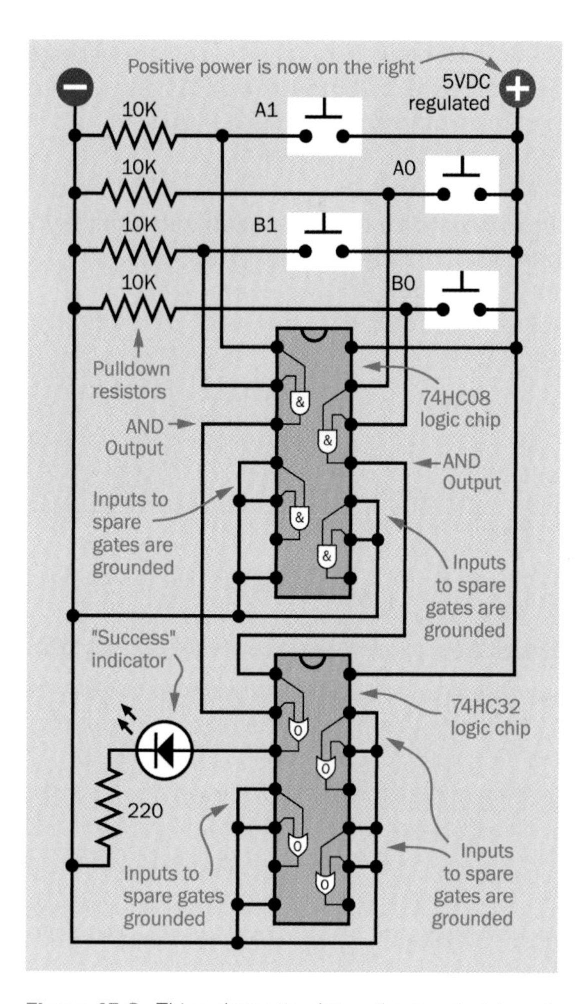

Figure 15-8. *This schematic shows the simplest breadboardable version of the ESP test.*

The breadboarded version of this circuit is shown in Figure 15-9.

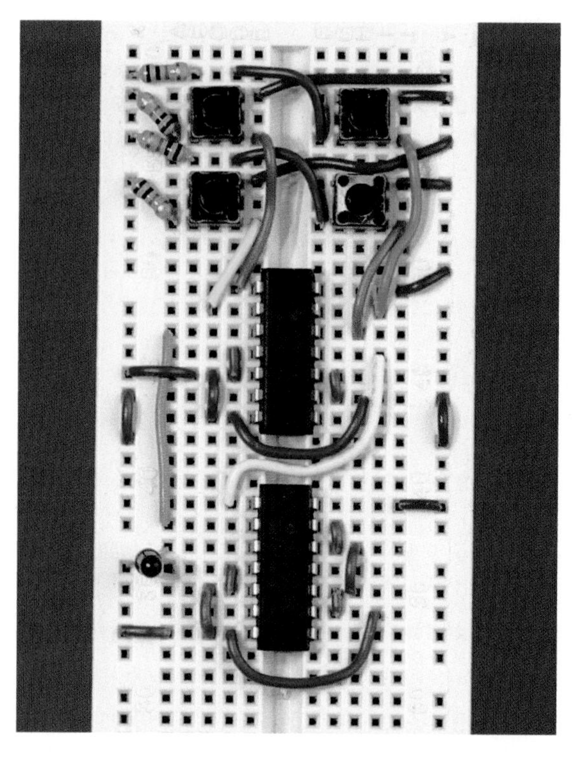

Figure 15-9. *The simplest and most basic demonstration version of the Telepathy Tester. Pushbuttons for the participants are included as four tactile switches at the top. The LED at lower left is the only output.*

I have not shown the 7805 voltage regulator and the two capacitors, which are always required with 74HC00 chips. Any time you see "5VDC regulated" on a schematic, you know that the voltage regulator and its capacitors are necessary.

Unused chip inputs must be tied to negative ground to prevent them from responding to stray electromagnetic fields. Unused outputs from each gate can be left unconnected.

You should find that if you press the pair of buttons labelled A0 and B0, or the buttons labelled A1 and B1, the LED lights up, while the other combinations of buttons will not do anything.

So far, so good. But now that you have the circuit in front of you, I think I can convince you that it badly needs some enhancements if it's going to become seriously usable.

Making It Better

Generally, I thinking that we need to add user-friendly notifications and cheat-proofing:

- In an ESP test, Annabel must not be able to see which button Boris is pressing, and Boris must not be able to see which button Annabel is pressing. This creates a problem: how does either person know when the next match will begin? Really we need a "ready" prompt for Annabel, which will light up when Boris is pressing a button but she hasn't pressed hers yet; and a "ready" prompt for Boris, which will light up when Annabel is pressing a button but he hasn't pressed his yet.
- As I mentioned earlier, even sincere people may be tempted to cheat if they believe that their ESP powers need a little bit of help because they aren't working well on this particular day. Unfortunately, in our Telepathy Test, cheating is really easy. A or B can guarantee a hit just by pressing both buttons at once!
- Currently we only have one indicator that lights up when Annabel and Boris are successful. We should really have another indicator which lights up if they fail.

In the next experiment, I'll implement these improvements—with some surprising results.

Experiment 16: Enhanced ESP

16

Before providing you with a revised schematic, I'm going to start by spelling out the requirements in words. I think a verbal description should always be the first step in developing a logic diagram.

The second step is to convert the logic diagram into a schematic that uses actual components.

Are You Ready?

Adding "ready" prompts to the Telepathy Tester seems easy enough. Perhaps I can express the task like this:

"If button A0 OR A1 is pressed, then an indicator tells Boris that Annabel is ready. Alternatively, if button B0 OR B1 is pressed, then another indicator shows Annabel that Boris is ready."

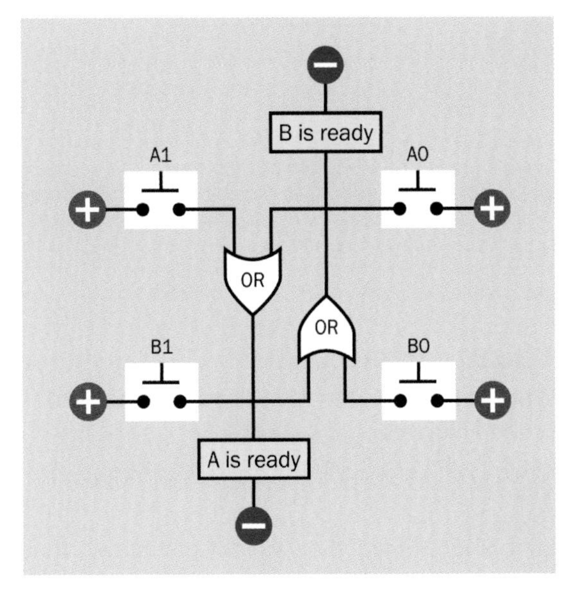

Figure 16-1. *Two OR gates can be used to show each person if the other person has already pressed a button. The "ready" indicators do not reveal which button has actually been pressed.*

Figure 16-1 translates these two sentences into a logic diagram. For the sake of clarity, I have shown this separately from the previous logic diagram in Figure 15-2. However, the two diagrams could be combined, as shown in Figure 16-2, because you can take the output from a single pushbutton and share it among two or more logic inputs. (In Figure 16-2 I have showed the connections from Figure 15-2 in gray, to distinguish them from the new connections.)

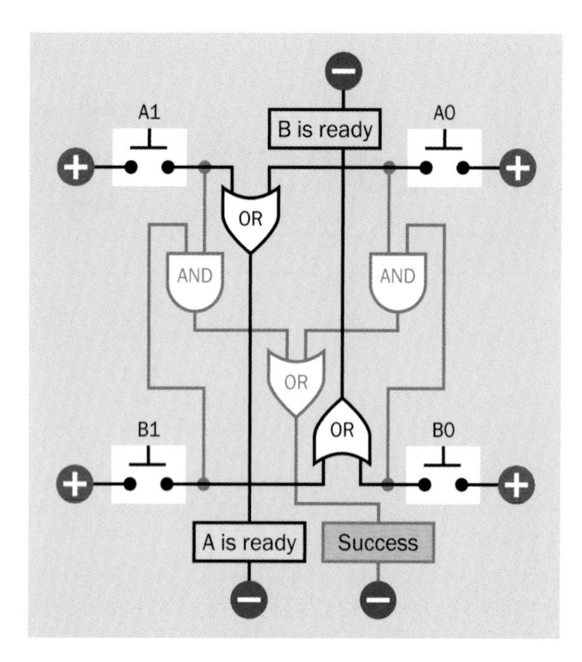

Figure 16-2. The previous two logic diagrams are shown here combined, sharing each pushbutton output between inputs on two separate logic gates. Unfortunately, combining multiple logic diagrams quickly results in complexity, which is difficult to interpret. In an effort to provide some clarity, connections from the previous logic diagram are shown here in gray.

Cheating Revealed

What about the "Cheat" indicator? It can be described like this:

"If A0 AND A1 are being pressed, then an indicator shows that Annabel is cheating. If B0 AND B1 are being pressed, then an indicator shows that Boris is cheating."

Figure 16-3 shows this as another separate logic diagram. Once again, this can be combined with the other diagrams by sharing the output from each pushbutton among multiple logic gates.

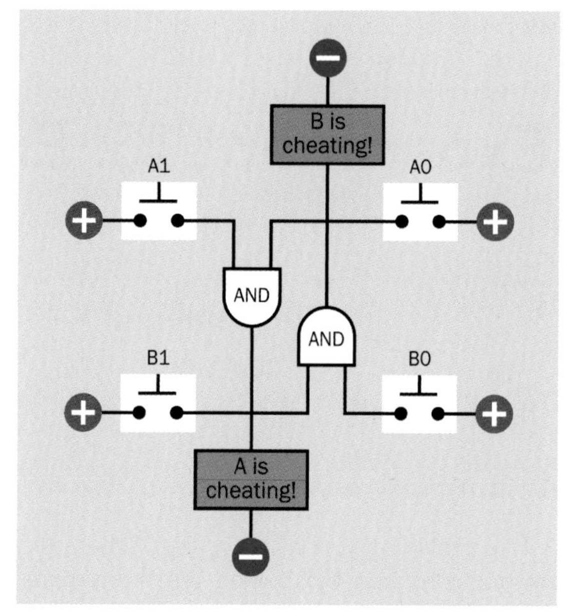

Figure 16-3. By adding two AND gates, we can see if either player has cheated by pressing two buttons at once.

Failure Indicated

Lastly, the "Fail" indicator. If the players press buttons that are not opposite each other, the test is a failure. Well, this can be displayed easily. We can describe it like this:

"If A0 AND B1 are pressed, OR if A1 AND B0 are pressed, then the players fail."

Figure 16-4 shows a logic diagram representing this statement.

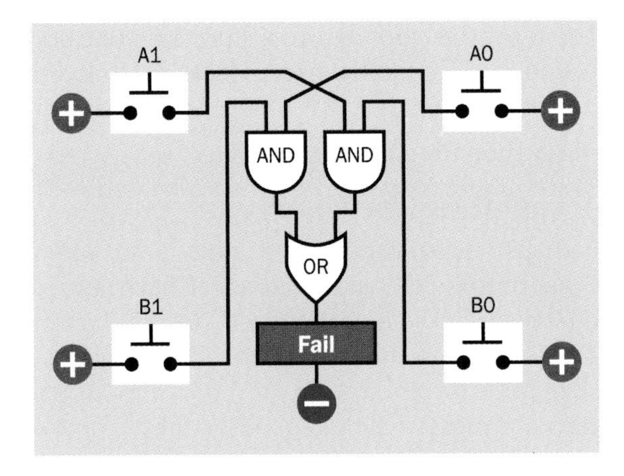

Figure 16-4. *A couple more AND gates, feeding into an OR gate, will show if either of the incorrect button combinations is pressed.*

Would you like to add this functionality to the breadboarded circuit in Figure 15-8? You may have noticed that there were several gates in the chips in that schematic that were not being used. Because every gate operates independently of the others in a multigate chip, we could make use of unused gates to provide the initial features that I have described above. For instance, the "ready" test requires just two OR gates, and there are three OR gates inside the 74HC32 chip that are not being used.

I'm tempted to say, "Sure, go ahead, add the new features!" But maybe you'd better not, because I don't want you to get mad at me if you take time to build something that doesn't work properly. And the fact is, if you add the features I just described, they won't work properly. I can prove this by asking you to try a little thought experiment.

Conflicts

Suppose that Boris cheats by pressing B0 and B1 simultaneously. At the same time, Annabel is playing by the rules, so she presses only one button, A0. What happens?

Buttons A0, B0, and B1 are all being pressed. Because B0 and B1 constitute cheating, the "Cheat" indicator comes on, as it should.

But wait—Annabel is pressing A0. Because A0 and B0 are both being pressed, the "Success" indicator comes on!

This is not all. Because B1 and A0 are also being pressed, the "Fail" indicator comes on. And because each player has pressed at least one button, both of the "ready" indicators lights up.

This is a catastrophe. *All* the indicators come on!

What went wrong? The problem was that my descriptions were inadequate. I only thought about the buttons that have to be pressed to create a logical output. I didn't consider which buttons must *not* be pressed to create that output. For example, I began describing the "ready" logic like this:

"If button A0 OR A1 is pressed, then an indicator tells Boris that Annabel is ready."

But I should have written it like this:

"If button A0 OR A1 is pressed, AND neither B0 NOR B1 has been pressed yet, then an indicator tells Boris that Annabel is ready."

In other words, I have to make sure that Boris hasn't pressed one of his buttons yet, before the system tells him that Annabel is waiting for him to press one of his buttons.

Similarly, the "Success" indicator or the "Fail" indicator should only light up if neither A NOR B has cheated.

Hmm, that NOR word has turned up twice, now. Evidently I am going to have to use a NOR gate here, somehow, and this is all starting to look extremely complicated. Who would have thought that such a simple game could create unexpected issues? I think a chart will help to untangle the confusion.

The Untangling

Take a look at Figure 16-5. This now takes into account the buttons that are not being pressed, as well as the buttons that are being pressed. A0, A1, B0, and B1 represent the four buttons, with red meaning that a button is pressed, and black meaning that a button is not being pressed. A gray button with an X in it means that its state doesn't matter because it is irrelevant to this particular logic test. On the right are the messages that should be created by each combination of buttons. Remember, the black (unpressed) button states are just as important now as the red (pressed) button states.

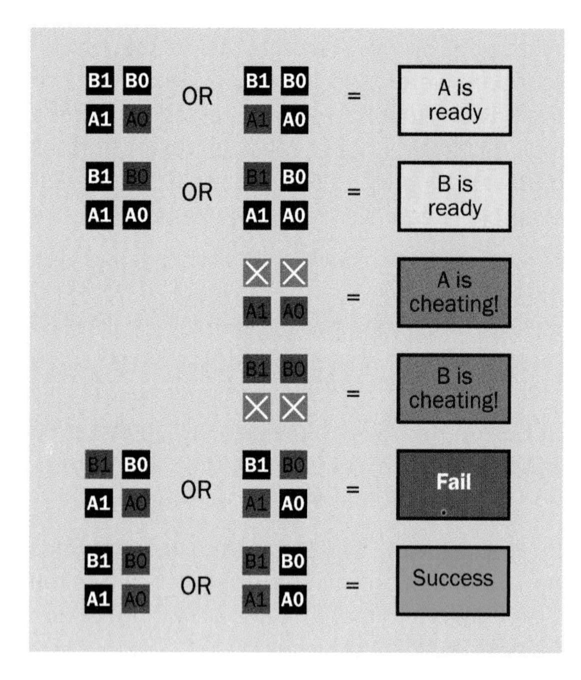

Figure 16-5. *In this chart, A0, A1, B0, and B1 indicate a button that is pressed (red) or not pressed (black). A gray X means that in that particular test, the state of the button is irrelevant and can be ignored. The colored boxes on the right represent the indicators that should light up in response to each combination of buttons.*

On the first line, Boris should only see a "ready" light for Annabel if Annabel has pressed one button (not both, which would be cheating), AND Boris has pressed neither B0 NOR B1.

The second line uses this same logic for Annabel's "ready" light.

In the third and fourth lines, if someone is cheating by pressing both buttons, we don't care what the other person is doing.

The "Fail" indicator and the "Success" indicator will light up only if those specific red-colored buttons are pressed, but other buttons are not being pressed.

Translating the Chart

Perhaps you're wondering if there is still some other button combination that will have consequences that I haven't predicted. No, the chart in Figure 16-5 specifies an outcome for every possible button combination. (If three or four buttons are pressed simultaneously, these combinations are all covered by the "Cheating" test.)

Now I can translate each line of the chart into a logic diagram, and this time, I feel confident that it will work. However, I will require two additional types of gates, to do this: NOR and XOR. In case you don't recall how they function, you can check back to Figure 15-5.

Describing them in words, the NOR and XOR gates work like this:

- The NOR gate has a low output if either or both of its inputs are high. The output only goes high if both of its inputs are low.
- An XOR gate has a low output if both of its inputs are high or low. The output only goes high if either of its inputs is high while the other is low. (XOR stands for "exclusive-OR," in case you were wondering. It is pronounced "ex-or.")

In Figure 16-6 you can see how these gates are used to imitate the tests shown in Figure 16-5.

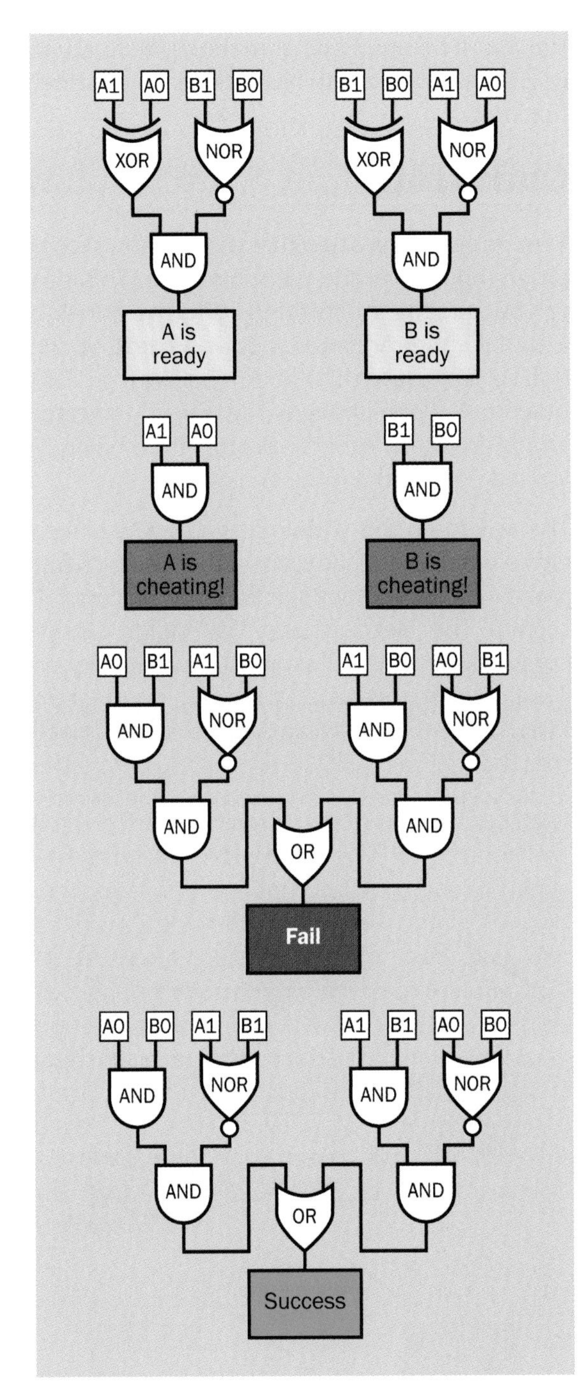

Figure 16-6. *Logic diagrams that emulate the relationships shown in the previous figure. Each of the logic inputs (A0, A1, B0, B1) represents a connection with a pushbutton that can be pressed or not pressed and will provide a high input when it is pressed, while a pulldown resistor (not shown) will provide a low input when the button is not pressed.*

In Figure 16-7 I have selected four possible button combinations as samples to see how they work in the logic diagram that decides whether the "A is ready" indicator should light up. In the lower two combinations, the "ready" indicator is not illuminated.

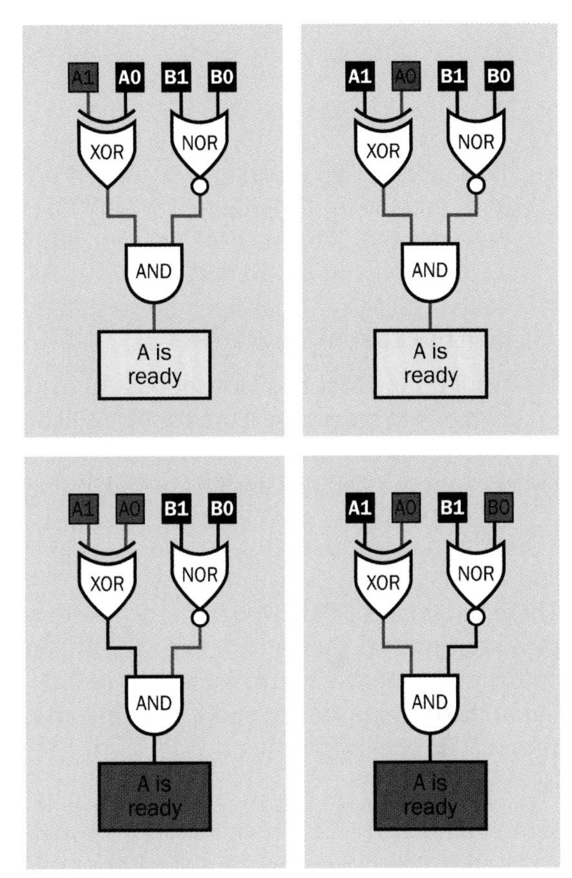

Figure 16-7. *The results of four arbitrary combinations of the pushbuttons, for the "ready" test for Player A.*

The diagrams in Figure 16-6 can be summarized in words like this:

- Testing for "Player A is ready": if either A0 or A1 is high (not both), AND neither B0 NOR B1 has been pressed yet, the "A is ready" indicator lights up.
- Testing for "Player B is ready": if either B0 or B1 is high (not both), AND neither A0 NOR

A1 has been pressed yet, the "B is ready" indicator lights up.

- Testing for "Player A is cheating": if A1 AND A0 are both pressed, the "A is cheating!" indicator lights up.
- Testing for "Player B is cheating": if B1 AND B0 are both pressed, the "B is cheating!" indicator lights up.

Now things get slightly more complicated:

- Testing for a "Fail" situation: if A0 AND B1 are both pressed, AND neither A1 NOR B0 has been pressed, then the "Fail" indicator lights up. OR, if A1 and B0 are both pressed, AND neither A0 NOR B1 has been pressed, then then the "Fail" indicator lights up.
- Testing for a "Success" situation: if A0 AND B0 are both pressed, AND neither A1 NOR B1 has been pressed, then the "Success" indicator lights up. OR, if A1 and B1 are both pressed, AND if neither A0 NOR B0 has been pressed, then the "Success" indicator lights up.

These statements are derived from each pattern of button presses shown in Figure 16-5. If you read the statements aloud to yourself while looking at the button-press symbols, you'll see that they match, one-for-one.

Now you could build the circuit. You would assemble enough chips to provide all the necessary gates and carefully link each button with the various appropriate inputs of the gates. Remember, it's quite okay to connect one button to several gate inputs.

The trouble is, this has turned into a much larger project than you might have expected. To provide all the logic gates shown in Figure 16-6 you would need three quad two-input AND chips, one quad two-input OR chip, two quad two-input NOR chips, and one quad two-input XOR chip. That's seven, altogether. They would not all fit onto a breadboard.

Hm. Can it be simplified, somehow? Well, actually, yes, it can. We can think of this as "optimizing" the logic.

Optimizing

Remember earlier I said that the "cheating" condition should override everything else? This idea can be used to simplify things. What I'm thinking is this: if either Annabel or Boris is cheating, we don't even have to think about "Success" or "Fail" outcomes. We just light up the "Cheat" indicator and prevent any other indicators from being illuminated, and that's the end of it.

The way to do this with logic gates is to have a cheat-detection section, and if there is no cheating, it sends an "okay" signal to the success/fail section. The "Success" and "Fail" lights cannot come on without the "okay, no cheating" signal. I've illustrated this concept using a flow diagram in Figure 16-8. In this diagram, we have to make sure that neither A NOR B is cheating before the "okay" signal leads us down to see whether the trial is a success.

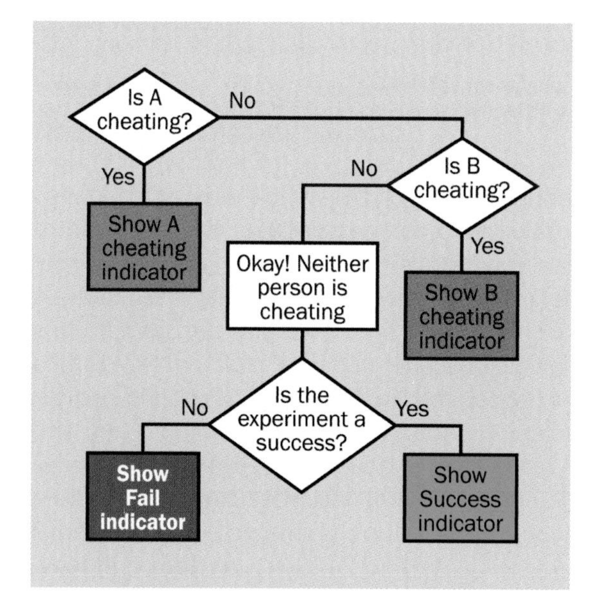

Figure 16-8. *A flow diagram shows the concept of verifying that no cheating has occurred, before we determine whether the outcome is a success or a fail.*

How can we translate this into logic gates? Well, a NOR gate gives a high output if, and only if, both of its inputs are low. If we use a low input to mean that a person isn't cheating, the output from the NOR gate will be high so long as neither person is cheating.

In the lower section of Figure 16-9 you see a NOR gate added, tapping in to the "cheating" indicators for A and B. So long as NEITHER Annabel NOR Boris is cheating, the NOR gate gives a high output. This is added to some three-input AND gates, which require all their inputs to be high before a "Success" or "Fail" indicator can light up. This is the same as saying that if either player cheats, the "Success" and "Fail" indicators won't light up.

To accommodate the three-input AND gates, I also had to rework the logic at the bottom of the diagram, using four XOR gates. I leave it to you to figure out why they should work. Just refer back to Figure 16-5.

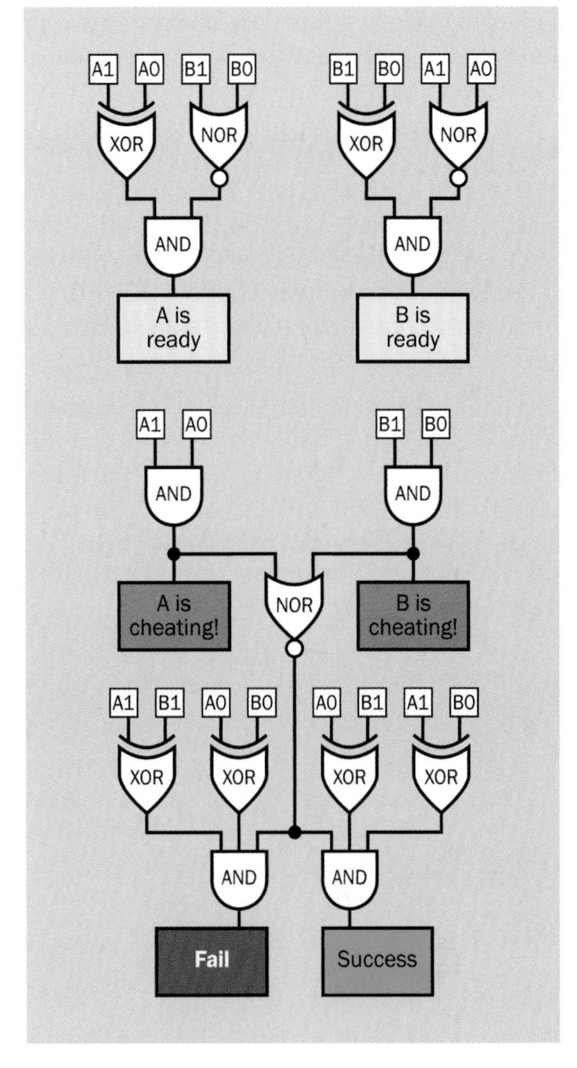

Figure 16-9. *A simplified version of the previous logic diagram adds a NOR gate, which must give a high input, meaning that neither A NOR B is cheating, before the "Success" or "Fail" indicator can light up.*

Perhaps you are wondering—what is the system for simplifying or optimizing a logic diagram? The formal way to do it is by learning proper Boolean notation and looking for functions that duplicate or contradict each other. There's an example of this kind of thing on Wikipedia (*http://bit.ly/1jJBThe*), but I find it difficult. I prefer just to stare at a logic diagram while imagining all of its possible states and trying to see other ways to satisfy the requirements. Once I think I have a simplified version, I check every combination of inputs to make sure that they all work. This is an intuitive approach, not the classical approach, but it works for me.

Building It

Now that the circuit has been optimized, you can build it. You will need one quad two-input AND chip, one triple three-input AND, one quad two-input NOR, and two quad two-input XORs. That's just five chips instead of seven. (I managed to get rid of the OR chip entirely.)

The pinouts of the additional chips are shown in Figure 16-10 and Figure 16-11. Notice that the NOR gates, inside the chip, are inverted compared with AND, OR, and XOR gates inside their chips. Be careful when making connections.

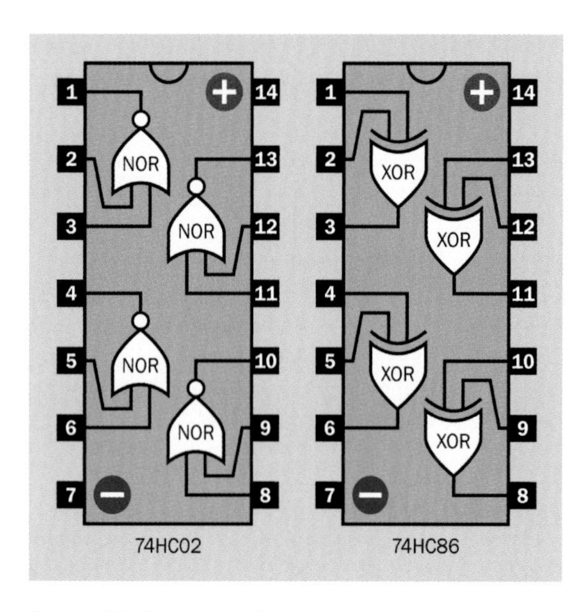

Figure 16-10. *A 14-pin logic chip can contain four two-input NOR gates or four two-input XOR gates in the configurations shown here. It's important to remember that the inputs and outputs of the NOR gates are inverted compared with the gates inside other quad-two-input chips.*

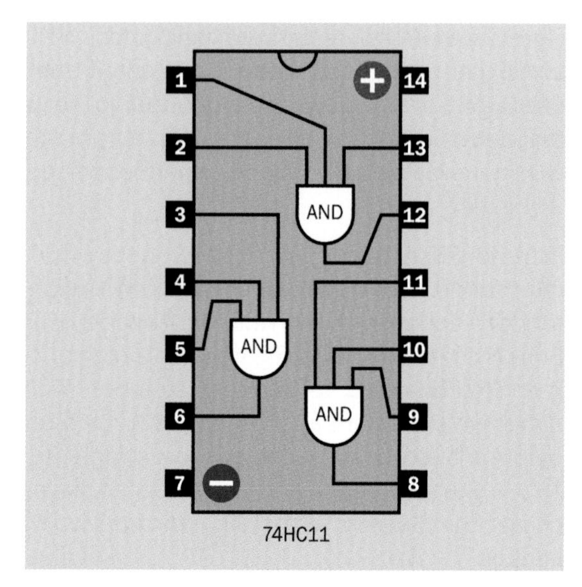

Figure 16-11. *A 14-pin logic chip can contain three three-input gates, as shown here.*

You can see how everything is wired together for a breadboard layout in Figure 16-12. XOR gates have an X in them, AND gates have an & sign, and NOR gates are represented by letter N. (These are not standard abbreviations.)

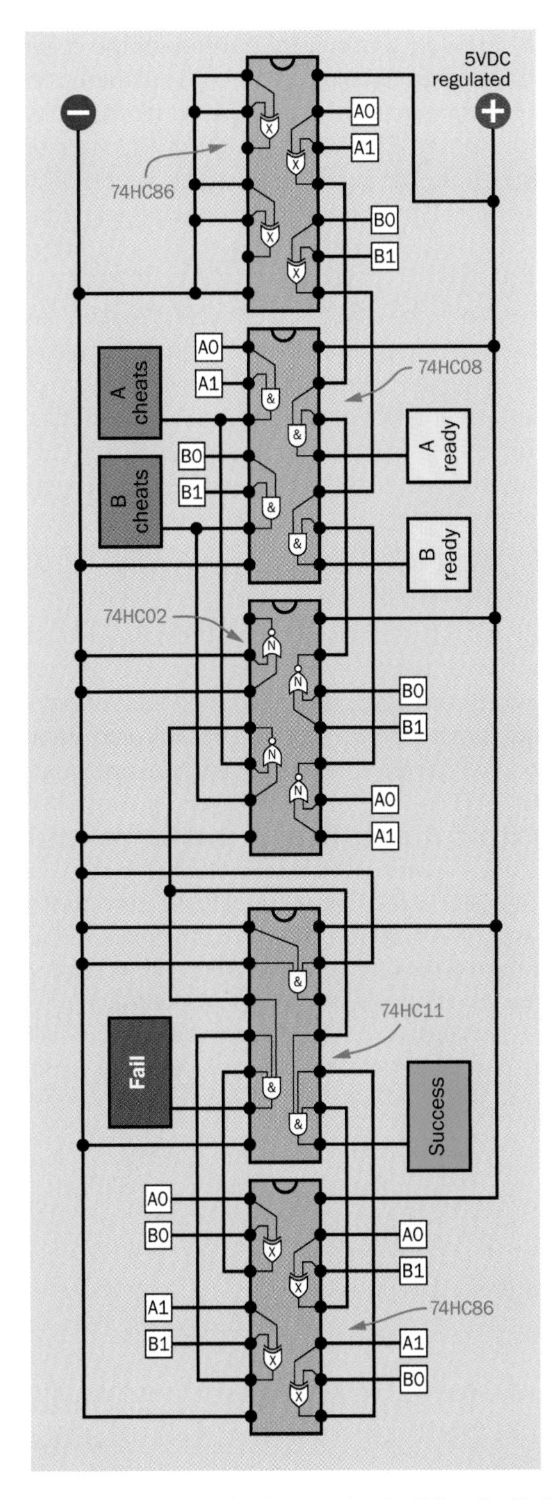

Figure 16-12. *The final schematic for the Telepathy Test, using five logic chips.*

Because the schematic in Figure 16-12 barely fits within one printed book column, I had to leave out the pushbuttons (which you are likely to mount separately anyway). They are shown in Figure 16-13. The output from each button goes to five button labels. You need to connect each of these labels to an identical label in Figure 16-12. Thus, the first A0 label in Figure 16-13 can connect with the A0 label in the XOR chip at the top of Figure 16-12, and so on. The easiest way to accomplish this is with a multicolor ribbon cable.

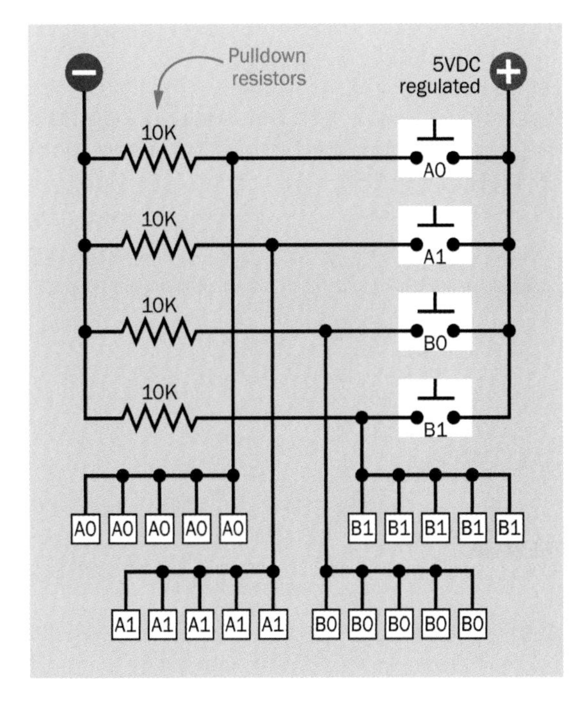

Figure 16-13. *Four SPST normally-open pushbuttons should be used in conjunction with the ESP Test circuit. The output from each of them is distributed among five chips. Because each group of five button-inputs is high or low simultaneously, one pulldown resistor will be sufficient for each group.*

Probably you would also mount the LED message indicators separately. Just be careful to add a 220Ω series resistor for each LED, to avoid overloading the outputs from the logic chips (unless you are using LEDs that contain their own series resistors).

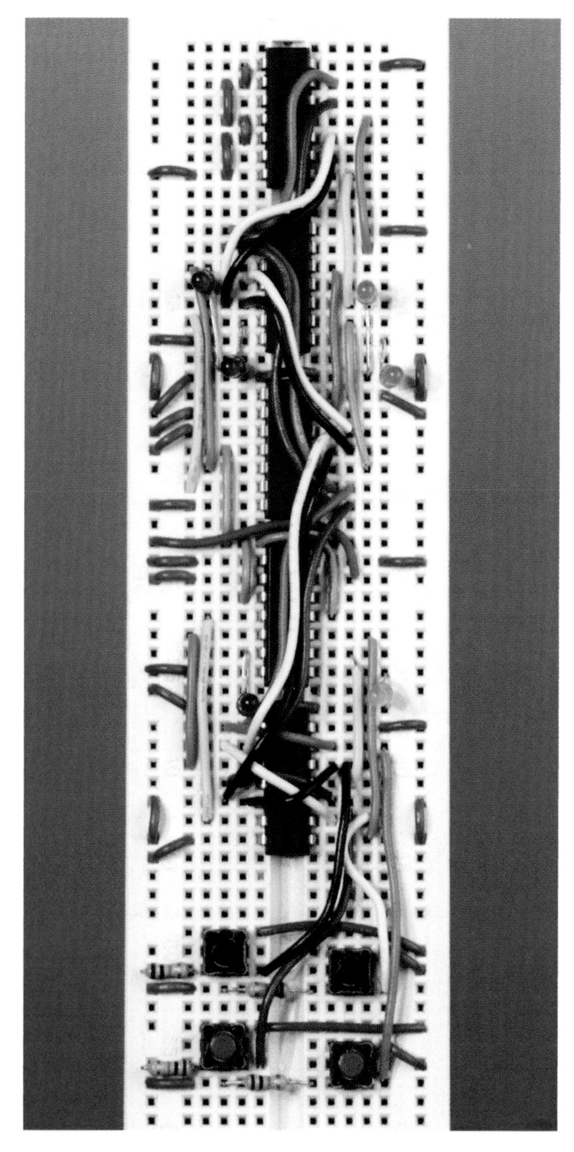

Figure 16-14. *A breadboarded version of the complete Telepathy Test, using five logic chips. Six LEDs provide outputs that are identified in the schematic of the circuit. No series resistors are shown because I chose the type of LED that contains its own series resistor. The two push-buttons for Player A are the red ones at the bottom. The buttons for Player B are the black ones, immediately above. This configuration of the circuit is only for demonstration purposes. For a functional Telepathy Test, the buttons would be located separately so that the players would be unable to see each other's choices.*

In Figure 16-14 you can see a breadboarded version of this circuit. After I made the connections between the five chips, I realized there was still just enough space at the bottom of the breadboard for four tactile switches, so I put them there and connected each of them to its respective logic inputs. There was no space to locate the wires from the switches along each side of the board, so I ran some of them down the middle.

Details

You should be careful that the LEDs that you use as "Cheat" indicators (the orange rectangles in the schematic) don't affect the voltage output from their AND gates too much. The outputs from these AND gates don't just supply the LEDs; they also connect with the inputs of NOR gates in the next chip. Each of these inputs must be at least 4V when there is a high output from the AND gate supplying it. An LED may pull down this voltage, so be sure to check it with your meter.

If you decide to build a fully-functional version of this circuit instead of just a breadboard demo, you will need two copies of each "Success" and "Fail" LED—one for each player. To control the load of two LEDs driven in parallel, you can increase the series resistor or amplify the chip outputs. The easiest way to amplify the outputs is to pass them through the transistor pairs in Darlington array such as the ULN2003 chip. Darlingtons that can amplify seven or eight inputs are readily available.

Alternatively, you can drive each pair of LEDs in series. The advantage of this plan is that none of the power from the logic chip is wasted in a resistor. Unfortunately the results are difficult to predict, because LEDs are more sensitive to current than to voltage, and different LEDs have very different specifications. Be sure to measure the current passing through LEDs when they are wired in series, and add a small resistor if you find that the LEDs are drawing more than their rated current.

I really encourage you to build this circuit. It's quite easy, because very few components are involved. The only requirement, really, is to be careful and precise in making connections. I think you can do that if you decide to give it a try.

If you actually use a version of this circuit, bear in mind that one player's actions may not be entirely random, and the other player may learn this pattern—even if the first player is unaware of it. This takes us into the realm of randomicity, which I will be dealing with later.

The Digital Difference

A superficial inspection of the schematic in Figure 16-12 shows that it looks very different from the schematics in the experiments that involved op-amps. For a start, there are no capacitors, and the only resistors are for pulldown purposes. No transistors, either! This is because logic chips have been designed to talk to each other without any help from other components. So long as you use chips from the same family, the output from one is guaranteed to be acceptable to the input of another.

You can also share the output from one gate among the inputs of several other gates. For instance, the output from the lower-left NOR gate in the 74HC02 chip in Figure 16-12 is shared by the inputs of two AND gates below. Driving multiple gates from one output is known as *fanout*, and the HC series should allow a single logic output to be shared among as many as 10 logic inputs.

Making It Even Better

Now, could the schematic be optimized still further? Probably. I'm thinking that instead of testing separately for "Fail" and "Success," I could use logic that basically says: if the central NOR gate says that neither player is cheating, and two buttons have been pressed, and we don't have a "Fail" condition, then we must have a "Success" condition. In other words, "Success" could be defined as "NOT Fail."

However, my head is starting to ache from trying to figure all this out, so I will not attempt to optimize the circuit any further. You should give it a try yourself, though, if logic interests you. If you succeed in reducing the chip count from five to four, please let me know. But don't try too hard, because after the next experiment, I'm going to show you how the number of logic chips can be reduced to just two, if we add one extra component called a decoder. Stay tuned!

Not So Simple?

When you started reading about the Telepathy Tester, maybe you thought it was too simple to be interesting. But now maybe you're feeling that it became too *complicated* to be interesting! Well, okay, once again, you don't have to wrestle with it if you don't want to. Hold on to the conclusions we reached, though, because they are recurring facts of life in digital circuits:

- A logic problem which looks simple at first can become extremely complicated when too many extra tests and conditions are added. Logic conflicts can occur. It's easy to imagine new features that would be nice to have—but think twice before you try to add them.
- User input is always a problem, because we have to imagine every possible quirk of human behavior that may occur, and process it appropriately.
- There are systematic ways to build a logic circuit, but they will not create the simplest possible circuit with the lowest chip count. Optimizing can almost always reduce the chip count—but on the other hand, it can create a circuit that is more difficult to understand, more likely to contain errors, and perhaps more difficult to modify subsequently.

Desktop computers have always used microprocessors, but logic chips were needed in addition, and the number of chips (the "chip count") was a big deal, because chips weren't cheap in those days. The Telepathy Test shows you the kind of design process that the computer pioneers went through. Even today, people who design CPUs are still dealing with logic states—but their task has become easier because they not only have better design tools, but powerful simulation software to test everything.

Could We Use a Microcontroller?

Well, could we? Of course the answer is yes, absolutely, yes! Each button would connect with an input to the microcontroller, and you would write program code to look for different patterns of button-presses. IF-THEN statements would then branch to different outcomes.

Logic errors would still be possible (actually, they would be probable) but overall, the design process would be much less of a headache, and the hardware would become trivial. Instead of fifteen logic gates in five logic chips, you would just need one microcontroller. If someone asked me to build a Telepathy Tester tomorrow, I have no doubt that I would use a microcontroller.

However, my purpose here is to show you how stuff works. Logic is absolutely fundamental in all digital devices, and the best way to learn about it (as always) is with hands-on involvement. For that purpose, there's no substitute for those old logic chips.

Although, actually—there are two possible substitutes. As I just mentioned, there is the possibility of using a decoder, which contains multiple logic gates, so that you don't have to wire so many gates together yourself. I'll get to that in experiments 19 and 20.

The other possibility is to replace each logic gate with nothing more than a pair of old-fashioned electromechanical switches.

I'm going to show you a game circuit, next, which can be built from switches—although you will see that it can still be built from chips, or even a combination of switches and chips, if you prefer.

Experiment 17: Let's Rock!

17

Rock, Paper, Scissors is a truly ancient, international game, but just in case you have somehow never played it, I will recap the rules. Two opponents face each other, and on a count of three, each of them makes a hand gesture. The gesture can consist of:

- A fist, to represent a rock
- A flat palm, to represent paper
- Two fingers to indicate scissors

Comparisons determine the winner. Rock blunts scissors, scissors cut paper, and paper wraps rock.

There's an obvious similarity, here, to the Telepathy Test, because two people are facing each other, trying to sense each other's intentions. However, there are some differences which will affect an electronic version of the game. First, each player has three choices instead of two. Second, if both players make the same choice, the game is a tie. Third, if they make different choices, one of them wins while the other loses.

Background: Probability

Let's assume for a moment that telepathy doesn't exist. Does this mean that Rock, Paper, Scissors is a game of pure chance?

No, because two human beings will be playing the game, and their choices will not be entirely random. In fact, many people have irrational ideas about randomicity.

For instance, it's a common belief that if you're tossing a coin, and heads comes up ten times in a row, tails will be more likely to come up next. This is known as the Monte Carlo fallacy, named after a casino in Monte Carlo where black came up twenty-six times in succession on one of its roulette wheels on August 18, 1913. A lot of people lost a lot of money midway through that streak by betting on red. After all, if black had come up ten times, or fifteen times, or even twenty times, red just had to come up next—or so they believed.

The belief is a fallacy because a roulette wheel has no memory. Nor does a coin. If you have thrown heads a dozen times in a row, the coin doesn't know that. Consequently, the odds of heads coming up on the next throw are still exactly the same as before.

People, however, are different. They do remember what they've done, and their memories influence their decisions. If someone playing Rock, Paper, Scissors knows that he has made a rock on three consecutive turns of the game, he's more likely to try something else on the next turn. He'll probably feel that to behave unpredictably, he shouldn't keep repeating himself. Therefore he becomes more likely to choose paper or scissors on the next turn.

You can take advantage of this by making scissors. This way, you'll either tie or win. Either way, you won't lose.

The trouble is, if you're playing against an experienced opponent, he may realize that you won't expect him to repeat himself. Consequently, he may repeat himself just to defy your expectations.

But what if you know him well enough to expect this? Once again, you can anticipate his behavior and modify your strategy accordingly—but if he picks up on that, he will modify his.

This recursive process in which people keep trying to second-guess each other is a common theme in the fascinating field of game theory, a branch of mathematics that became so influential in the 1960s, it affected US foreign policy and the nuclear arms race.

Background: Game Theory

Game theory was established as a discipline in 1944 when computer genius John von Neumann published *Theory of Games and Economic Behavior* with Oskar Morgenstern. The concepts were refined in the early 1950s and quickly became popular among theorists in RAND, the think tank in Washington, DC.

Game theory can describe any situation where two (or more) "players" are seeking a strategy to acquire an advantage but do not have complete information about each other, or cannot trust each other. In a game of poker, for instance, a player may bluff, while other players must try to figure out if the person is bluffing, and if so, how to respond. The response can then feed back to affect the player who chose to bluff, until finally the cards are revealed, and one player wins.

A military confrontation can likewise involve bluffing, challenges, and attempts to second-guess the opponent. This led RAND advisors such as Herman Kahn to argue that it might be "rational" under some circumstances for the Soviet Union to launch a first strike against the United States. Therefore both nations should build a capability to launch a second strike after suffering a first strike, since this would act as a deterrent to the first strike.

At *http://www.gametheory.net* you will find a simple description of some of the assumptions of game theory. One of them is that players will behave rationally to maximize their self interest. Yet would any statesman really decide that it was "rational" to kill hundreds of millions of people in another nation, and turn it into a radioactive wasteland, to gain an advantage?

Maybe not—but game theory said that you couldn't be sure of that. Therefore the United States spent billions of dollars to develop the hydrogen bomb and deploy missiles in silos that were designed to survive a preemptive attack.

This may seem a long way from a game such as Rock, Paper, Scissors, but really the only differences are the scale, the number of variables, and the seriousness of the outcome.

The Logic

How should we represent the logic for a Rock, Paper, Scissors game? If you followed the steps to develop the Telepathy Tester, this shouldn't be too hard. The basic logic for Player A is shown in Figure 17-1. Annabel and Boris are now playing this game under their previous identities as Player A and Player B, and each of them has a choice of three pushbuttons labelled Rock, Paper, and Scissors. For convenience I have abbreviated Annabel's pushbuttons as AR, AP, and AS, while Boris has BR, BP, and BS.

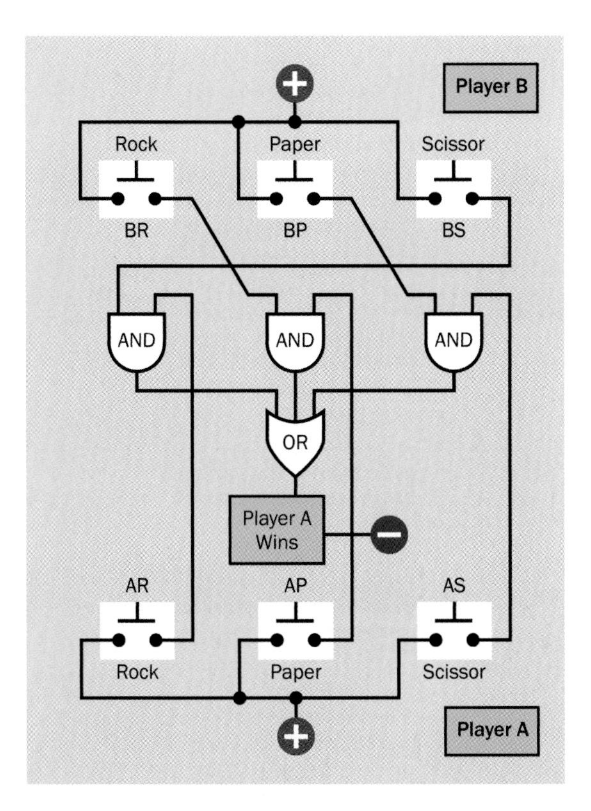

Figure 17-1. *This network of logic gates will display a "Win" message in response to any of the three winning combinations for Player A in the Rock, Paper, Scissors game.*

If Annabel presses the Rock button while Boris presses the Scissors button, Annabel wins, because rock blunts scissors. If A presses Paper and B presses Rock, A wins again, because paper wraps rock. If A presses Scissors while B presses Paper, A still wins, as scissors cut paper. These three winning options are shown in the figure, as you'll see if you trace out the connections.

Boris, of course, has his own three winning options, and then there are three possible ties if Annabel and Boris happen to press the same-labelled buttons. This is all summarized in Figure 17-2.

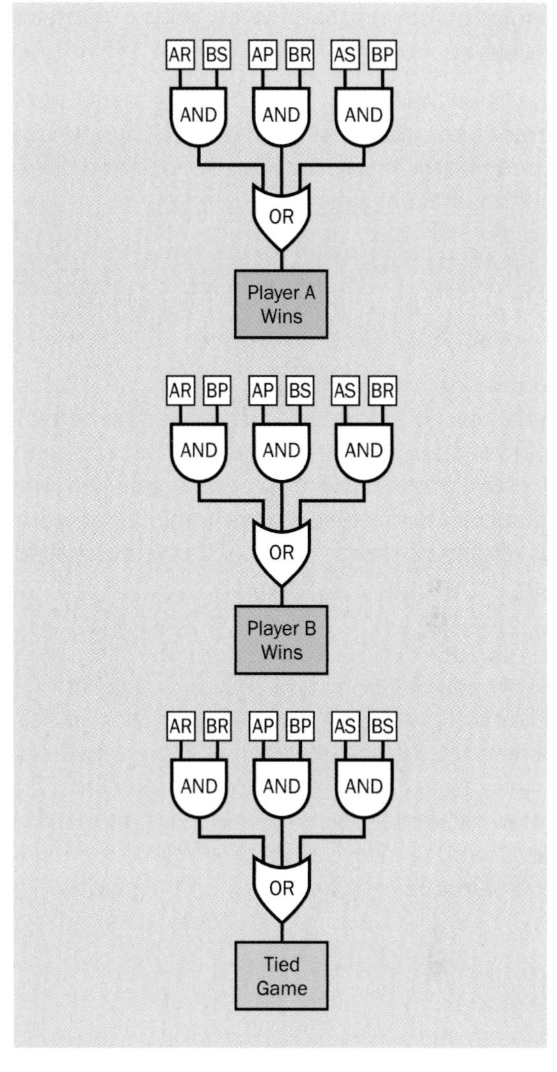

Figure 17-2. *Three logic networks to give appropriate messages for each of the winning switch combinations and the tie combinations in the Rock, Paper, Scissors game.*

Pretty simple, so far. But as always the problems begin when we consider issues such as playability of the game and cheating.

Who's On?

As in the Telepathy Test, each player's set of pushbuttons must be hidden from the other player. But this means that at the end of each round, there is no visual confirmation of which button each player actually pressed. We only have three indicators saying that "Player A Wins," "Player B Wins," or "Tied Game."

Really an LED should light up beside each button that is pressed—but only after both players have made their choices. How can this be done? Well, let's spell out the logic.

First we have to wait until the game has ended, with Player A winning OR Player B winning OR the game ending in a tie. If I use a three-input OR gate to represent this, I can tap into the three game outcomes shown in Figure 17-2. The result will look like Figure 17-3. The three game outcomes feed into the three-input OR gate, and when it has a high output, this signals that the game has ended one way or another. The output is ANDed with each button, as a prerequisite to allow the button to light an LED. The LEDs are shown as yellow circles to save space in the diagram. Each LED is identified with the same two-letter abbreviation as the button that lights it.

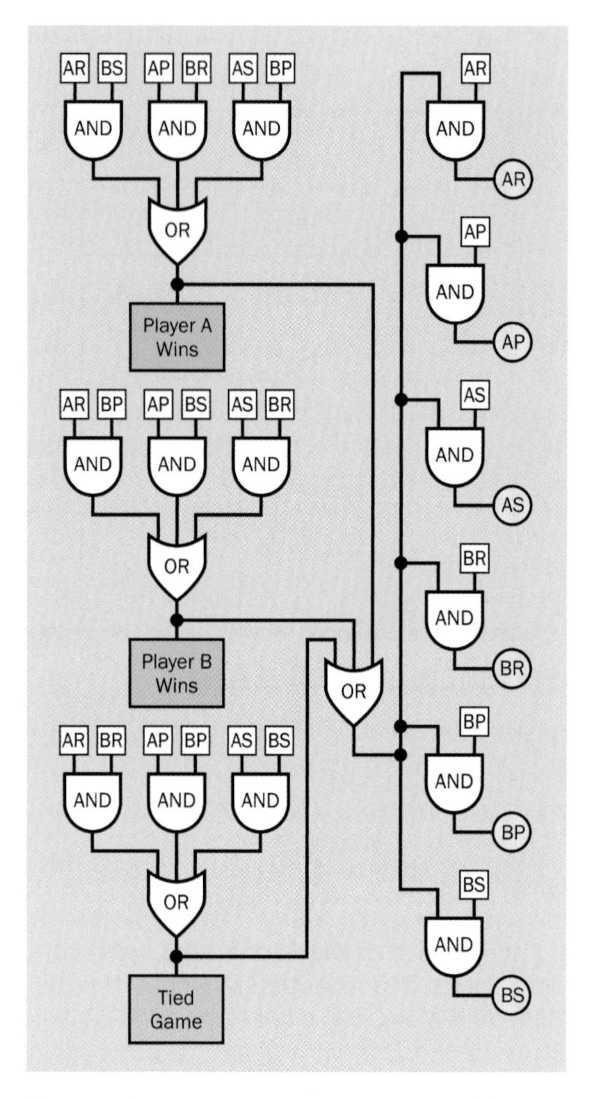

Figure 17-3. *A logic network that prevents an LED from lighting beside any switch until after a game has reached an end point.*

Who's Cheating?

This is beginning to look complicated (again). But the worst is still to come (again) because of the cheating problem. How can we signal if someone breaks the rules by pressing two buttons at once? In the Telepathy Tester, an XOR gate was helpful for this purpose because it has a high output if one input is high and the other is low, but it has a low output if both inputs go high. In the Telepathy Tester, a high input from an XOR meant that a player had made a move and was not cheating.

But XORs are usually restricted to two inputs, and in Rock, Paper, Scissors, we have three buttons for each player. This is a problem!

Spelling it out always helps. Annabel is cheating if AR AND AP are both pressed, OR if AP AND AS are both pressed, OR if AS AND AR are both pressed. (She is also cheating if she presses all three buttons, but we don't need to test for that, because pressing three buttons must always entail pressing two buttons, and we are already testing for every combination of two buttons.)

Cheating should light an alarm indicator and should also suppress any normal outcome from the game. The same thing should occur if Boris cheats. In other words, if Annabel cheats OR Boris cheats, the outputs to the "Win" indicators and the "Tied Game" indicator should all be suppressed.

It's certainly possible to implement this feature. We just need three more AND gates for each player to test each pair of buttons; and an OR gate for the outputs from the ANDs; and a NOR gate that gives a high output only if both players are *not* cheating; and then the NOR gate supplies another AND gate for each legal outcome of the game. But this doesn't sound very easy to build.

Background: Gate Arrays

As early as the 1970s, manufacturers were marketing chips that contained programmable arrays of logic gates. All the designs had the same basic goal of allowing a "generic" array of gates to be programmed with connections for a specific application to create a customized logic circuit on a chip. Many methods were used to achieve this, in devices with acronyms such as PLA, PAL, GAL, and CPLD (all of which you can look up online, if you're interested). Ultimately, field-programmable gate arrays (FPGAs) were developed, containing not only logic gates but more sophisticated capabilities that can be chosen by the end user.

Unfortunately, programming an FPGA requires a hardware description language, appropriate software (usually licensed by the chip manufacturer), and some appropriate hardware. This is not the kind of thing to set up in a home workshop, and so, for our purposes, we're stuck with old-fashioned chips.

Or are we? I would really prefer a simpler option than a logic diagram with more gates than will fit on a page—and fortunately, that option exists.

Experiment 18: Time to Switch

18

Figure 18-1 introduces a new concept. It shows how each logic gate can be emulated by a pair of plain, ordinary switches, and each switch can be considered as an input to the gate.

When a switch is pressed, it's the same as a logic input being high. When it is not pressed, this is the same as a logic input being low. Thus, in the illustration, the top pair of switches gives a high output only if the left switch AND the right switch are both pressed. Just like an AND gate!

Although—not quite. An AND gate output is either high or low. When the switches are open in Figure 18-1, the output is an open circuit. It will be floating, with no defined voltage. We need to add a pulldown resistor to control the state of each floating output. I omitted the resistors for simplicity.

There is another, more important difference between a logic gate and a pair of switches: the switches will conduct electricity in either direction, while the logic gate will not. You cannot feed current into the output of a logic gate and have it emerge from the input. This difference can create new headaches—although in some ways, it can also make things simpler.

Background: An XNOR Made from Light Switches

Just a brief digression—did you know that your home is very likely to have an XNOR gate in the wiring for its lighting? Typically this occurs where there is a flight of stairs. You have a switch at the top of the stairs, and another switch at the bottom. When the light is on, either of the switches will turn it off. When the light is off, either of the switches will turn it on.

Check the switch logic for XNOR in Figure 18-1. Imagine how the power flows for each combination of switches—and how it will flow if the position of either switch is changed.

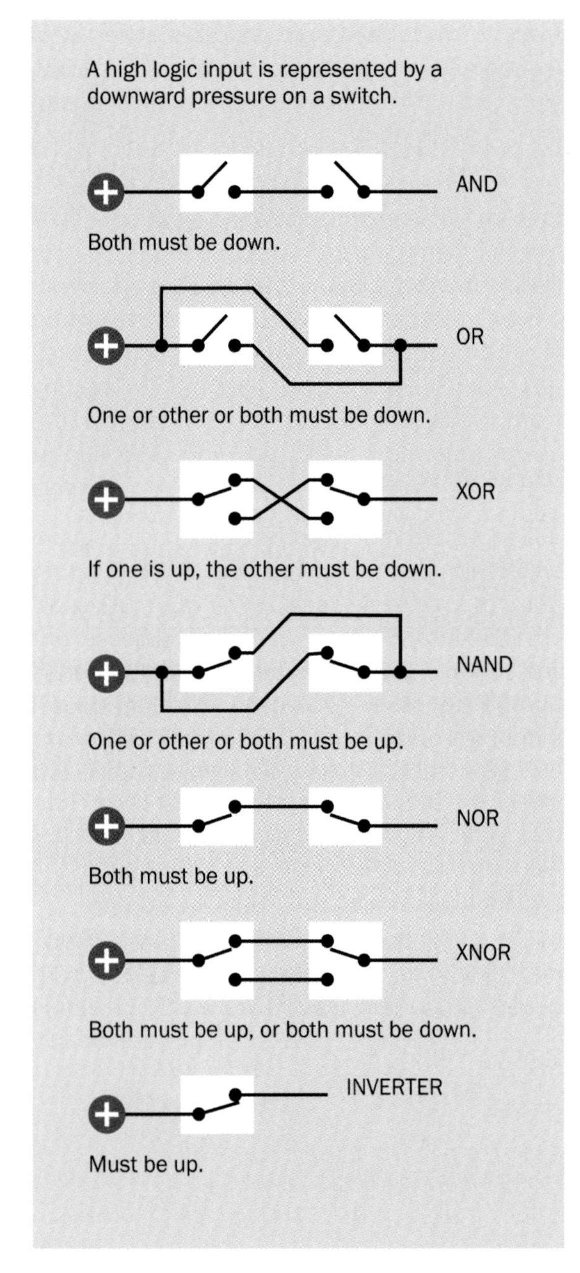

Figure 18-1. *Each of the switches emulates an input to a logic gate if pressure on the switch is seen as being equivalent to a high input. To avoid the output floating when switches are open, a pulldown resistor can be added.*

Can you imagine a logic circuit that would provide the same functionality for three switches? That is, any of the switches must always reverse the state of a single light bulb. I'll give you a hint: the center switch must have two poles.

Back to the Rock

Take a look at Figure 18-2 and compare it with Figure 17-1. They both have the same function, but the new version needs no logic gates. Each pair of switches that will create a win for Player A is wired in series, which is the same thing as ANDing them. All three pairs are tied together, which is the same thing as ORing them.

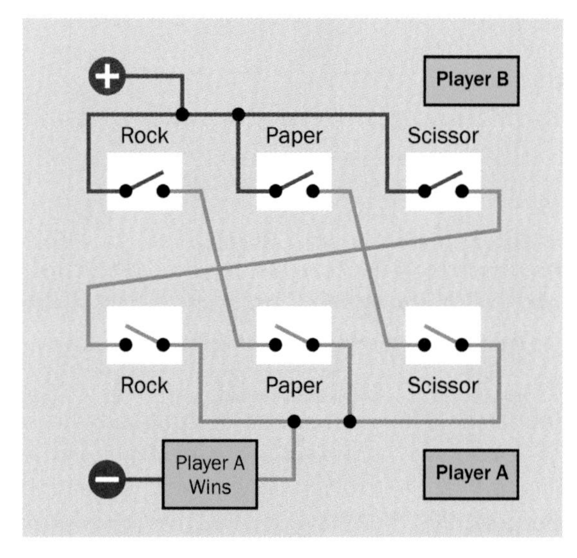

Figure 18-2. *Using only switches, a circuit can produce a "Win" output for any of the three combinations favoring Player A in a Rock, Paper, Scissors game. The wire colors are included to provide clarity and can be seen as being like colored insulation.*

If you're wondering why I am showing switches in the schematic now, instead of pushbuttons, it's because this circuit is going to end up with multiple-pole, double-throw switches; and although multiple-pole, double-throw pushbuttons do exist, they are more difficult to represent in schematics and take up more space. Just imagine that the switches are spring-loaded to return to their open state when they are not being pressed.

The conductors in Figure 18-2 are colored just for clarity because more are going to be added in the next few steps. Think of the colors as being like colored insulation on hookup wire.

Now if you look at Figure 18-3, you see the switch-and-wiring logic that responds if Player B wins. Can we combine the two circuits? We can't connect the output from one switch to multiple other switches, in the way we connected the output from one switch to multiple logic gates. The hardwired connections will allow electricity to circle around and flow back, causing erroneous results. We have to use multiple-pole switches to keep each circuit separate, as shown in Figure 18-4.

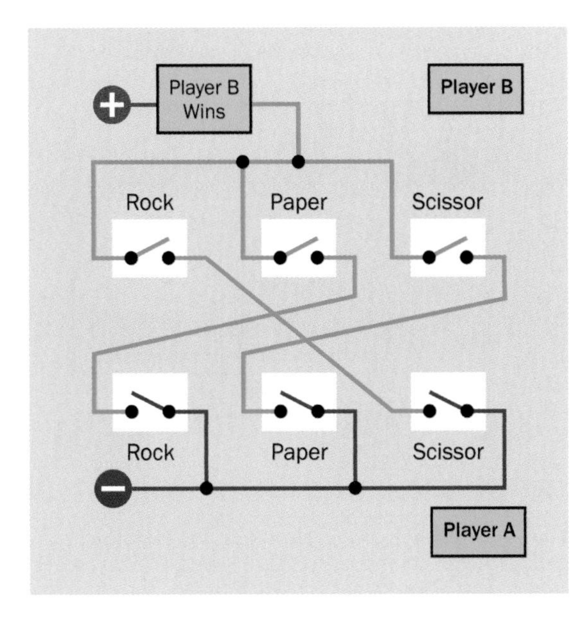

Figure 18-3. *A switched circuit that produces a "Win" output for any of the combinations favoring Player B in a Rock, Paper, Scissors game.*

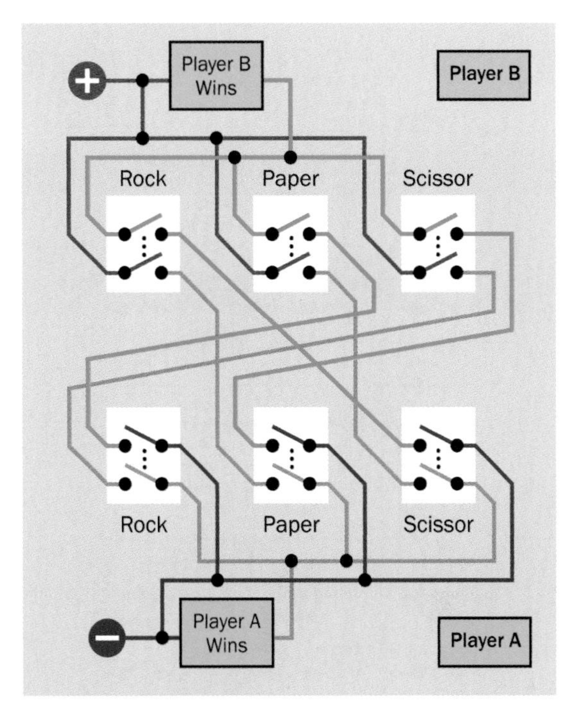

Figure 18-4. *When the two previous schematics are combined, double-pole switches must be used to keep the two circuits electrically separate.*

This circuit requires two-pole switches, but as I add extra features, still more poles will be required. Fortunately, pushbutton switches and push-twice switches (which use one push for "on" and a second push for "off") are available with two, four, six, and even eight poles. (They are often used in relatively cheap stereo equipment.)

Showing Which Button

In Experiment 17, I complained that trying to do everything with logic gates was giving me a headache. This was just after I showed you a logic diagram in Figure 17-3 that would illuminate an LED beside a switch that was closed, but only after both players had closed their switches.

Is there an easier way to do this with wires instead of logic gates? I think so. Figure 18-5 shows how. Because I put a positive power supply at the top for Player B, and negative ground at the bottom for Player A, the LEDs can be grafted into the middle of the circuit.

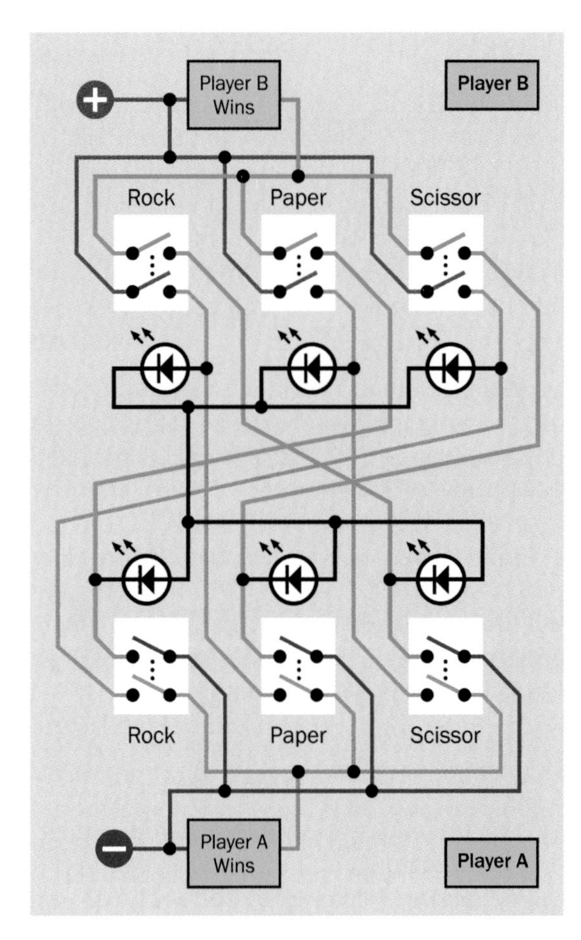

Figure 18-5. *This configuration enables an LED to indicate which switch has been pressed, but only after both players have made their choices.*

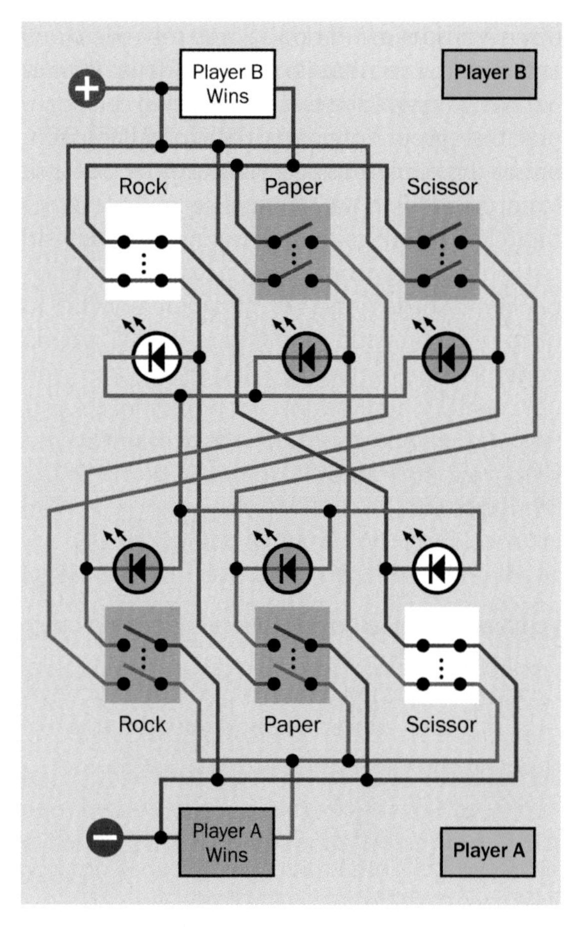

Figure 18-6. *In this example showing how two switches illuminate the appropriate LEDs, Player B has chosen "Rock" while Player A has chosen "Scissors."*

Figure 18-6 shows an example of how this circuit works, in a situation where Player B presses the "Rock" switch while Player A presses "Scissors." The conductors that are now directly connected with positive are shown in red, while the negative connections are shown in blue. Inactive wires, switches, LEDs, and indicators are grayed out. The magenta wires carry a reduced current that has passed through one LED but has not yet passed through a second LED.

A component needs to have positive on one side and negative on the other to be activated. Therefore the "Player B Wins" indicator lights up, while the "Player A Wins" indicator does not.

The current has to pass through one LED to get to another LED. Between LEDs, the wires are shown in magenta, as they have an intermediate voltage. If you think of the two LEDs as being in series, with positive at one end and negative at the other, you'll see why the two LEDs adjacent to closed switches will light up, while the others remain inactive. Remember that because LEDs are diodes, they block voltage of the wrong polarity.

If you trace other switch combinations, I think you will find that each of them lights only the correct LEDs and indicators.

In an actual circuit, if you put two LEDs in series, and each LED is of the type that has a resistor built into it, you may find that the light output is unacceptably dim. This is because the current is passing through two LEDs and two resistors in succession. You can try using two generic LEDs wired in series without any resistors, and use your meter to check that the current is below the maximum specified for each component. Most likely, for optimum performance, you should use generic LEDs with a small-value resistor substituted for the vertical magenta wire on the left of the schematic, linking the two groups of LEDs. Choose the value of the resistor by trial and error, working from 220Ω downward, until the current is correct. Remember that while most LEDs are happy with a current of 20mA, some have a lower limit.

Cheat Proofing

Now for the hard part: cheat prevention. The way I've chosen to do this is by using the existing circuit with an additional switch configuration that controls the availability of power to the circuit for each player.

The principle of the thing is shown in Figure 18-7, where three double-pole switches are shown in different combinations. Note that all of these switches are normally closed. When you press a switch, you open it. To draw your attention, I have colored a switch green when it is being pressed to open it.

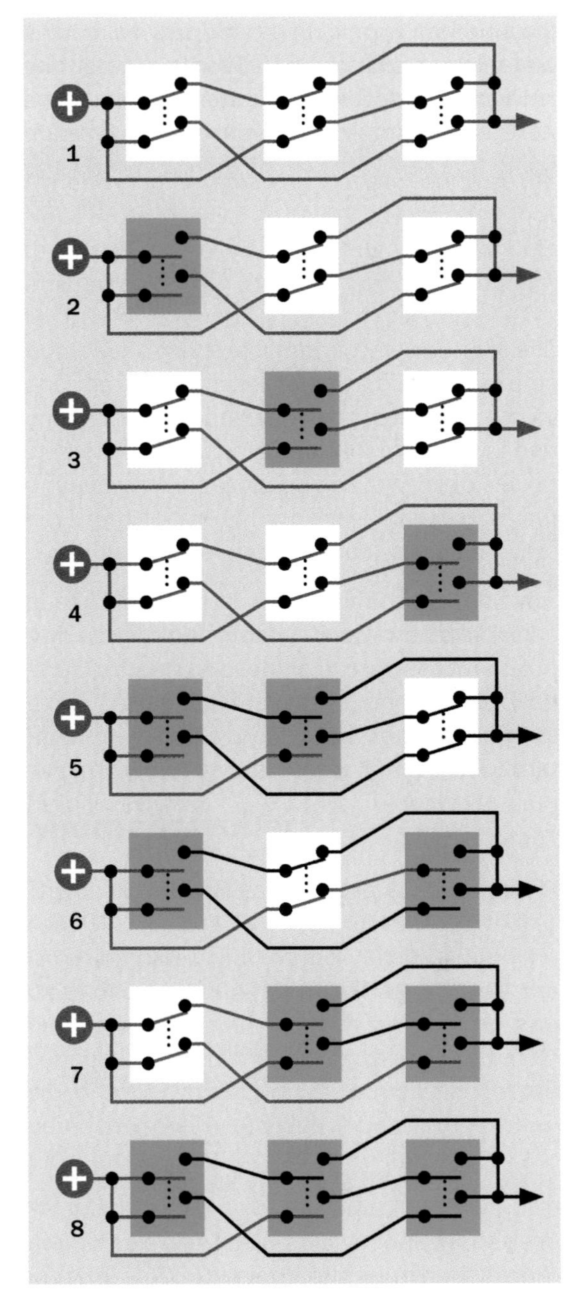

Figure 18-7. *Wired in this way, three normally-closed double-pole switches will allow current to pass through when no switches or a single switch is pressed, but will block current when two or three switches are pressed. The green highlight indicates switch(es) that are pressed.*

At the top, none of the switches is being pressed, so they all remain closed, and the electricity passes through. In the next three examples from the top, one switch at a time is being pressed, but the other two remain closed, allowing at least one pathway for electricity to pass through. In the next three examples, two switches at a time are being pressed (remember, green means the switch is being pressed), and the remaining unpressed switch is insufficient to connect power through the chain. In the final example, with all three switches pressed, the electricity is blocked.

The wires carrying current are shown in red. You can see that if no switches are pressed, or any one switch is pressed, power is connected, but if any two switches or all three switches are pressed, power is blocked. This provides an anti-cheating system that can be used in the Rock, Paper, Scissors game.

In Figure 18-8 the anti-cheating system has been added to the previous circuit, using an extra two poles on each switch. For the sake of clarity, I have only shown contacts that are being used in the circuit. I have not shown any unused switch contacts. Thus every switch shows two normally-open contacts and two normally-closed contacts.

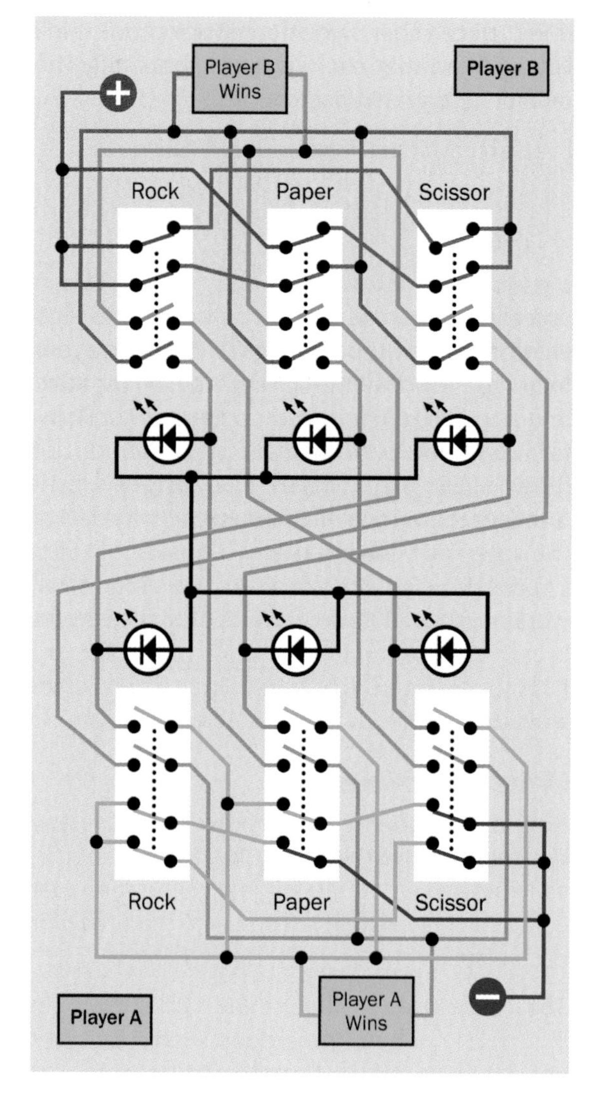

Figure 18-8. *The previous circuit has been expanded to include an anti-cheating system, so that power is shut off when either player presses two or three switches instead of one. See text for details.*

I have tried to clarify it with wire colors. Any wire that is permanently connected to the positive side of the power supply is colored red. Wires that normally carry positive voltage, through normally-closed switches, are magenta. If two or three switches are pressed simultaneously, the magenta wires no longer carry positive voltage.

Similarly, at the bottom end of the circuit, dark blue wires are permanently connected to the negative side of the power supply, while pale blue (cyan) wires will no longer have a connection to the negative side if two or more switches are pressed.

In this way, the system shuts off power to the circuit if either person cheats.

Fit to Be Tied

Lastly, do we need an audible signal if Annabel and Boris have a tied game by pressing buttons opposite each other? They'll see from the pushbutton LEDs that the game is tied. Still, if you want a beep, it can be added. To do this without requiring an extra pole on the switches, you need three beepers. Fortunately they're very cheap (some costing less than $1). You can graft them into your circuit as shown in Figure 18-9. The circular symbols show where the beepers go. They should be the polarized type, passing current in one direction, and they should be the kind that makes a sound when you apply a DC voltage. You don't want a loudspeaker or the kind of beeper that requires you to feed an audio frequency into it.

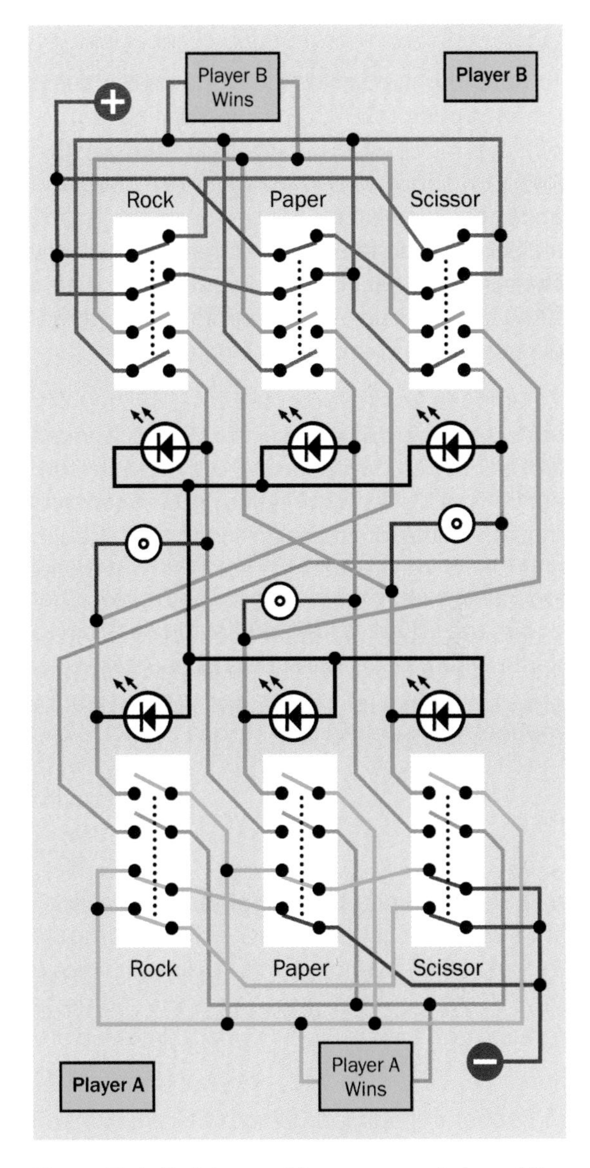

Figure 18-9. *Each beeper (shown as a symbol consisting of two concentric circles) will sound if the two switches connected with it are closed, indicating a tied game.*

Why do you need more than one beeper? Because if you only had one, you would have to run three wires to each side of it from existing switch contacts, and electricity would flow in through one wire and then out through any of the other wires. Consequently, any pair of switches would sound the beeper.

Wiring It

To build this circuit, you'll have to use soldered point-to-point wiring. The pin spacing on the switches may be 0.1", but if you inserted a switch into a breadboard, each pair of pins would end up sharing one lateral conductor on the board, making it impossible to power them individually. The switches that I've seen are narrower than a through-hole chip, and therefore cannot straddle the central channel in the board.

A computer rendering of a typical 4PDT pushbutton switch is shown in Figure 18-10. The red labels that begin with letter P are the pole connections of the switch. The four sets of contacts that match the four poles are labelled C1, C2, C3, and C4. NC means that the contact is normally closed, while NO means that it is normally open. Whether you use a spring-loaded pushbutton or one that you press twice is up to you. This type of switch is often referred to as a "slider," because the internal contacts slide.

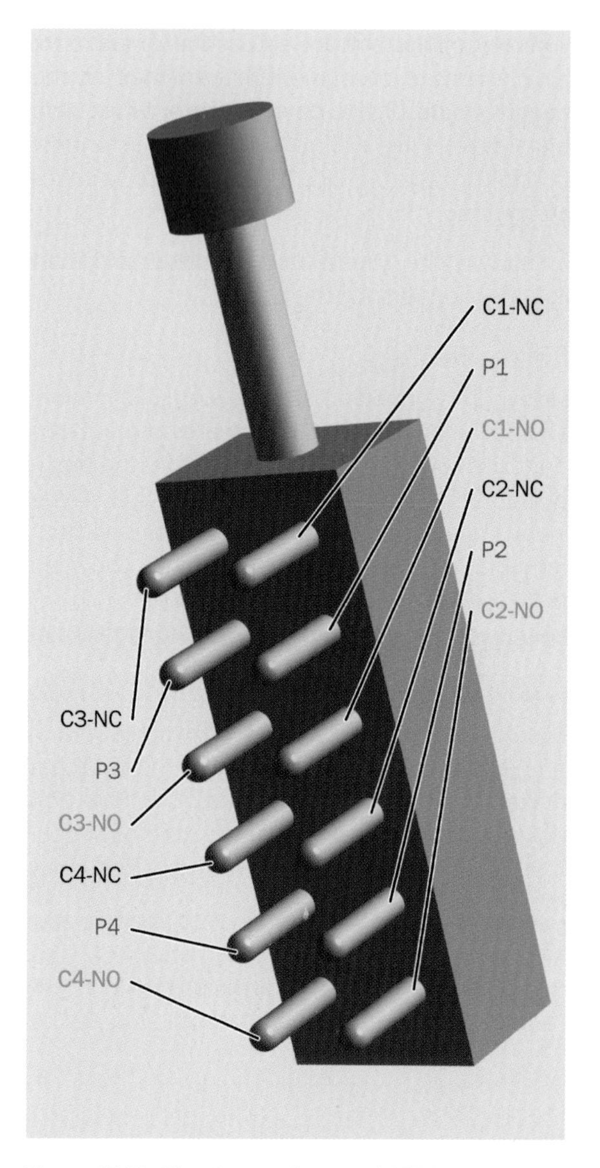

Figure 18-10. *The pinouts of a typical 4PDT slider switch. The poles are numbers P1 through P4, while contacts are numbered C1 through C4. Normally-closed contacts are labelled NC, while normally-open contacts are labelled NO.*

Two typical schematic symbols representing a 4PDT slider switch are shown in Figure 18-11. The labels match the labels on the 3D rendering of the switch. On the righthand side, the vertical gray bar consists of an insulating material with conductive sections added to it, shown in black. These black sections short together the contacts adjacent to them. When the bar is pushed down by the plunger of the switch, it shorts a different pair of contacts. You may find either of these types of symbols being used in manufacturers' datasheets.

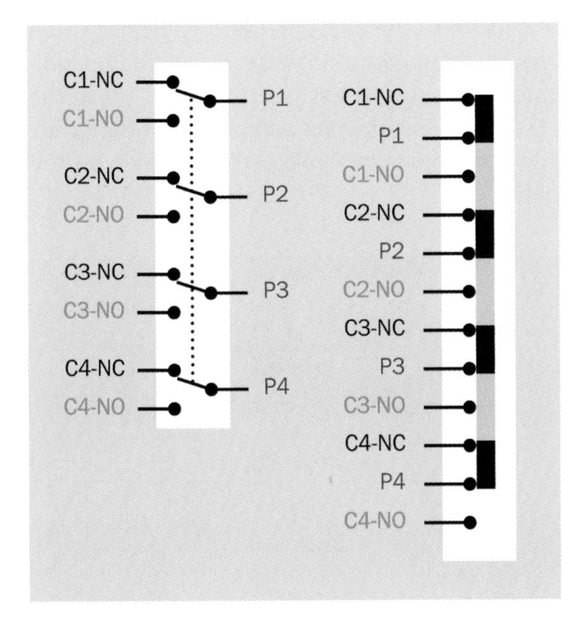

Figure 18-11. *Two commonly used schematic symbols to represent a 4PDT slider switch.*

Figure 18-12 shows how to hook up the switches to make the Rock, Paper, Scissors game. Pins that have soldered connections with wires are shown solid black, while unused pins are white. A few of the wires have been moved around relative to the schematic shown previously, to minimize crossovers and avoid any locations where three wires meet on a pin. Such locations are more difficult to solder.

This circuit lacks the cheat-detecting feature, in case you want to keep things relatively simple (initially, at least). The wire colors match the colors that I used in the schematics, but of course any colors will do. Notice that the switches at the bottom are upside down relative to the switches at the top. This is because you may want to mount them on opposite sides of a box or board, in which case the switches will be rotated relative to each other.

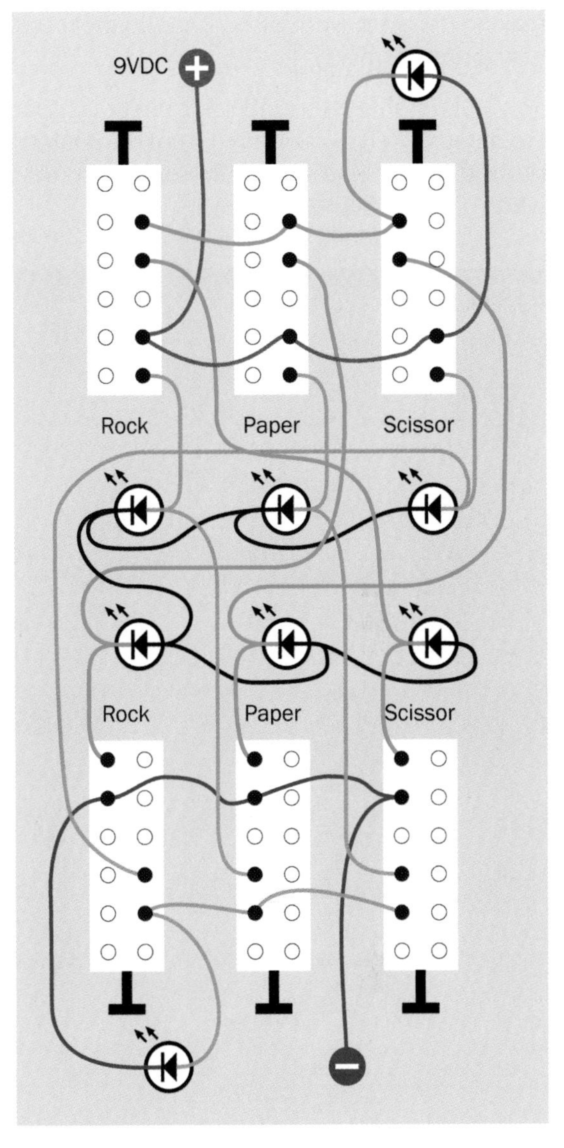

Figure 18-12. *Wiring six 4PDT slider switches for the simpler version of the Rock, Paper, Scissors game, which does not include cheating prevention. Series resistors for the LEDs are not shown.*

Soldering wires to the switch pins requires some care to avoid solder bridges that will cause short circuits. I think you'll have an easier time if you use wire that is thinner than the 24 gauge that I have recommended for breadboarding work. Personally I like to use rainbow-colored ribbon cable, which can be pulled apart into separate colored conductors. The stranded wire is flexible, making it easy to work with, although the result may look a bit messy.

Figure 18-13 shows the underside of a board that I soldered myself. My soldering is not the neatest, but the object here is to build something that works, and the board does work.

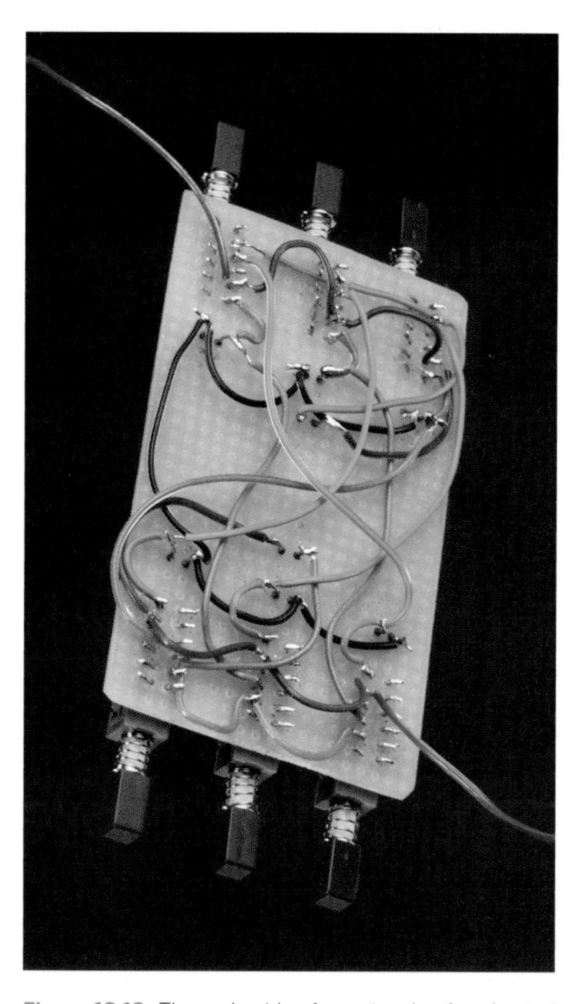

Figure 18-13. *The underside of a rectangle of perforated board showing the slider switches wired without the "no cheating" feature.*

I used 12V LEDs that contain their own series resistors, which is why you don't see any resistors in the wiring. I was pleased to find that I could power two of these LEDs in series, and they would still be bright enough, even though the power inevitably goes through two of the internal resistors in addition to two of the LEDs.

Figure 18-14 shows the top side of the same board. I left the LED leads long, because the idea is to push the LEDs up through holes drilled in the lid of an enclosure, while the pushbuttons will be accessed through holes in the side of the enclosure. At the time of writing, I haven't gotten around to building this box. The red LEDs indicate which switch has been pressed, while the LEDs that look gray are actually blue when they light up, indicating which player has won the game.

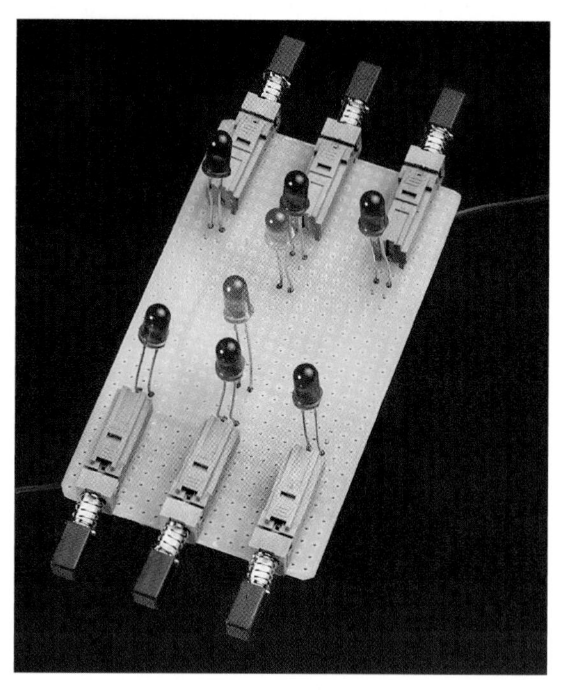

Figure 18-14. *The top of the rock-paper-scissors circuit. The LED leads were left long so that they could be pushed up through holes in the lid of an enclosure.*

I'm suggesting a 9VDC power supply for this project, because it will use no electricity at all until two buttons are pressed, and even then the LEDs will not use much power. If they are rated for 12V, they will look okay with a 9V battery.

If you use generic LEDs that require external series resistors, you can use 470Ω resistors for the single LEDs that show who wins a game, but I don't know what value resistors you will need with the LEDs that are wired in series. If you reduce the series resistor to make the LEDs brighter, be careful not to exceed the maximum forward current recommended in the datasheet.

A version of the game that uses two pieces of perforated board, on opposite ends of a box that has LEDs mounted in the top, is shown in Figure 18-15. This configuration hides each person's switches from the other person. A 9V battery is dimly visible inside the box, clamped under a piece of plastic. Replacement will require removing some of the screws, but the battery should be good for a couple of years if the game isn't played too often.

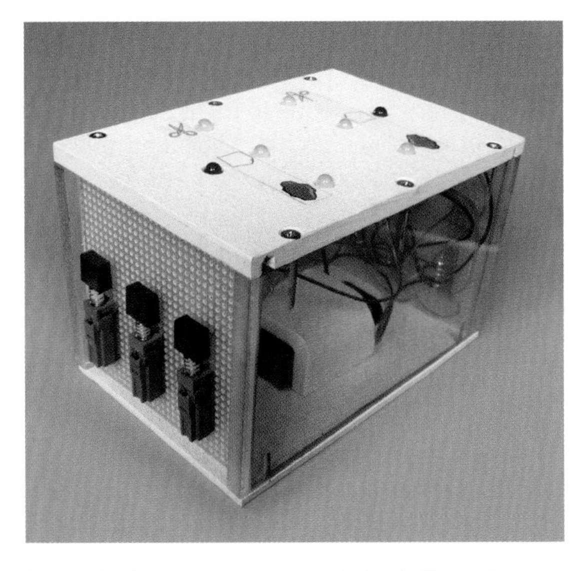

Figure 18-15. *A finished version of a Rock, Paper, Scissors game using switches only.*

Cheat-Proofed Wiring

Figure 18-16 shows the switches hooked up including the cheating-prevention feature. It really doesn't add much to the soldering time and is a nice feature to have. If you already wired the uncheat-proofed version, the only wires you will have to change are the blue and red ones that provide power. Everything else is additional.

If you also want to add beepers, I'm going to leave that to you.

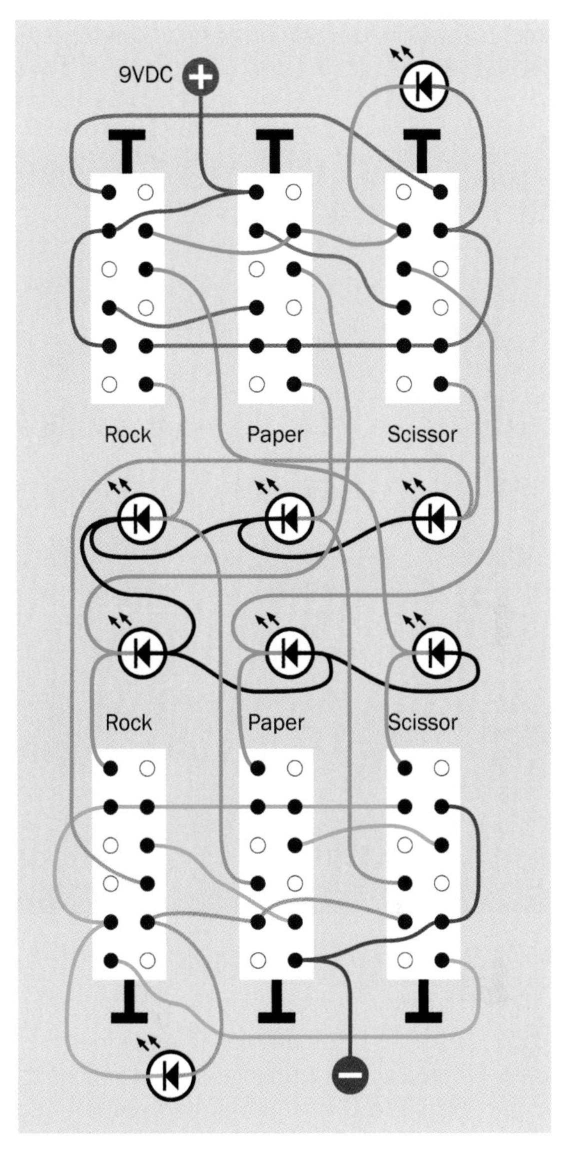

Figure 18-16. *Wiring diagram for the enhanced, switched version of the Rock, Paper, Scissors game, including cheating prevention.*

The underside of the perforated board with cheat-proofing wiring added to it is shown in Figure 18-17.

Conclusion

Switches looked simpler than logic gates to begin with, but in the end, they turned out to be complicated in a different way.

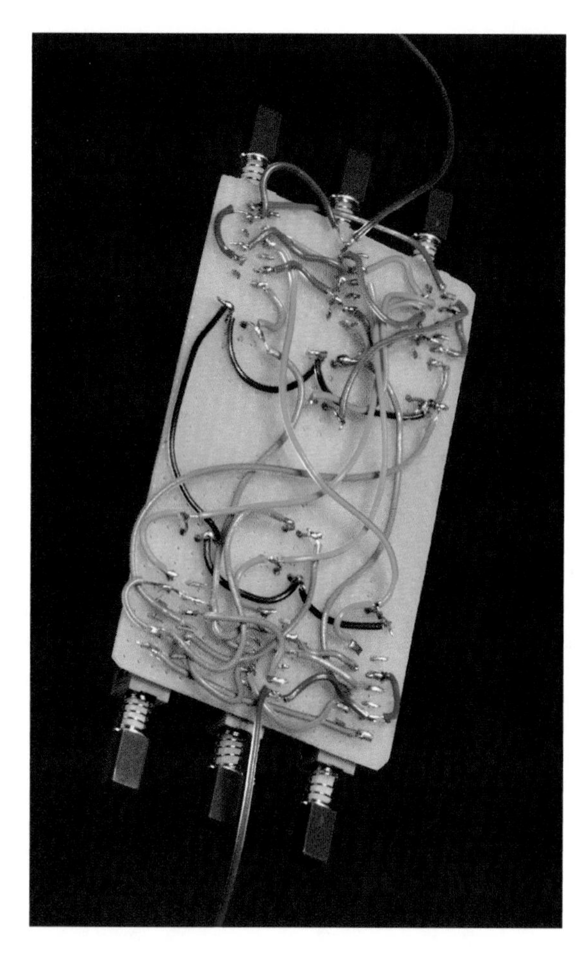

Figure 18-17. *The same board shown previously, now with cheat-proofing added.*

Logic gates can be assembled in a manner that is—well, logical. You write down what you want, using ANDs and ORs, and then represent it in a diagram with AND and OR gates (and other gates, if necessary). Finally you choose some chips containing the appropriate gates and link them together in the same way that you drew your diagram. This is not a big challenge.

The problem with logic gates is that because each one is not very powerful, you need a lot of them. Consequently, the circuit requires a lot of components, becomes confusing, and can encourage wiring errors.

Switches have several advantages. They're simple, they're easy to understand, they don't require any power when they're off, and they can conduct a lot of power (relatively speaking) when they're on. I like the idea of a circuit that doesn't need an on-off switch because the switches *are* the circuit!

The problems start when you want the circuit to perform multiple functions. This inevitably leads to multiple-pole switches, and the tangle of wiring may have unpredictable results unless you're really careful. If you want to optimize the circuit, you have to be even more careful.

Most problematic, you need a strange kind of intuitive ability to design a circuit entirely from switches. There's no equivalent of the logical system that enables a circuit of chips to be derived from a verbal description using ANDs and ORs. Really, it's rare for anyone to try to build a useful circuit using only switches. I included this one just so that you could get a sense of what's involved.

Are there any options other than logic gates and switches? Absolutely! There are three possibilities:

1. You can use relays. Many years ago, entire telephone systems were relay based. You might assume that a logic circuit using relays can be very similar to one using transistors, because the components behave similarly. But don't forget that a relay can have multiple poles, like a switch, enabling one input signal to energize multiple separate outputs.
2. You could introduce a decoder, which I mentioned before. It contains a whole bunch of prewired logic gates so that you don't have to deal with them individually. I'll deal with decoders in detail in the next two experiments.
3. Yet again, you can use a microcontroller. In the Rock, Paper, Scissors game, as in the Telepathy Test, a microcontroller would make things much easier. But, as in the Telepathy Test, you wouldn't learn much about logic.

Experiment 19: Decoding Telepathy

19

Take a look at Figure 19-1, where I have redrawn the Telepathy Test logic diagram using a decoder chip. Although this circuit now incorporates every possible feature that I talked about earlier, it requires only three chips. A circuit doesn't get much simpler than this.

Decoder Testing

So what is this mysterious component that makes everything so simple? Before you build the circuit, let's start by bench-testing a decoder. The one I'm recommending is the 74HC4514. I'm specifying the HC family because whenever I buy a chip, I'm thinking that I might reuse it for some other purpose in the future, which may require it to power LEDs (at least while testing the circuit). Remember that the 74HC00 chip series can source or sink up to 20mA, while the output from a 4000 series chip is much more limited.

However, the 74HC4514 is relatively expensive (more than $3 when bought singly), and you should know that its cousin, the old CMOS version, which is simply numbered 4514, is more widely used and cheaper. For this experiment, because the decoder isn't going to drive any LEDs, either of the chip versions will work. They have the same pinouts, as shown in Figure 19-2.

Connect the chip as shown in Figure 19-3. At the top of the schematic, I'm specifying switches instead of pushbuttons to test the chip, because you may find it inconvenient to hold down two or three buttons simultaneously. However, either switches or pushbuttons will work. The breadboarded version is shown in Figure 19-4.

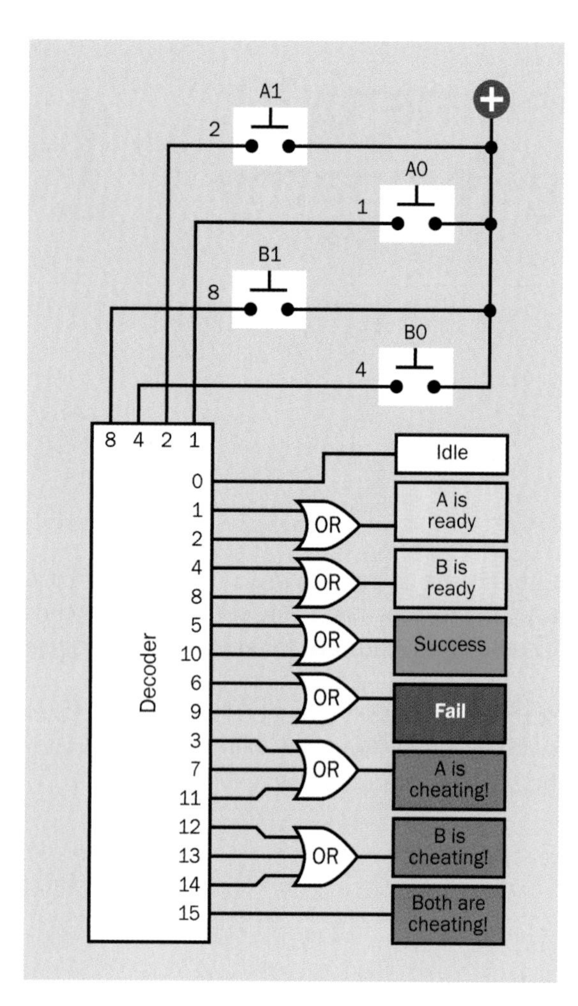

Figure 19-1. *Using a decoder chip, the number of logic gates in the Telepathy Test project has been radically reduced. Pin locations on the actual chip have been reshuffled here to reduce wiring crossovers in the diagram.*

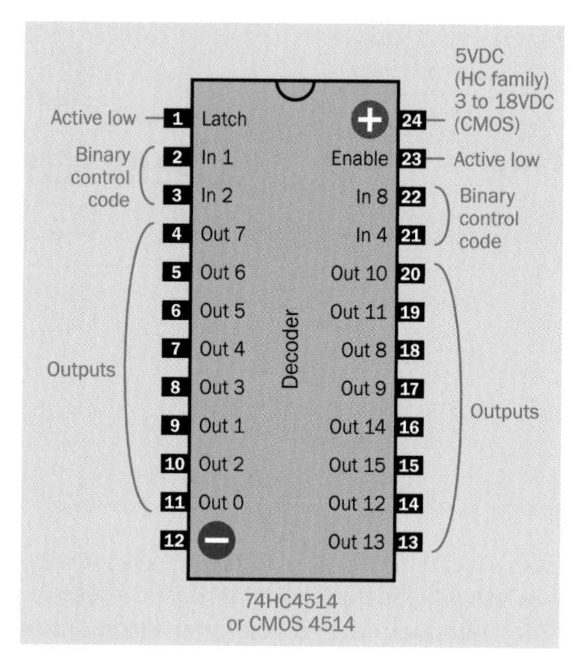

Figure 19-2. *Pinouts of the decoder chip that can simplify the Telepathy Test project.*

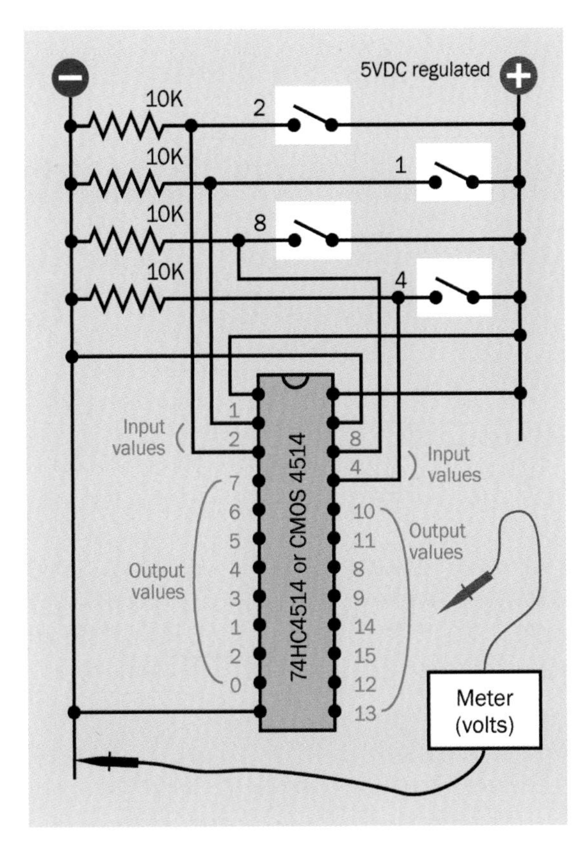

Figure 19-3. *Setting up the enclоder chip for exploratory testing. The meter probe can be touched against any of the unconnected pins. See text for additional details.*

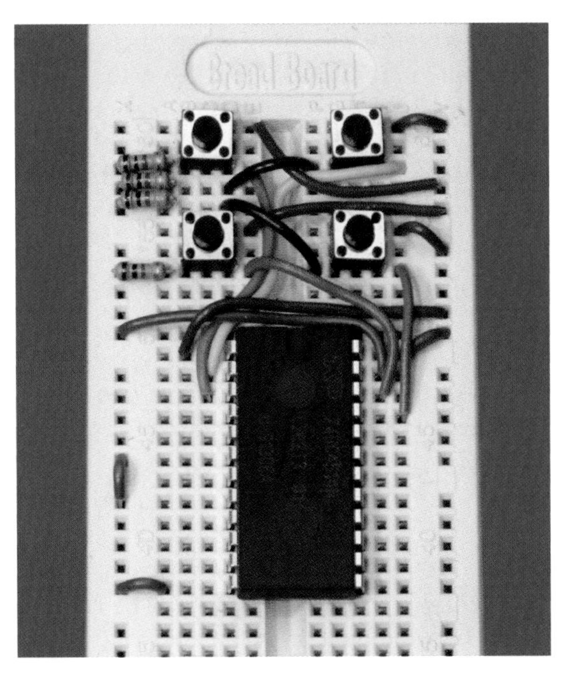

Figure 19-4. *The breadboarded version of the decoder test circuit. Decoder chips from some manufacturers may be narrower than the one shown here, but the functionality is the same.*

Set your meter to measure DC volts, and connect the negative probe to the negative bus, using a jumper wire with alligator clips at each end or a minigrabber on your meter probe. You will be touching the other, positive meter probe to the pins of the chip that are shown as outputs in the diagram.

Because the chip performs simple arithmetical calculations, its pins have values asssigned to them, as shown in the figure. The input pins have values 1, 2, 4, and 8. These numbers are also shown beside the switches that supply power to the inputs. The output pins have values 0 through 15. You'll see that they are not in numerical order (this is not a typographical error in the figure). To avoid confusion, you may want to write the numbers on adhesive labels that you stick to the breadboard beside the chip (if you can write small enough, with a fine-tip pen).

Begin with all of the switches open. Run the meter probe along all the output pins of the chip, and you should find that the output pin with value 0 is high, while all the others are low. Now close switch 1, and the output pin with value 1 will be high, while all the others are low. Now close switches 1, 2, and 4 simultaneously, and keep them closed. You will find that output pin with value 7 is now high, while all the others are low.

The decoder is functioning like a little adding machine. The values of input pins that have a high state are summed together. The chip then energizes the output pin that has the corresponding value.

Here's the most important feature of a decoder chip:

- The 1, 2, 4, and 8 input values are chosen so that each output value can only be created by one unique combination of inputs.
- Therefore, the value of the output can tell us which switches have been closed.

In case this isn't entirely clear, take a look at Figure 19-5, which shows four possible states of the chip, chosen at random. I have simplified the figure by showing inputs at the top of the decoder, in sequence, and outputs down the right-hand side of the decoder, in sequence. A red input indicates a high value, created by a closed input switch, while a red output also indicates a high value. If you add up the high values at the top, that gives you the value of the output pin that will be high.

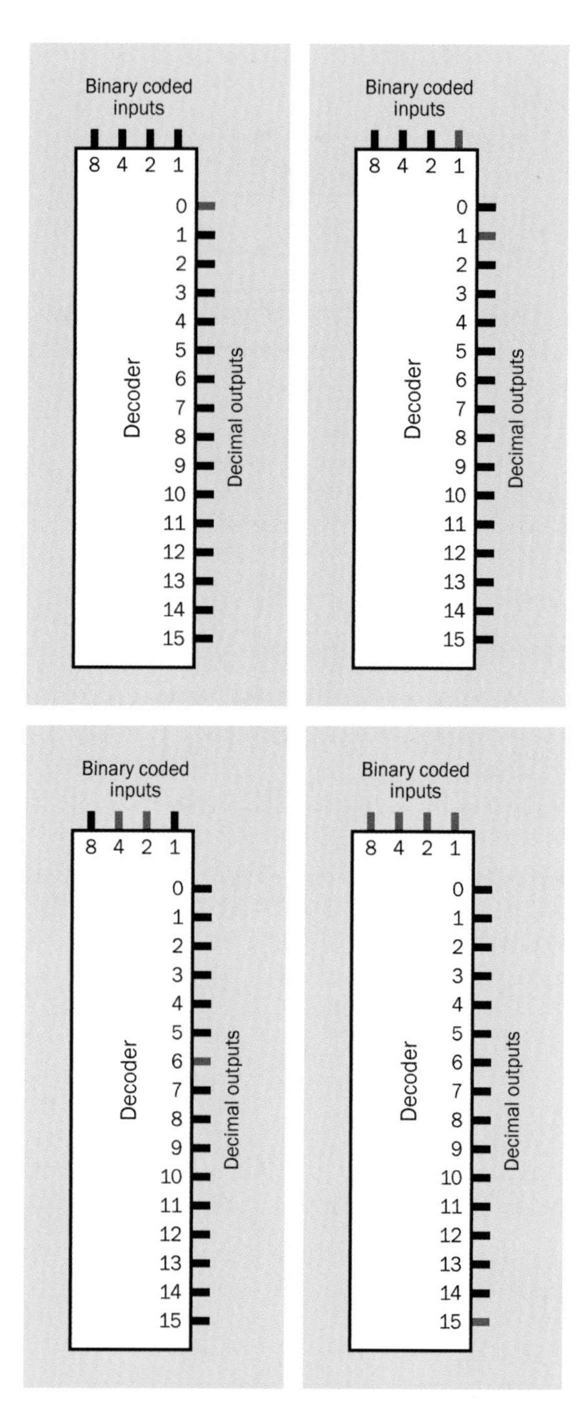

Figure 19-5. *Four examples of the behavior of a decoder chip, which creates a high output (shown in red) on a pin whose assigned value is the total of the values of the input pins that have high states. Pins in this diagram have been sorted into numerical sequence to clarify the process. Pins on the actual chip are not arranged sequentially.*

Remembering Binary

Now go back to Figure 19-1. This is almost the same as the diagram in Figure 19-5 except that I have moved and grouped the numeric values of the output pins so that they align conveniently with the logic gates.

In the Telepathy Test, we can now see easily which switches have been pressed or not pressed by Boris, and which switches have been pressed or not pressed by Annabel. For instance, if output pins on the decoder with values 1 OR 2 are high, switches A0 OR A1 (with values 1 OR 2) are closed, and all the other switches are not closed. Therefore, in the Telepathy Test, A has made a selection, while B has not done so yet.

Suppose the output pin with value 6 has a high output. This can only mean that switches A1 and B0 (with values 2 and 4) are closed, while the other switches are open. Similarly if the output pin with value 9 has a high output, this can only mean that switches A0 and B1 (with values 1 and 8) are closed, while the other switches are open. Therefore, if pins with value 6 OR 9 have a high output, each player has closed a switch, and the switches are not opposite each other, so A and B failed that particular telepathy test.

You can go through all the combinations and check the outputs.

In Figure 19-1, you will see that I added an extra output indicator, which says that both players are cheating when all four pushbuttons are closed. I also added an output which states that the circuit is "Idle" when no switches at all have been closed.

The inputs to the decoder are described as "binary" inputs because their values are the same as the place values in a binary number. I wrote about binary code briefly in *Make: Electronics*, so I'm just including one quick-reference diagram in Figure 19-6 to show the relationship between binary numbers and pin values on the decoder chip. It is called a "decoder" because it takes a number in binary code and decodes it to provide a single high state on a pin that can be identified with decimal notation.

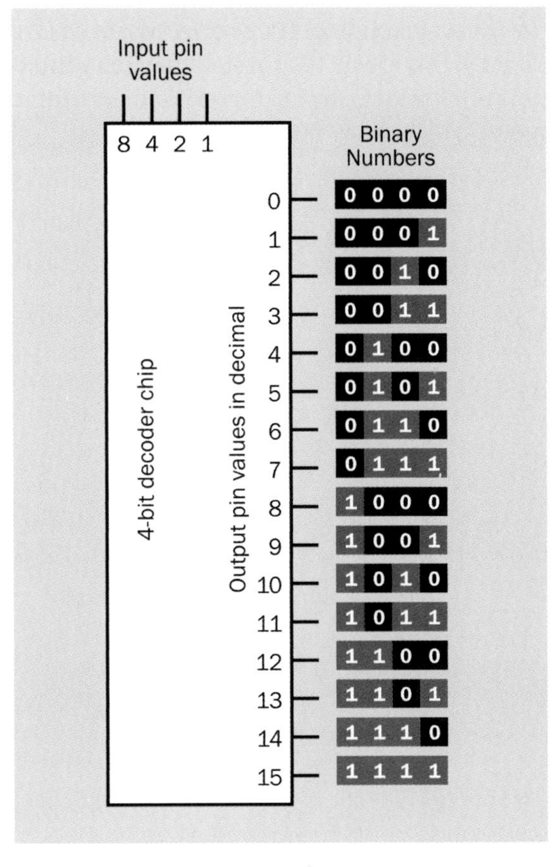

Figure 19-6. *The high and low states of decoder input pins can be thought of as 1s and 0s in a binary number. The output pins of a decoder have the values of equivalent decimal numbers.*

- Because a binary digit is often referred to as a "bit," and because the 74HC4514 decoder (or its older 4514 cousin) has four binary inputs, it is often referred to as a "four bit" decoder. It may also be described as a "4/16 decoder," as it has four inputs and sixteen outputs.

Boarding It

The Telepathy Test is now very easy to build. The full schematic is shown in Figure 19-7. In this circuit, you can use either the cheaper CMOS version of the decoder (the 4514) or the more expensive HC version (the 74HC4514, which was developed to emulate the earlier 4514 with the same pinouts). Note that the old CMOS chips don't have enough output power to drive LEDs directly. This is why I suggested that you should perform the initial test in this experiment with a meter, which draws negligible current.

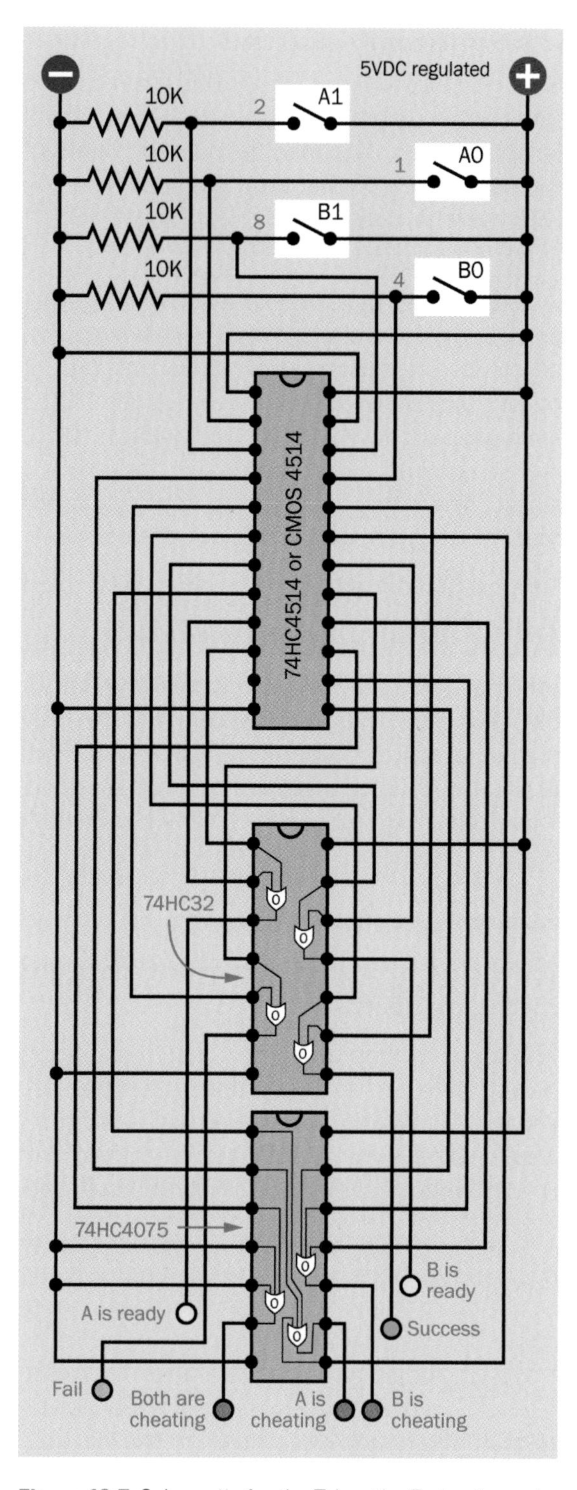

Figure 19-7. *Schematic for the Telepathy Test using a decoder chip and two logic chips.*

In the circuit shown here, you'll need a quad two-input OR chip—the 74HC32, which you have used previously. Its pinouts are shown in Figure 15-7. The other logic chip is a triple three-input OR chip—the 74HC4075, which you have not dealt with before. Its pinouts are shown in Figure 19-8. The internal connections of this chip are similar to those of the triple three-input AND chip, which was shown in Figure 16-11. They are not exactly the same, however, and you should follow the schematic carefully when making connections.

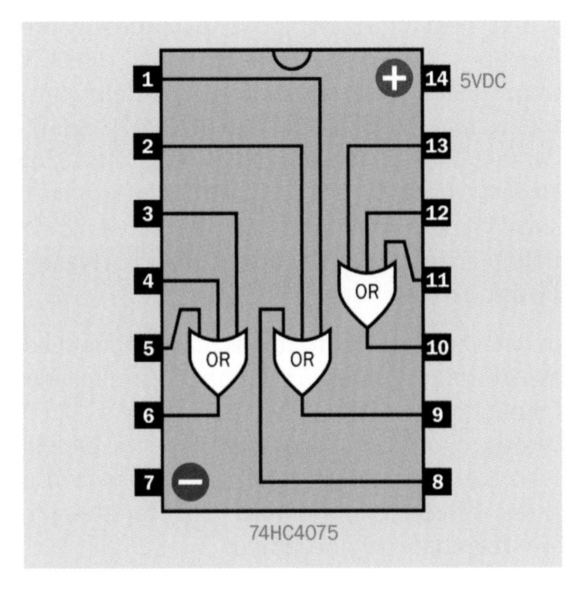

Figure 19-8. *Internal connections of the three OR gates inside a 74HC4075 chip.*

If you refer back to Figure 19-1, you'll see that almost all the decoder outputs pass through logic gates before lighting an appropriate LED. If we use 74HC00 series logic gates, these will be powerful enough to light the LEDs. But what if you choose to use the cheaper 4000 series decoder chip? Outputs with values 0 and 15 are shown connected directly to indicators.

You can get around this by omitting the "Idle" indicator that would be connected to output value 0, powering the "Both are cheating!" output from output value 15 by passing the signal through a third three-input OR gate (not shown in Figure 19-1). If you tie two inputs of a three-input OR gate to negative ground and apply a signal to the third input, the OR gate just passes the signal through. It acts as a "buffer," with an output state that is always the same as the state of its one active input. This is shown in Figure 19-7.

All of the LEDs share a single series resistor. This is acceptable and sufficient because only one of the LEDs will be powered at a time.

When placing jumper wires, I suggest it's a good idea to print a paper copy of the schematic in Figure 19-7 and apply colored ink to each line after you establish the corresponding jumper on the breadboard. This will reduce the risk of wiring errors, which tend to be common where multiple conductors are parallel and close together.

A breadboarded version of the circuit is shown in Figure 19-9. In it you will see that the decoder that I used is an extra-wide chip. Some versions of the 74HC4514 look like this, while others are normal width. The functionality is the same.

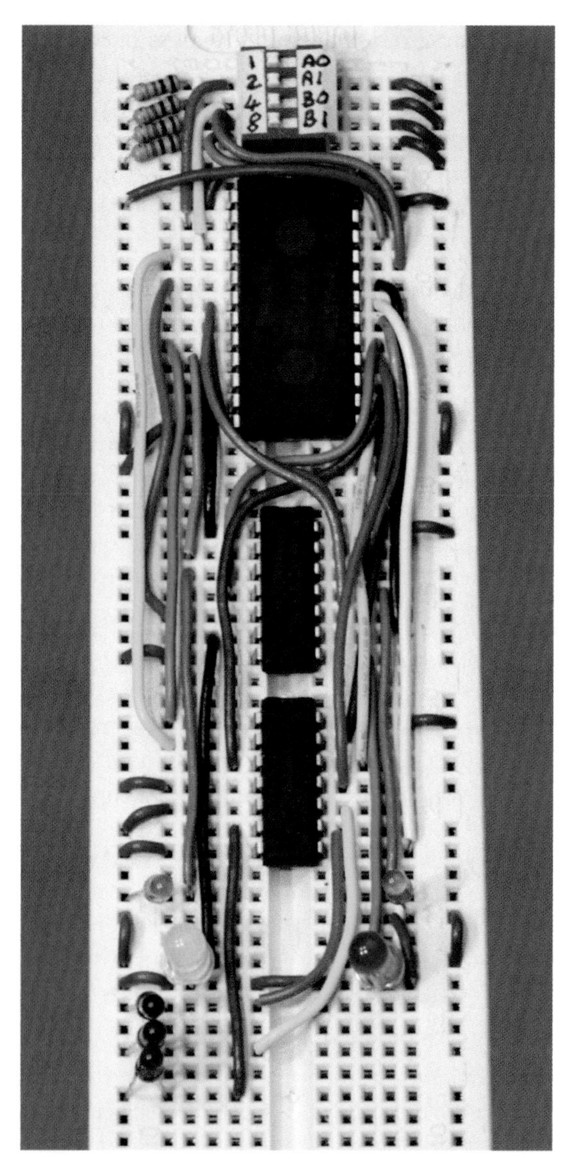

Figure 19-9. *The Telepathy Test wired using a decoder to reduce the chip count and the complexity of the circuit.*

Although the 4514 chip should work acceptably in this circuit, be cautious about mixing it with HC chips if you reuse it in some other circuit in the future.

Decoder Pinouts

Getting back to the pinouts of the 4514 and the 74HC4514, shown in Figure 19-2: there are two numbering systems here. The white numbers in the black rectangles are the standard pin numbers, which are applied to any chip. They always start at the top-left pin and proceed counterclockwise around the edge of the chip. I will always refer to these numbers as "Pin 1," "Pin 2," and so on.

The numbers that I have added inside the body of the chip are the values that the pins represent when they have a high state. For instance, when Pin 22 has a high state, this represents an input of value 8. When Pin 20 has a high state, this represents an output of value 10.

A datasheet for this component will identify the functions of the pins in some other way, which is unfortunately not standardized. The input pins, for instance, may be labelled A0, A1, A2, and A3. Or, they may be identified as A, B, C, and D, or DATA1, DATA2, DATA3, and DATA4. Since there is such a lack of standardization, I felt it would be helpful to make the functions of the pins clearer by using my own scheme.

In datasheets, the output pins may be identified as Y0 through Y15, or S0 through S15, or similarly. I thought it was clearer to identify them as Out 0 through Out 15. (I often wonder why the people who create datasheets are so reluctant to use three letters—as in the word "Out," for example—rather than just one letter.)

Going back to the pin-numbering scheme, there are a couple of chip features that I have not explained so far. Pin 1 is the Latch Enable pin, which is active low. In other words, when this pin is held low, it latches the output state and tells the chip to ignore any inputs until further notice. This is not a feature that we need in the Telepathy Test circuit, so the breadboard schematic shows Pin 1 tied to the positive side of the power supply. In datasheets, the Latch Enable pin is often identified as LE, but may alternatively be identified as Strobe. Why, I don't know.

Pin 23 is a more general Enable pin, and it, too, is active low. This means that while Pin 23 is low, the functions of the chip are enabled. If Pin 23 is pushed high, the chip is disabled, and since I don't want it to be disabled, Pin 23 is shown tied to negative ground in the breadboard schematic. In datasheets, the Enable pin is often identified as E, but can alternatively be identified as Inhibit.

Decoder chips are not used so often these days, but they can still be useful for small projects. Assigning place values to a set of switches, and then decoding the input from them, is a simple and powerful way to assess user input.

Can you think of another application where we could do this? Well, how about the Rock, Paper, Scissors game? There are six switches in that game, so what if we wire them to a 6-bit decoder?

This idea suffers from two snags. First, there's no such thing as a 6-bit decoder chip. And second, if such a chip existed, it would have 64 output pins. That's a lot to deal with, especially since most of the outputs would be irrelevant, as they would involve a player cheating by pressing more than one switch. We would have to OR literally dozens of outputs, all of which would illuminate a "cheating" indicator.

However, there are 3-bit decoders. Maybe we could use one of them for one set of switches, and other for the other set. Would that be helpful? My answer to that is "maybe." I'll explain why in the next experiment.

Experiment 20: Decoding Rock, Paper, Scissors

20

I invite you to inspect Figure 20-1. This circuit is the simplest way I could find to create a fully-featured version of the Rock, Paper, Scissors game, using two decoders to simplify the logic while retaining switches to activate the LEDs and a beeper. Because it uses switches as well as logic gates, you can think of it as a hybrid circuit.

The Logic

I'll deal with the logic first. You'll remember that at the end of Experiment 17, I decided that logic gates were inconvenient and complicated for the task of detecting if either player had cheated. Well, this difficulty has now disappeared.

In the schematic, each switch now has a binary-code place value of 1, 2, or 4, and each set of three switches feeds the inputs on a three-bit decoder. All the possible switch combinations can now be expressed with just eight outputs, numbered 0 through 7. I have shuffled the sequence of the numbers to minimize wiring crossovers in the schematic, and output 0 is currently unconnected.

Let's look at just one of the decoders, since their connections are so similar. Output value 3 from Decoder A can only be created if Player A presses switch 1 in combination with switch 2. Output value 5 can only be created by pressing switch 1 with switch 4—and so on. Consequently, we now know that if we have output 3 OR output 5 OR output 6 OR output 7 from Decoder A, Player A has cheated. We simply combine these outputs through a quad-input OR gate, and send them to a "cheat" indicator, and that job is done.

The remaining logic is similar to the logic I used before. I have used diagonal lines to connect the logic gates, in an effort to make the connections easier to follow. For example, suppose Player A presses switch 4 (meaning "scissors"). If Player B presses switch 2 (meaning "paper"), scissors cut paper, so Player B wins. This is shown in Figure 20-2 where the red lines are carrying positive power.

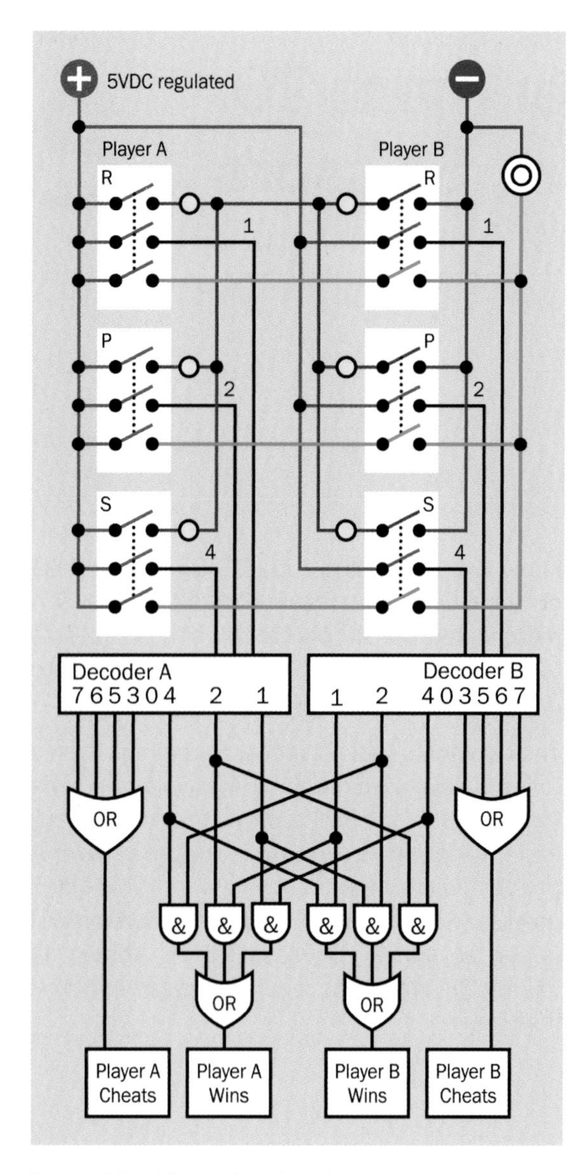

Figure 20-1. *Mixing decoders, logic gates, and multipole switches may be the simplest way to emulate the Rock, Paper, Scissors game.*

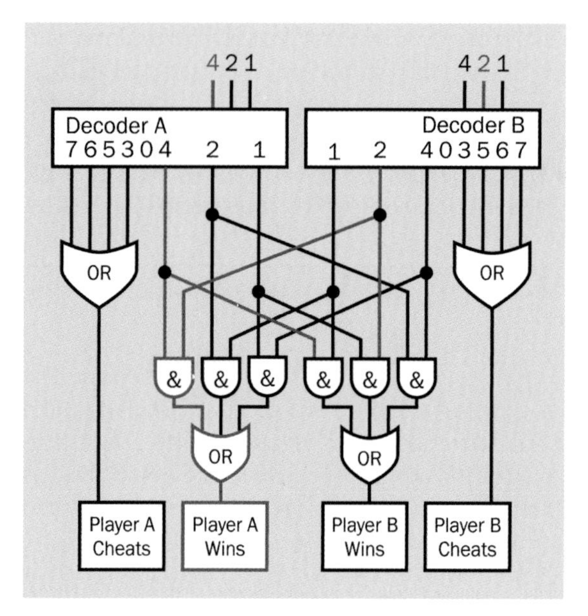

Figure 20-2. *In this example, Player A chooses "rock" and Player B chooses "scissors," using the array of switches shown in the previous schematic.*

You can follow all the possible combinations this way.

But what if both players press the same-numbered switches, creating a tied game? The tie combinations do not activate any of the AND gates, so the "who wins" indicators stay dark. I chose to use the switches themselves to deal with this possibility, because the wiring is so simple. Arbitrarily, in Figure 20-1, I colored the relevant wires green. You'll see that if any two switches opposite each other are pressed, this activates a beeper, shown at top-right. Now that the wiring diagram has been simplified, one beeper will do the job.

Meanwhile, the LEDs that show which switch has been pressed are also activated by the switches. Each LED is shown as a yellow circle to save space in the schematic, and they are connected by brown wires. The LEDs are wired in series, as before, so that both players must press switches before any LEDs light up. In the event of a tie, the beeper sounds, and the players can see two LEDs illuminated opposite each other.

I think this is the simplest way to represent the game electronically, although I'm always open to other ideas.

The Specification

If you come up with a different system, using either logic gates or switches, I'd love to see it. Just make sure it satisfies these criteria:

- A "cheat" indicator must light up if either player presses more than one button.
- Two other indicators must show which player wins, and they must only work if no one cheats. Note that my schematic satisfies that requirement, because each of pins 1, 2, and 4 on Decoder A and Decoder B will only have a high output if a single switch is pressed. Multiple switches will create a different decoder output.
- A beeper should sound if there is a tie.
- Each switch should light an LED showing that it has been pressed, but not until both players have made their choices.

An Unobtainable Or

The version of the game in Figure 20-1 now requires six switches or pushbuttons, two decoders, two quad two-input AND gates, one triple three-input OR gate, and one dual four-input OR gate. Add it all up, and the chip count is significantly lower than if I had tried to accomplish all this just with logic chips. And, the circuit is easier to build and understand.

If you want to wire this circuit, it should be fairly simple.

But wait a minute. Why am I using that "should" word? Because although it should be simple, it isn't! The problem is a matter of availability:

- A dual four-input OR gate is not available (in DIP format) in the HC family. It can only be obtained in the old CMOS 4000 series, which is not really powerful enough to drive an LED.

This is annoying, but it's not an unusual event. Anyone who designs a circuit may discover that the exact part that he wants does not exist, or has reached what the suppliers refer to as "end of life."

What about the other parts that I want to use in this particular circuit? Are they available?

The 74HC237 is a three-bit decoder in the HC family; no trouble there.

We have used triple three-input OR chips and quad two-input AND chips in the HC family, so we know they exist.

It's just the dual four-input OR chip that's a problem.

What to do? Well, you can always emulate a logic gate by using other logic gates, and OR gates make this particularly easy. Figure 20-3 shows how three two-input OR gates can be combined to perform the same function as a single four-input OR gate. So, I can replace each four-input OR gate in this way.

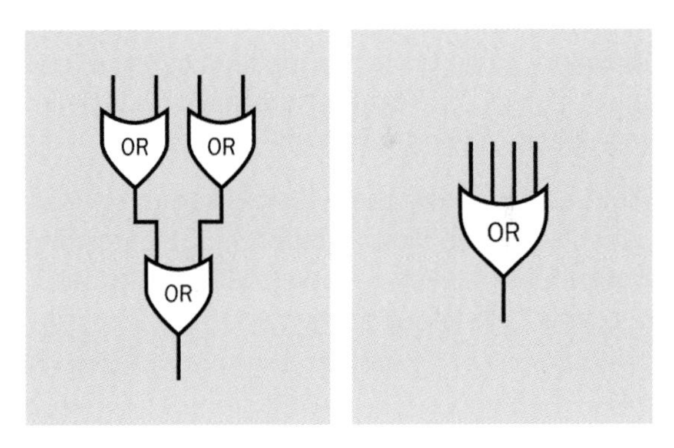

Figure 20-3. *Three two-input OR gates have exactly the same function as a single four-input OR gate when they are combined in this configuration.*

It's annoying, though, that this will require two quad two-input OR chips instead of one dual four-input OR chip. I want all the chips to fit on my breadboard. Is there another option?

Saved by NOR

Each of the (unobtainable) four-input OR gates was saying, "If any of the decoder pins 3, 5, 6, or

7 is high, this means two or more buttons have been pressed, which is cheating." But let's think what this means. If 3 OR 5 OR 6 OR 7 is high, this must mean that neither 0 NOR 1 NOR 2 NOR 4 is high—because those are the only other pins available. In other words, instead of looking for a high output on one of the output pins that indicates cheating, we can look for an *absence* of a high output from any of the other pins. It amounts to the same thing.

A four-input NOR gate on output pins with values 0, 1, 2, and 4 will serve exactly the same function as an OR gate on output pins with values 3, 5, 6, and 7. And guess what: a dual four-input NOR chip does exist in the HC family. It's the 74HC4002.

Why is there a four-input NOR chip, but not a four-input OR chip? I have no idea. And how do I know which types of chips and gates are manufactured in each family? Because Wikipedia has a page listing every logic chip that exists. Just search for "List of 7400 series integrated circuits." Very convenient! And after verifying that the chip exists in theory, I can go to a supplier such as *http://www.mouser.com* to make sure that it is still being manufactured and sold in the real world.

Figure 20-4 shows the wiring diagram revised, substituting the NOR gates for the OR gates. The output pins in each decoder that have values 3, 5, 6, and 7 are left unconnected.

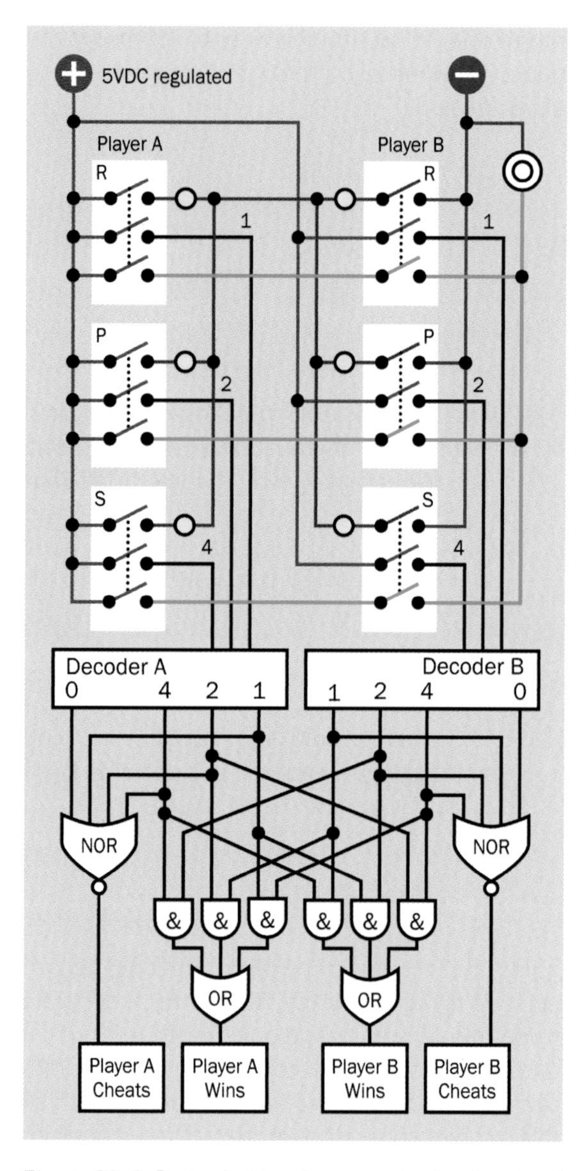

Figure 20-4. *Revised wiring diagram using two four-input NOR gates instead of two four-input OR gates, which are unavailable in the HC family, in through-hole packaging.*

Boarding the Simplified Scheme

Now that the logic is settled, it's time to put the chips together. You'll need two 74HC08 quad two-input AND chips, of the same type which you have used before. Their internal workings were shown in Figure 15-7. (Two of the AND gates in these chips will be left unused.) You will also need one 74HC4075 triple three-input OR chip, which you ran across in Experiment 19 (see

Chapter 19). Its internal connections were shown in Figure 19-8. (The third OR gate in that chip will be left unused.) And the last of the logic chips will be be the 74HC4002 dual four-input NOR chip. Its pinouts are shown in Figure 20-5.

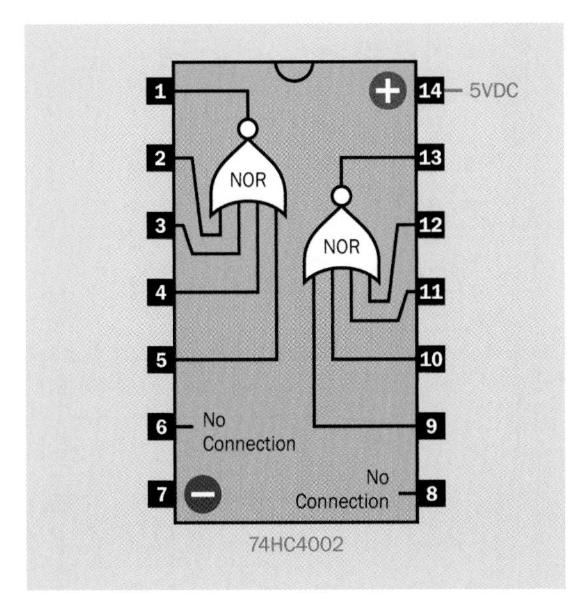

Figure 20-5. *Pinouts of the 74HC4002 dual quad-input NOR chip.*

Of course you will also need two 74HC237 decoder chips. Figure 20-6 shows the pin functions of this component. Note that it has two output-enable pins, one of which is active when high, the other being active when low. Both of them must be active to enable the output. Therefore, Pin 5 should be tied to ground, while Pin 6 should be tied to the positive bus of the breadboard.

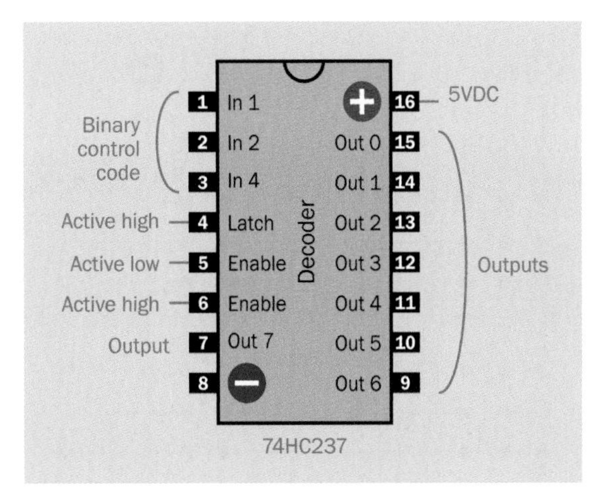

Figure 20-6. *Pin functions of the 74HC237 decoder chip.*

Pin 4 will latch the output, causing the chip to ignore changes in any of its inputs. Since Pin 4 is active high, it should be grounded to allow the decoder to respond.

I have used the same pin-naming vocabulary as in Figure 19-2. The outputs are labelled Out 0 through Out 7. The inputs are identified with the binary place values, as In 1, In 2, and In 4.

With this information, it's now relatively easy to breadboard the hybrid version of Rock, Paper, Scissors. Once again I have placed the switches in a separate schematic. You'll find it in Figure 20-7. The hookup wiring of the switches is shown in Figure 20-8.

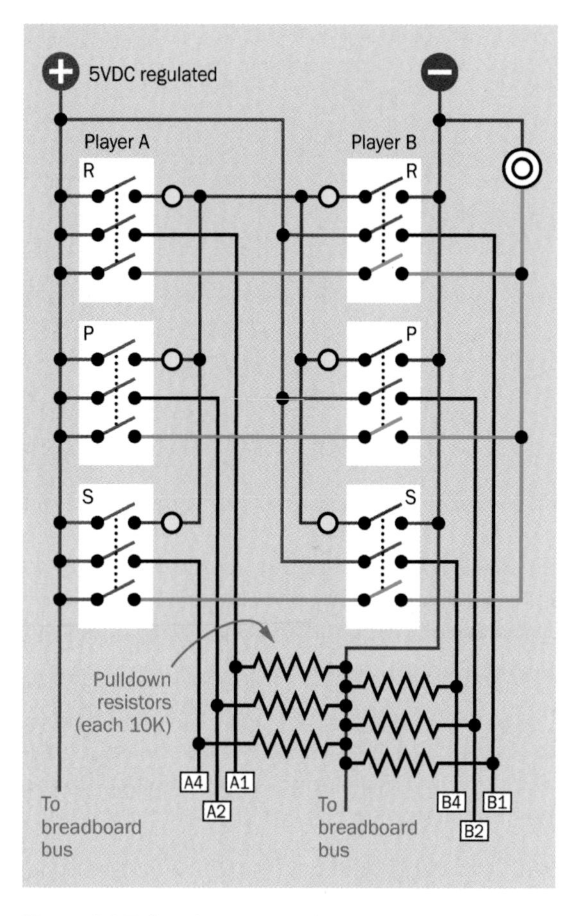

Figure 20-7. Switch outputs to be connected with equivalent labels in the breadboard schematic.

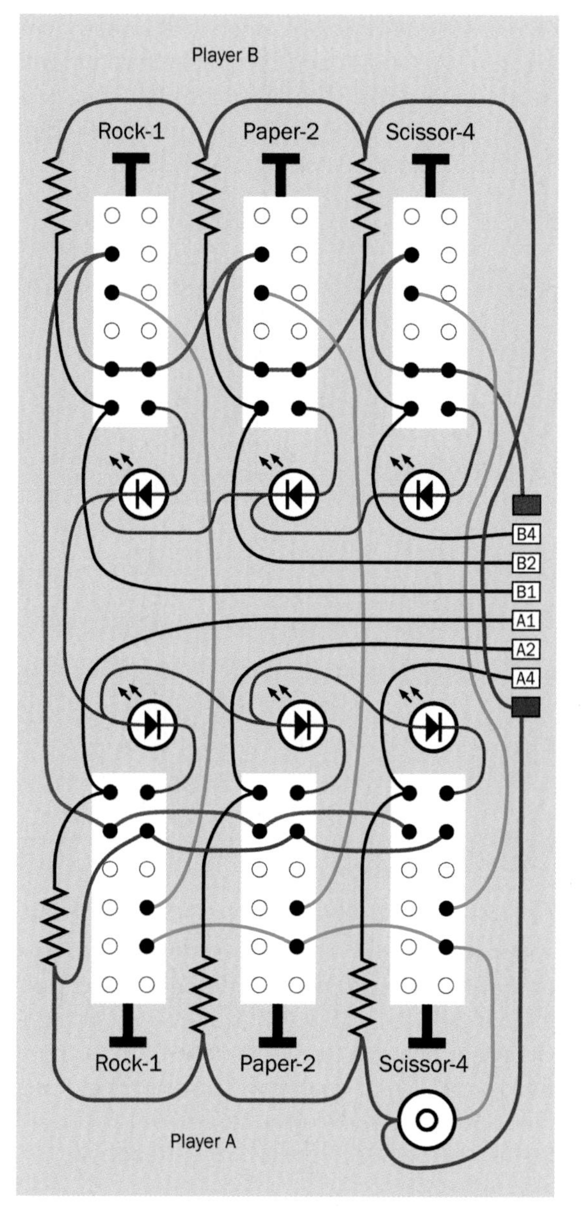

Figure 20-8. *Wiring the slider switches shown in the previous schematic. An edge connector, at right, can be used to link this circuit with the main circuit on the breadboard, including positive and negative sides of the power supply.*

Actual slider switches mounted on perforated board are shown in Figure 20-9.

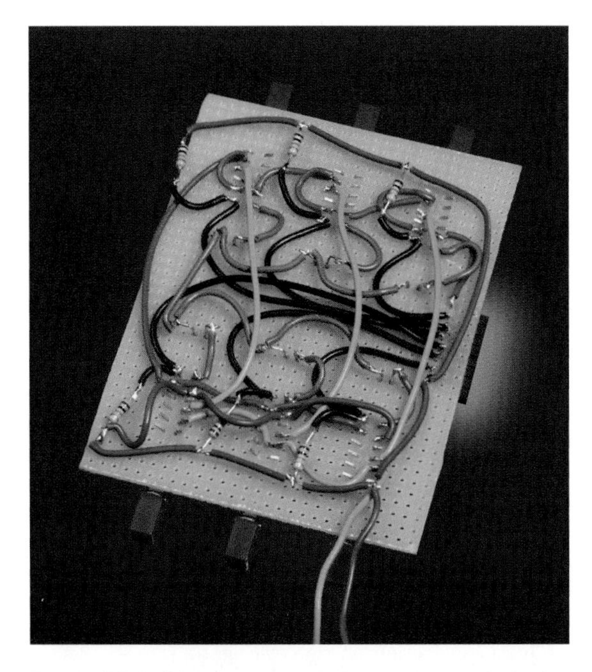

Figure 20-9. *Slider switches for connection with the hybrid Rock, Paper, Scissors game. An edge connector to link this board with the main breadboard for the game is visible in the pale area of the background color at right.*

The switch outputs (labelled A1, A2, and so on) connect with the chip inputs that have the same labels, in Figure 20-10.

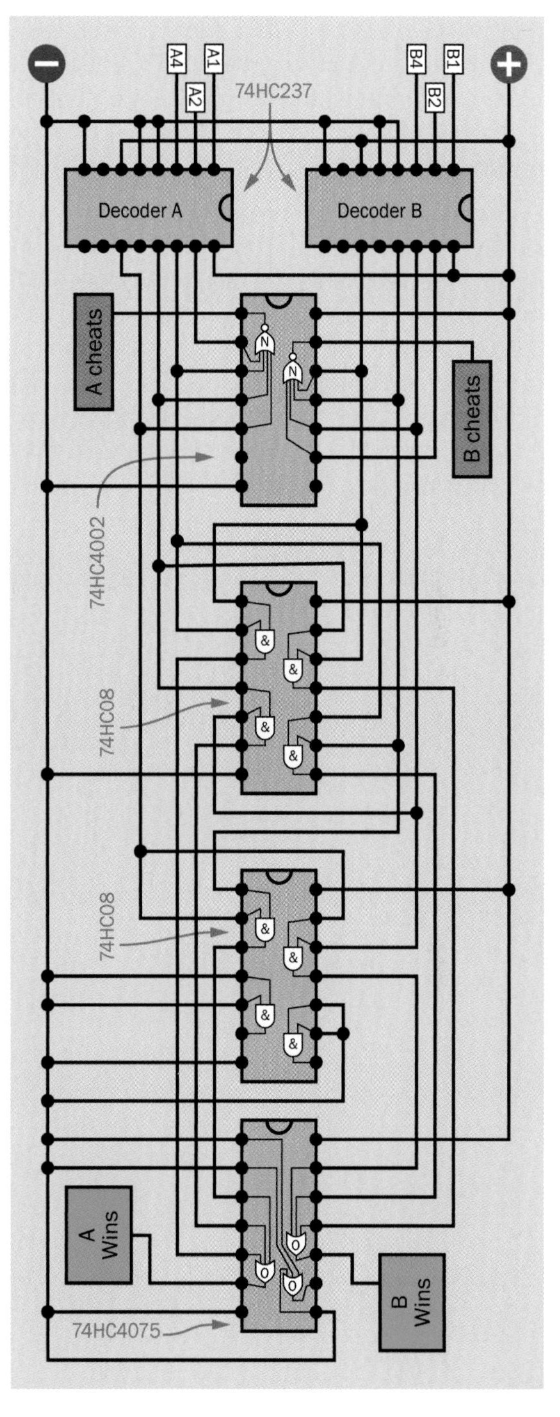

Figure 20-10. *Hybrid Rock, Paper, Scissors breadboard schematic.*

Normally, as I think you know, I draw schematics that show components approximately where they will be placed on a breadboard. In this case, I had to violate that rule, because there wasn't room on a printed page to stack all the chips in a vertical column. Therefore, I had to rotate the decoder chips.

Everything will fit on a breadboard, though. The circuit is shown in Figure 20-11. For testing purposes, I used six miniature tactile switches, each of which fits in one column on the breadboard. You can see their little yellow and red buttons at the top of the photograph. Because these are SPST switches, they don't enable wiring to show a tie or display an LED. You need the multipole switches in Figure 20-9 to do that.

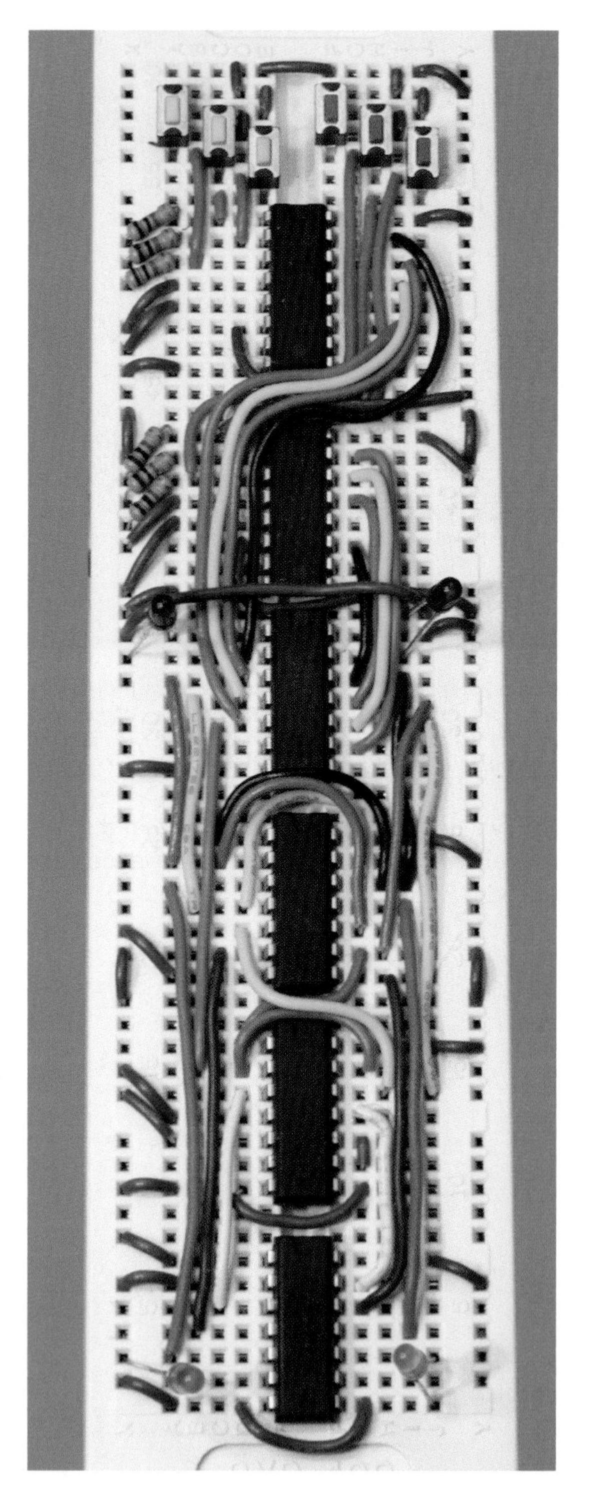

Figure 20-11. *Breadboarded version of the hybrid Rock, Paper, Scissors breadboard schematic.*

Make Even More

Because I can never resist thinking of little extra features to add to a project, it occurred to me that the switches in this circuit could have an additional function. Triple-pole slide switches or pushbuttons are relatively unusual. Almost certainly, you'll be using four-pole pushbuttons, leaving one pole in each switch unconnected. Is there anything useful and interesting that you could do with those spare poles?

Well, yes. You could use them to switch on the power for the rest of the circuit. In the all-switches version of Rock, Paper, Scissors, I found it pleasing that there was no separate on-off switch. The pushbuttons connected power at the same time they determined the outcome of each game.

You could do the same thing with the hybrid version. Wire the unused contacts of Player A's switches to the positive side of the power supply in parallel so that if any switch is pressed, it connects power through to a separate power supply bus feeding the decoders and logic chips. Player B's switches can be wired the same way. As soon as either person presses a switch, the logic chips are powered up and ready to interpret switch inputs. As soon as the switches are released, the system automatically powers down.

Have I built this enhanced version? No, I didn't get around to it. That's why I'm just sketching an outline of it, leaving you to add this feature yourself.

Undecoding

If you look up "decoder" in an online parts supplier, you'll probably run across three other kinds of chips as well: encoders, multiplexers, and demultiplexers. This can be quite confusing, as the same chip can be described as both a decoder and a multiplexer. What, exactly, is a multiplexer, anyway?

My job here is to eliminate confusion (or at least reduce it to a tolerable level), so this is the topic that I'm going to tackle next.

Experiment 21: The Hot Slot

21

Previously, I mentioned that because a coin has no memory, it has the same chance of showing heads or tails every time you flip it. (See Chapter 17.) The odds are always the same. Likewise, the traditional "wheel of fortune" doesn't remember where it stopped on the previous spin, and provided it is turned randomly each time, the odds of any number coming up never vary.

Not all games are like this. If you've ever played Battleship, for instance, you know that your odds of success will vary during the game, especially if you reach a point where you have eliminated a lot of empty squares, leaving your opponents' ships nowhere to hide.

I decided to create a two-player coin game with variable odds. Initially the stake would be low, and the odds of winning would be low—but as each player added more coins, the jackpot would get bigger and, at the same time, the odds of winning would gradually improve. I thought this should create some tension and drama in the game, even if it was just played for tokens rather than money.

This is how the Hot Slot game came to exist.

If you read *Make* magazine, you may remember that one of my columns discussed a game in which coins were used to complete an electrical circuit and win a prize. The Hot Slot game shares that basic concept, but in every other respect, it's very different.

Muxing It

This game enables me to keep my promise to introduce you to a multiplexer. "Mux" is the colloquial term that people often use for this component.

Many variants exist, but the one I chose for this experiment is the 4067B. It's an old-school CMOS chip, although unlike many CMOS components of its era, it is not restricted by low output current. You can think of it as merely serving as a channel through which you pass current.

Plug the 4067B into your breadboard, and add four switches and pulldown resistors as shown in Figure 21-1. This is very similar to the test circuit for investigating the 4514 decoder (see Figure 19-3), except that instead of sixteen output pins, we now have sixteen "channel" pins that can function as inputs.

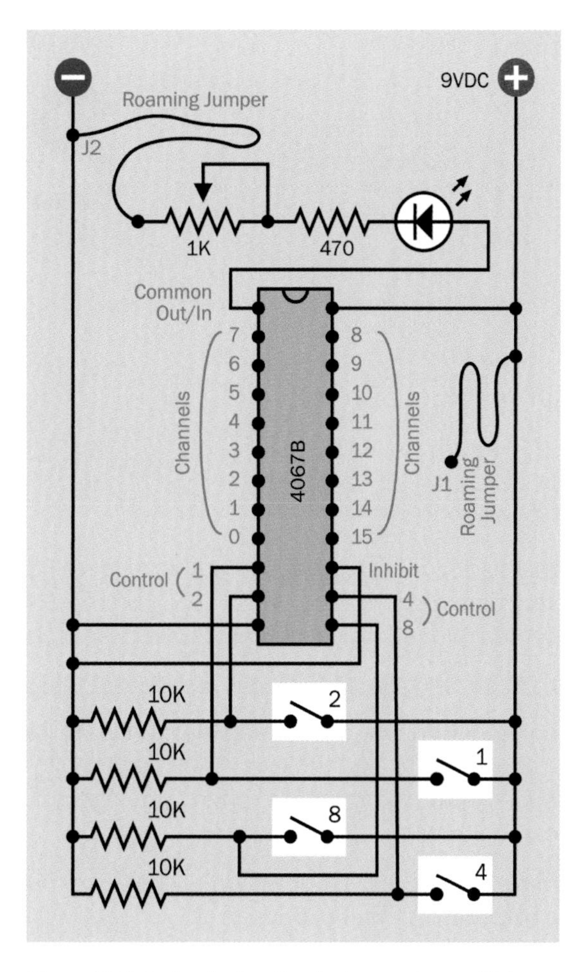

Figure 21-1. *A simple circuit to test an analog multiplexer.*

The switches are at the bottom because the pins that are associated with the switches are also at the bottom end of the chip. This keeps the schematic as uncluttered as possible.

You don't need a meter to test the multiplexer, because it's quite happy to drive an LED. This is attached to pin 1, known as the "Common Out/In" pin for reasons which I will explain in a minute. A series resistor protects the LED from excessive current, and a trimmer potentiometer will demonstrate that we can vary the current.

The wire attached to the trimmer has wiggles in it to suggest that it is a flexible jumper wire—the type that you should generally avoid, with a little plug on each end. I'm calling it a "roaming jumper" because I want you to be able to move it around. Right now, the end of it labelled J2 is plugged into the negative-ground bus. If you don't have a jumper wire of this type, you can use a length of solid-core 24-gauge hookup wire, with a quarter inch of the insulation stripped at each end.

At the right side of the schematic, you see another "roaming jumper." I want you to be able to touch the end of it, labelled J1, to any of the channel pins that have values 0 through 15.

You can power this circuit with 9VDC from a battery. The voltage does not have to be regulated, and you don't need any smoothing capacitors. However, CMOS chips are sensitive to static electricity, so take the usual precaution of grounding yourself before handling it.

Jumping and Roaming

Now, let's see what happens if you close the two switches labelled 2 and 4. If they work the same way as the switches that were used with an encoder, they should activate the channel that has value 6, because the channel number is determined by adding the values of the switches that are closed.

Sure enough, if you touch jumper J1 on channel pin 6, you should see the LED light up. Positive power is flowing from the bus, through J1 and into the channel with value 6. An internal connection in the chip takes the power to the Common Out/In terminal, at pin 1. From there it travels through the LED and the series resistors, and out of J2 into the negative bus.

Adjust the 1K potentiometer, and the LED brightness will vary. Note that you are not just adjusting the output power; you are adjusting the power flowing through the whole chip. You can add a meter, set to measure mA, in series with J1, and you'll find the current is varying on the input side as well as the output side.

You do pay a small penalty for this service. The chip deducts a percentage of the voltage for its own purposes. You can prove this by setting your meter to measure volts and applying it between pin 6 and the Common Out/In pin.

Now here's the interesting part:

- Touch J1 on any of the other channel pins, and the LED should stay dark.
- If you change the switch combination, a different channel pin will become active.

Turn on all the switches, for instance, and the sum of their values (8 + 4 + 2 + 1) will activate the pin that has a value of 15.

What have you learned so far? And what else do you need to know?

Quick Facts About Muxes

- This mux has four binary-valued control pins, like a decoder.
- But it seems to be the opposite of a decoder. Instead of delivering power out of a pin with value 0 through 15, it accepts power into a pin with value 0 through 15. These pins are connected inside the chip through "channels."
- The pattern of high/low states on the control pins selects one of the channels.
- The selected channel will pass an input current through to a common pin.
- The channels which are not selected will not pass current. (Actually there is a little leakage, but it is negligible.)
- Maximum current input is 25mA, but this value becomes lower as the voltage goes up. Each passthrough transistor is rated for a maximum of 100mW.
- The power supply can range from 3VDC to 20VDC.
- The voltage that you apply to a channel cannot be higher than the power supply voltage or lower than negative ground.
- A good way to think of a multiplexer is to imagine it as a solid-state rotary switch, as suggested in Figure 21-2. In this figure, control pins with values 2 and 8 have been arbitrarily selected, making an internal connection between channel 10 and the common output/input.

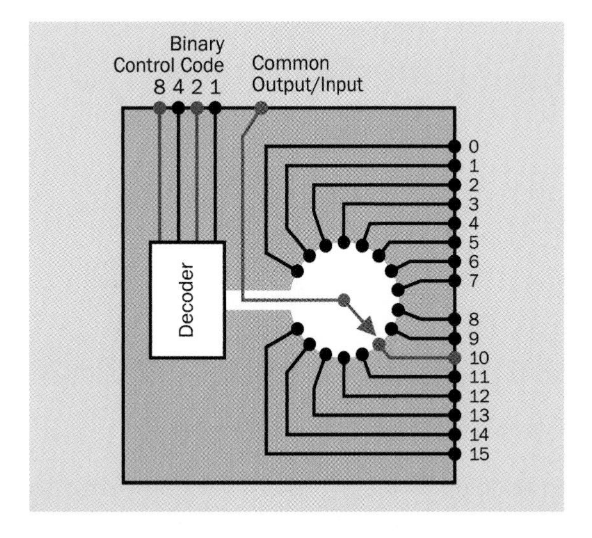

Figure 21-2. *A multiplexer works like a solid-state rotary switch. The purple connection is established by the binary code applied to the control inputs.*

Mux Pinouts

Pinouts for the 4067B are shown more formally, along with pin numbers, in Figure 21-3. The control pins are labelled 1, 2, 4, and 8, as these are their binary place values. (In a datasheet, these pins may be labelled A, B, C, and D.) The channel pins are labelled Chan 0 through Chan 15. (A datasheet may refer to them as Y0 through Y15, or similar.)

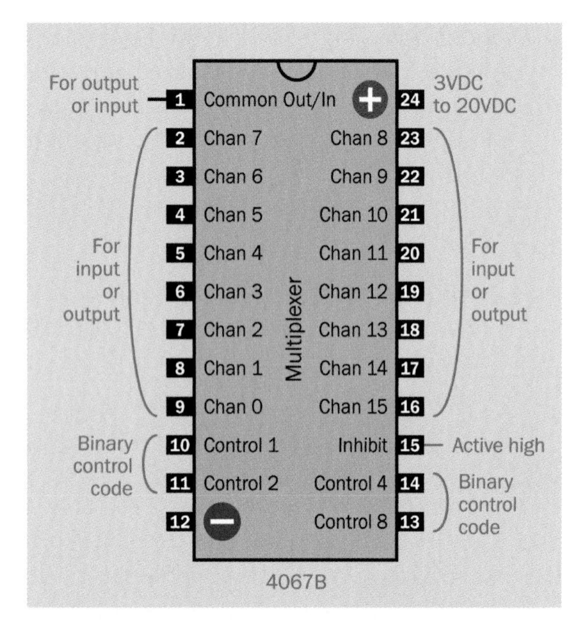

Figure 21-3. *Pinouts for the 4067B multiplexer.*

In addition to the pins that I have described so far, there's an inhibit pin, located at Pin 15. This is active high, meaning that if you connect it to positive voltage, the chip will be inhibited from responding to the control pins. I don't want this to happen, so in the test circuit, the inhibit pin is tied to ground. You can experiment by connecting it with the positive side of the power supply, at which point the chip will switch off its internal transistors.

Mux Applications

I'm going to use the multiplexer in the coin game that I have in mind. But what is it normally used for?

Consider its capabilities. Depending on its control input, it can select one of up to sixteen inputs and send the signal out through a common pin —very quickly. In fact it works so quickly, it can take two or four or more separate telecommunications signals on the channel pins and sample them one at a time, in sequence, combining them into a single output channel. Thus, one wire can carry two, four, or more signals. Of course at the other end, you need to split the incoming stream back into its original channels. For this purpose you need . . . a demultiplexer!

The same multiplexer that I've been talking about, the 4067B, can also function as a demultiplexer, because it is a bidirectional chip. This is why the channel pins in the pinout diagram are labelled "For input or output," and is why the Common pin is labelled "Out/In." The mux doesn't care which way the current flows through it.

You can prove this for yourself by modifying your test circuit slightly, as shown in Figure 21-4. Roaming jumper J1 is now plugged into the negative bus, while jumper J2 is plugged into the positive bus, and the LED has been turned around so that it responds to current flowing in the opposite direction.

You may wonder why a multiplexer isn't called a multiplexer/demultiplexer, since that's what it does. Well, if you look at a datasheet, it usually is called a multiplexer/demultiplexer—but this is such a mouthful, people just call it a mux.

A mux is also used where we need to switch more slowly between multiple inputs. Inside a computer, for example, it can select one of two or more video outputs.

In a stereo system, it can select among inputs from a CD player, a DVD player, an MP3 player, or some other audio source. Depending on the code applied to the control pins, the mux will choose one input and connect it with the common output pin. This used to be done with an electromechanical rotary switch or pushbuttons, but a solid-state switching device is more reliable and shouldn't introduce noise of the type that is characteristic of switch contacts.

A stereo system has two audio channels (more, if you have a home theater system with Dolby 5.1). To address this requirement, you can buy multiplexers that contain two or more switches, both governed by the same set of control pins. These multiplexers can be compared with a rotary switch that has several decks, all governed by the same control shaft and knob.

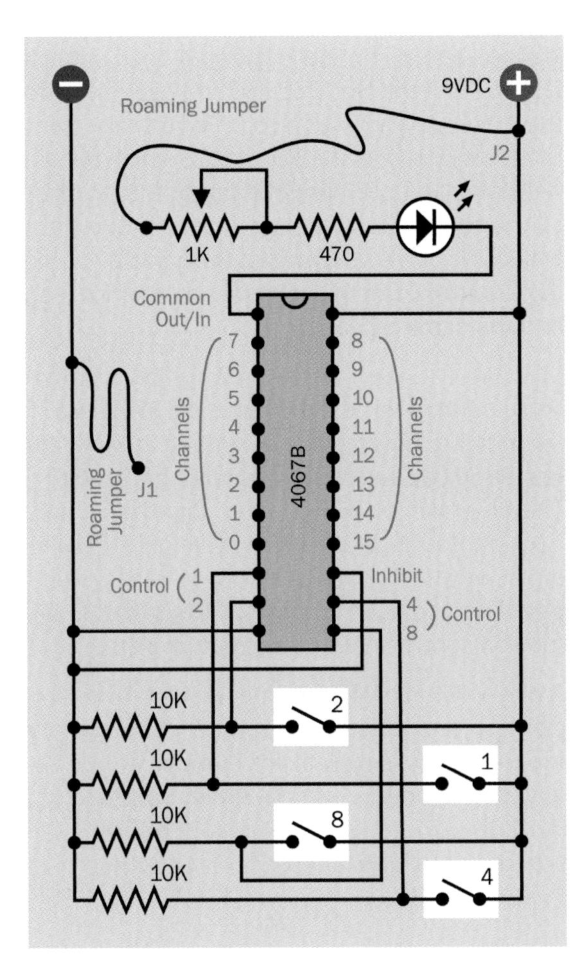

Figure 21-4. With Jumper J2 plugged into the positive bus, you can demonstrate the bidirectional capability of the multiplexer by using J1 to sink current into the negative bus. Don't forget to reverse the LED so that its polarity matches the current flow.

Analog Versus Digital Mux

The 4067B is an analog multiplexer. This means that it can pass and preserve AC signals, which oscillate above and below a neutral value. For example, you could hook up a multiplexer to a simple audio intercom and distribute its output to a variety of rooms in your house.

Digital multiplexers are also available. Their inputs must be within the usual high/low range for logic chips. The digital multiplexer then uses a binary code on its control pins to select one channel (just like an analog multiplexer), and senses whether it is high or low. But the chip generates its own output signal, to make sure that the output conforms with digital specifications.

A digital multiplexer is not reversible. The channels receive inputs, the Common pin provides an output, and that's the end of the story. AC signals cannot be used.

Since a digital mux is not reversible—what do you do if you want to decode its output at the other end? Well, you use a decoder! The control inputs on a decoder can be synchronized with the rate at which data packets are flowing in so that it chops the stream into segments of the correct size.

A decoder is really the same thing as a digital demultiplexer, although that term is seldom used. Meanwhile a digital multiplexer is similar to an encoder. This is why electronic parts suppliers often put encoders, decoders, multiplexers, and demultiplexers all in the same category. The supplier doesn't try to disentangle all the different varieties. It just shows you all the options and says, "You figure it out."

To assist you in this, here's a summary.

Quick Facts About Mux Variants

- A *decoder* can have two, three, or four control pins. A binary code on these pins selects one output pin from a series that have values from 0 upward. The selected pin has a logic-high output. The other output pins are logic-low. The decoder is a digital device using high and low logic states. Its input/output direction is not reversible.
- An *encoder* is the opposite of a decoder. It has multiple inputs, each of which is assigned a decimal value, from 0 upward. Only one pin at a time should have a logic-high input. The encoder converts the value of this input into a two, three, or four-bit binary

number which it outputs through two, three, or four output pins. It is a digital device using high and low logic states. Its input/output direction is not reversible. (We have not encountered an encoder in this book, yet. But we will.)

- A *digital multiplexer* is similar to an encoder, in that it receives an input from one of a sequence of pins, each of which is assigned a decimal value, from 0 upward. However, it has a single common, or output, pin. The binary states of two, three, or four control pins determine which input pin is connected with the output pin. It is a digital device using high and low logic states. Its input/output direction is not reversible.
- An *analog multiplexer* is similar to a digital multiplexer but does not create its own output voltage. This is the device that you tested in the experiment above. It merely passes the status of a selected channel pin through to a Common pin, via an internal connection. In this way, it functions as a digitally controlled, solid-state rotary switch. It will work with AC or DC signals over a wide range of voltages, and its current flow is reversible.
- An *analog demultiplexer* is usually just an analog multiplexer which is used in reverse so that the Common pin is used as an input and a Channel pin is selected as the output.
- A *digital demultiplexer* is usually the same thing as a decoder.

Game Design

I had to go through a lot of explanation before I could get back to building the Hot Slot game. Anyway, here's the way I'm imagining the game. I'll have a box with sixteen slots in it, each slot large enough to hold a coin. When a coin is inserted in a slot, the coin will create an electrical connection between two internal contacts.

One of the slots will be chosen at random to have live contacts. This will be the Hot Slot. It will be energized via a multiplexer, before the game begins, and the two people who are playing the game will have no way of knowing which it is.

A player's turn consists of placing one coin in a slot. If it isn't the Hot Slot, nothing happens, and the other player places a coin. Players take turns until someone hits the Hot Slot, at which point a beeper will sound. The winning player then takes all the coins that have been placed in the slots. One of the players then resets the game by pressing a button. The mux chooses a new Hot Slot, and the game repeats.

In Figure 21-5 you will see a simple flow diagram showing how this can be done. A 555 timer will be running in asynchronous mode, generating maybe 50,000 pulses per second—that is, 50KHz. The timer will drive a counter chip that counts from 0 through 15 over and over again, delivering its output as a series of four-bit binary numbers. These four-bit numbers will be wired to the four-bit control input of the multiplexer.

A player will allow the counter to run by pressing a button, and will stop the counter at an arbitrary moment. This is how a slot number will be selected randomly. I used a similar approach in *Make: Electronics*, so if you read that book, the scheme should be familiar.

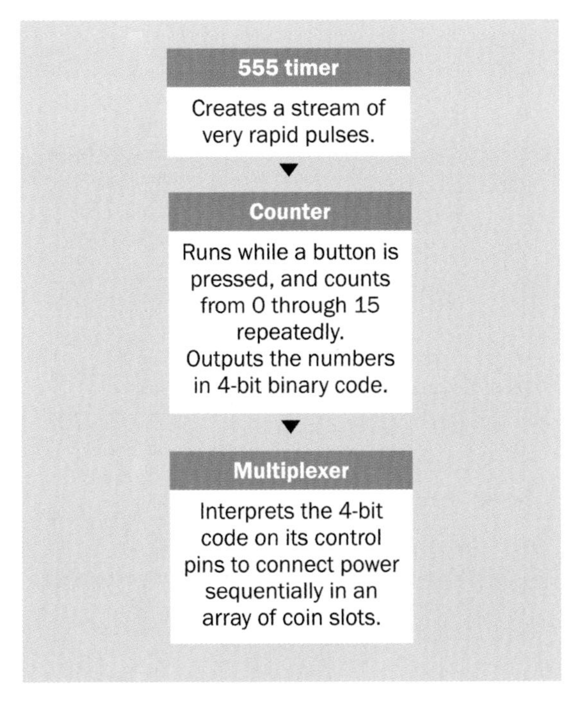

Figure 21-5. *The system to choose a random number which will be interpreted by a multiplexer, to choose one of sixteen coin slots.*

Slot Counting

I want to use a 9VDC power supply for the multiplexer to minimize the effect of its internal resistance. Is there a counter chip that can be powered by 9V? Yes, indeed. The 4520B counter is another old-school CMOS component which is still being manufactured in large quantities. Its pinouts are shown in Figure 21-6.

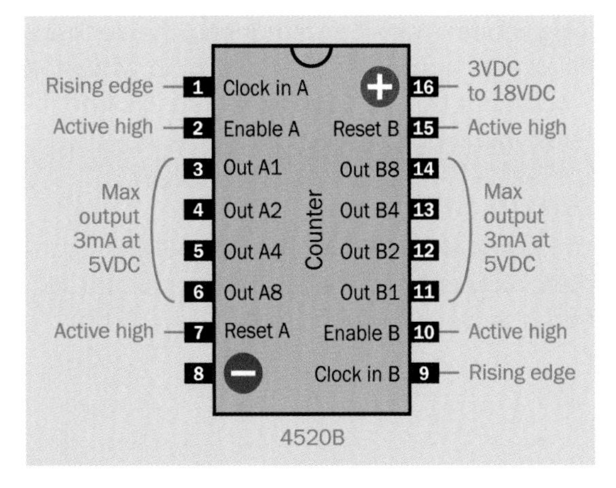

Figure 21-6. *The pinouts for a CMOS 4520B chip, which counts repeatedly from 0 to 15 decimal and expresses the running total as a four-bit binary output.*

The 4520B actually includes two 4-bit counters, which can be chained together to make an 8-bit counter capable of counting from 0 through 255. We only need four bits, so I won't use the second half of the chip.

In Figure 21-6 I have arbitrarily designated one counter as "A" and the other as "B." Manufacturers' datasheets may identify them this way, or may label them differently. I have also designated the outputs according to their binary place value (1, 2, 4, and 8). Datasheets may commonly designate them as Q1, Q2, Q3, and Q4, or some similar numbering scheme. There is no standardization.

The Reset pin (which is active high) sets all the outputs to 0, which is of no interest to us, because we want a random, arbitrary output. Therefore the Reset should be tied permanently to the negative side of the power supply in the Hot Slot game, to disable it.

The Enable pin is also active high, meaning that the counter will run when positive power is applied to Enable, and will freeze when Enable is connected to negative ground. This feature will be used in the Hot Slot schematic to stop the counter, to choose a slot at random.

The Clock pin advances the counter when the signal on the pin transitions from low to high. (If a high-to-low transition is required, the clock signal can be applied to the Enable pin while the Clock input is held low. This feature is of no interest in the Hot Slot game.)

Circuit Design

You can now build the circuit for picking a random number from 0 to 15. The breadboard section is shown in Figure 21-7.

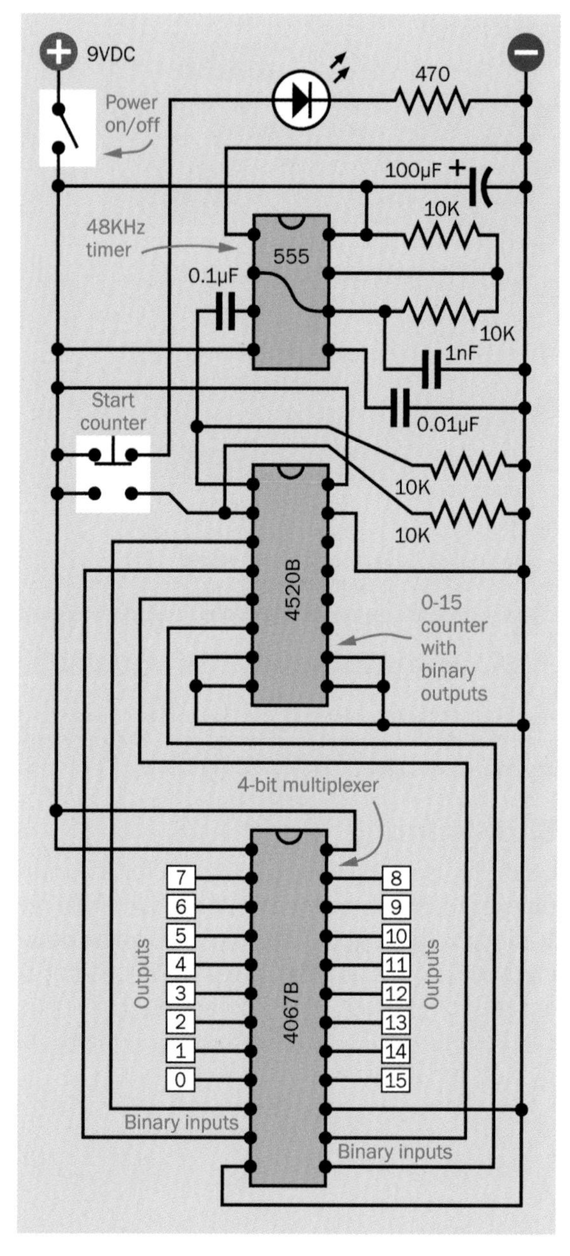

Figure 21-7. *The breadboard section of the circuit for the Hot Slot game.*

The counter and the multiplexer are both CMOS chips of similar vintage, so they should talk to each other without any problems. They should also be compatible with the output from the 555 timer. The only potential source of trouble is the tendency of a bipolar 555 timer to create voltage spikes, which can be picked up by the counter and misinterpreted as clock cycles. Placing a 100µF capacitor between the power supply pin of the timer and negative ground should deal with that issue. Make sure the capacitor is as near to the power supply pin as possible, and keep the leads of the capacitor as short as possible.

When the power switch at top-left is turned on, the LED is illuminated, and the 555 timer immediately starts sending clock pulses through the coupling capacitor to the 4520B counter. However, a 10K pulldown resistor on Pin 2 of the counter (the Enable pin) prevents the counter from counting.

To select a random number, the pushbutton must be pressed. This connects Pin 2 of the counter directly to the positive bus of the power supply, overriding the pulldown resistor and enabling the counter, which responds to the stream of pulses from the timer. When the pushbutton is released, the counter is disabled again, stopping at a randomly selected number. The LED comes on, indicating that the game is ready.

The number at which the counter has stopped is supplied to the 4067B multiplexer, which makes an internal connection between the corresponding channel pin and the common out/in pin. This receives power from the positive bus. Power from the bus goes out through the Channel pin (so the multiplexer is really behaving as a demultiplexer, if you want to be precise about it).

The sixteen channel pins are connected with the sixteen coin slots, which are depicted in Figure 21-8. The numbers down the left edge of this schematic indicate connections with the same-numbered channel pins of the multiplexer.

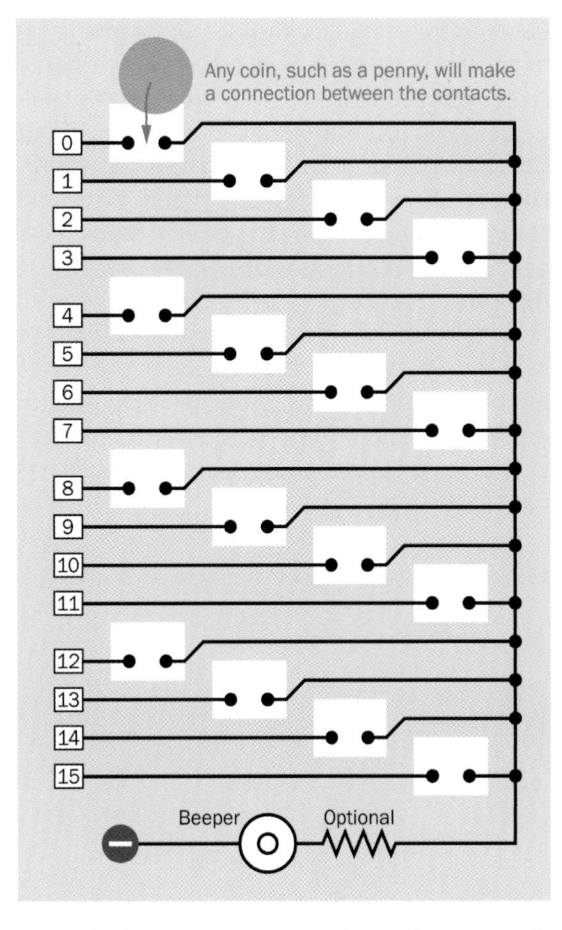

Figure 21-8. *In this schematic, each pair of contacts indicates a coin slot in the Hot Slot game. A coin makes an electrical connection between the contacts.*

Because only one channel at a time receives power from the multiplexer, only one slot is ever active. When a coin is inserted in that slot, it connects current through the beeper to negative ground. This ground must come from the negative ground on the breadboard to complete the circuit. A resistor may be placed in series with the beeper if necessary to limit its power consumption to around 15mA to avoid overloading the multiplexer.

When the coins are removed, the beeper stops. The button on the breadboard must now be pressed again, to choose a new random number; otherwise, the slot that was energized in the previous game will still be the same.

Building the circuit for the Hot Slot game is actually the easy part, as it only contains three chips. Fabricating the coin slots is more of a challenge. If you just want to test the circuit without going to this trouble, you can use two eight-contact DIP switches instead of coin slots, and close them one at a time to simulate the process of adding coins. This is shown in the breadboarded version of the game in Figure 21-9. I've used flexible jumper wires to connect the multiplexer with the DIP switches, because I regard these as temporary connections. Ideally a 17-conductor ribbon cable should be used (the 17th conductor would be a ground connection from the switches back to the main circuit).

Figure 21-9. *The breadboarded version of the Hot Slot game, using DIP switches as a substitute for coin slots.*

Slot Design

In my *Make* magazine feature using coins in slots, I suggested the slot design shown in Figure 21-10, using small sections of thin aluminum angle cut from a length that I bought in a hardware store. Naturally, for sixteen slots, you would need to make four copies of this assembly.

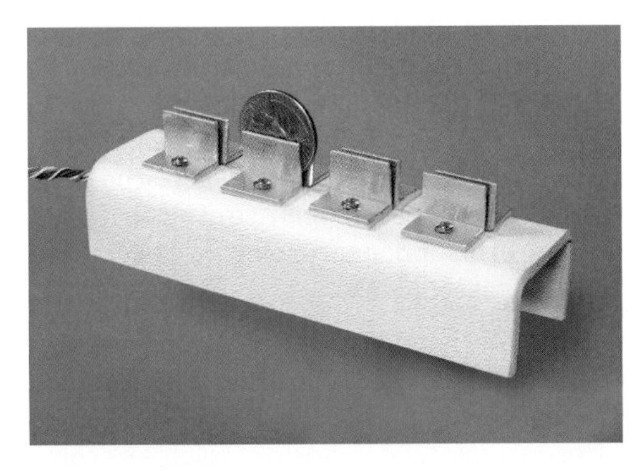

Figure 21-10. *A simple type of coin slot for the Hot Slot game that can be made relatively easily while providing a reliable contact with the coin.*

You can substitute a small section of wood for the ABS plastic that I used, and the slots can be installed in a box with corresponding apertures in the top. If you can come up with a more elegant slot design that does not require the coin to be jammed into place, I encourage you to do so. Eventually I'm going to modify this project with infrared sensors that will detect the presence of coins, but that won't be until Experiment 31.

Hot Slot Testing

When you test the circuit, I suggest you begin with a larger value, such as 47µF, for the timing capacitor connected with the 555 timer to make it run slowly. Add LEDs (sharing a 1K series resistor so they don't disturb the voltages in the circuit) to the output of the timer and the four outputs from the counter.

When the DPDT pushbutton in the schematic is not being pressed, use a meter to find which channel from the multiplexer is energized, and make sure its value matches the value of the binary input.

Once you are sure that the game is functioning correctly, substitute the correct capacitor to increase the timer speed, remove the LEDs, and you're ready to play.

Who Wins?

Assuming that the circuit does what it's supposed to do, there's still one very important unanswered question: when two people are playing the Hot Slot game, does each of them have an equal chance of winning? Or does the person who goes first (or second) have an edge?

Figuring this out is like the process of designing a logic circuit. The first step is to describe the game very clearly:

- The person who places the first coin has sixteen slots to choose from, and only one of them will win the game.
- Therefore, he has a 1-in-16 chance of ending the game on the first turn.

Turning it around, there is a 15/16 chance of the first player *not* hitting it. In that case, the second player takes his turn. Fifteen slots are still available, and therefore the second player has a 1-in-15 chance of winning.

How likely is it for the game to end on the second turn? Two things have to happen:

- The first player has to miss the hot slot. There is a 15/16 chance of this.
- The second player has to hit the hot slot. There is a 1/15 chance of this.

To find the probability of both of these events happening, we must multiply the odds together. If C is the chance of the game ending with the second turn:

```
C = 15/16 * 1/15
```

If you remember your high-school math, the 15s cancel out, so:

```
C = 1/16
```

- In other words, there is a 1-in-16 chance of a game ending on the second turn.

I'll take this one step farther. The chance of Player 2 winning the game on the second turn is 1/15, so the chance of him not winning is 14/15. If that's the way it works out, Player 1 now tries again, and because fourteen slots are still open, he has a 1/14 chance of ending the game on the third turn. If C is the chance of this happening:

```
C = 15/16 * 14/15 * 1/14
```

The 15s and the 14s cancel out, so:

```
C = 1/16
```

In fact, the chance of a game ending on the first turn, or the second turn, or any turn up to and including the sixteenth turn, is always 1/16.

I'm not saying that the odds of hitting the hot slot always remain the same at the moment when you place a bet. On the contrary, they become more and more favorable as a game progresses, because there are fewer open slots remaining. What I'm saying is that if you play dozens or hundreds of games, you will find that about one game in sixteen will end with the first bet, one in sixteen will end with the second bet, and so on.

Intuitively this seemed odd to me, so I wrote a little BASIC program to simulate 1,000 games. The random-number function used in computer languages often shows some uneven distribution of values, so I ran the simulation several times. It confirmed that my math was correct.

Hold on to this nonintuitive fact while I consider how many coins each player may win.

The Payoff

The first player places the first coin. Let's suppose he is lucky and hits the hot slot with that first bet. The only coin he can win is the one that he just played. So, he gets his coin back. He makes no profit at all!

If the first player was unsuccessful, the second player places a coin. If he wins, he gets his coin back, *and* he gets his opponent's coin, so he doubles his money. Evidently, it's infinitely more rewarding to win on the second turn than on the first.

But suppose the first two bets are both unsuccessful. Now it's the first player's turn again. If he hits the hot slot this time, he takes back the coin he just placed and the coin he placed in the first turn and the coin that his opponent placed in the second turn. So, he recovers two of his own coins, and one of his opponent's coins. He makes a 50% profit on his total stake.

On the other hand, if no one wins during the first three turns, and the second player does win on the fourth turn, the second player picks up two of his own coins and two of the first player's coins. Once again he doubles his money. In other words, he makes a 100% profit.

It's definitely an advantage to be the second player! But if we average this out over a long series of games, how big an advantage will this be?

Figure 21-11 shows the outcome for every possible game. For instance, if a game ends with the first turn, Player 1 has placed one coin, which has hit the Hot Slot, so, he gets his coin back. His net winnings (additional to his stake) are 0. Player 2 hasn't placed any coins, so his total stake is 0 and his winnings are 0.

Number of turns when a game ends	Player 1		Player 2	
	Total coins staked	Net win or loss	Total coins staked	Net win or loss
1	1	0	0	0
2	1	-1	1	+1
3	2	+1	1	-1
4	2	-2	2	+2
5	3	+2	2	-2
6	3	-3	3	+3
7	4	+3	3	-3
8	4	-4	4	+4
9	5	+4	4	-4
10	5	-5	5	+5
11	6	+5	5	-5
12	6	-6	6	+6
13	7	+6	6	-6
14	7	-7	7	+7
15	8	+7	7	-7
16	8	-8	8	+8
Total	72	-8	64	+8
Average	4.5	-0.5	4	+0.5

Figure 21-11. *This table shows the winnings and losses for each player, in games that end with 1 through 16 turns, and on an averaged basis. On average, the second player will win half a coin per game.*

The extra two lines at the end of the table sum each column of numbers and then divide by 16 to find the average for each player, per game. Remember, it's equally likely for a game to last for one turn, sixteen turns, or any number in between. Therefore, the average will tell us accurately how much each player can expect to win or lose in an average game.

Here's the bottom line: Player 1 loses half a coin per game, while Player 2 wins half a coin per game—on average.

Understanding the Odds

The reason behind this strange outcome is that Player 1 has a disadvantage when he goes first: he risks a stake in the game before Player 2 risks any of his coins. This disadvantage continues each time Player 1 takes a turn. He's always risking another coin before his opponent risks one of his. Consequently, as you see in the table, when an average game ends, Player 1's average stake has been 4.5 coins, while Player 2's average stake has been only 4 coins.

Because Player 2 wins eight coins for every 64 coins that he bets, he can expect a profit on his investment of 8 / 64 = 0.125, or 12.5%. This may be comparable to the odds on a Las Vegas slot machine, and is more than twice as much as a casino will make from roulette, where it pays out 36 / 38 = 95% of the money placed by bettors, and therefore takes 5%. (Actually the calculation is a little more complicated, as most casinos pay different odds when 0 and 00 come up in roulette, but the percentages are close enough.)

Personally I never play games of chance, because as this game illustrates, the odds are seldom in favor of a player. No matter how lucky you think you are, the mathematics of probability will always beat you in the end.

Suppose you didn't know anything about the Hot Slot game, and a friend invited you to play it. Suppose your friend said, "Since you're a beginner, I'll let you go first." What a nice guy! He sounds as if he's doing you a favor. After all, you'll have the first opportunity to find the Hot Slot. But actually it's a disadvantage, as we've seen. Evidently he is not such a good friend after all.

Would it be worthwhile for someone to try to make money by playing the Hot Slot game? Suppose two people are playing with pennies, and each person starts with 100 pennies. Because Player 2 wins an average of half a penny per game, he or she would hope to win all of Player 1's money within 200 games.

That sounds time-consuming. Suppose it takes two seconds for each person to place a coin, and ten seconds to remove coins from the slots and reset the system for the next game. Since a game will be completed in eight turns on average, it will last for slightly less than 30 seconds. Therefore 200 games will take 1 hour 40 minutes. That's a lot of time, just to win $1.

However, if the stakes were quarters instead of pennies, Player 2 could hope to make about $15 per hour. And if the players were using tokens with a value of, say, $1 each, Player 2 would be expecting to rake in $60 of his opponent's money per hour. (Of course, he would have to persuade the opponent to make the first move in every game.)

Once again, the message is clear: before gambling, do the math!

Background: Alternative Game Arrays

If Player 2 has a 12.5% advantage in a sixteen-coin game, would it be the same in a game with fewer or additional coins and slots?

No, the advantage will be different. To see why, I'll take an extreme example. Suppose we have a game in which there are only two slots. In this game, the person who goes first gets his own coin back if he wins and loses it if the second player wins. So the first player never wins anything! On average he will win half the games and lose half the games, so in an average game, he should expect to lose half a coin, while Player 2 should expect to win half a coin.

In fact, no matter how many coins and slots there are (so long as it's an even number), Player 1 always loses an average of half a coin per game, and Player 2 always wins an average of half a coin per game. Extra coins and slots simply mean that each game lasts longer, and Player 2 has to put up a higher stake, to win that half a coin.

Extra coins and slots help to hide the fact that Player 2 has an advantage. In a two-slot game, the advantage becomes obvious. In a sixteen-slot game, it's not obvious at all.

And a Microcontroller?

Yes, a microcontroller could run this game, but I don't think it would be any simpler. Your microcontroller probably would have fewer than 16 outputs to activate a Hot Slot. You could use just four outputs to provide a binary code from 0000 through 1111, but then you'd still need a decoder, analog multiplexer, or demultiplexer.

Bearing in mind the additional cost of the microcontroller, and the additional time involved in writing the program code, I think this is one experiment where discrete components are simpler.

Experiment 22: Logically Audible

22

In this experiment I want to take a break from logic problems and probability. I'm going to show you how to build something that is fun, weird, and simple (although later in the book, I'm going to find a way to make it more complicated). I'm going to show you how to pass audio signals through logic chips.

Background: Neither Here nor Theremin

Long ago, in the earliest days of electronics, a gadget called a theremin created creepy noises on the sound tracks of horror movies. The theremin (pronounced "ther-a-min") was played in real time by a performer who waved his or her hands around two rods that were sensitive to fluctuations in capacitance between the theremin and ground. A skilled operator could make a theremin play recognizable melodies, although it sounded a bit like a violin bow moving to and fro across a tensioned hand saw.

Search online, and you'll find a complete explanation of how the theremin worked. You can even find MP3s of theremin performances, and a component kit that you can buy to build a theremin of your own.

When you used a phototransistor to control the frequency of a timer in Experiment 3, you created results that sounded a bit like a theremin. Now that you have become well acquainted with logic chips, you can use that knowledge to combine two (or more) audio frequencies, like a supertheremin.

Logical Audio

You might think a logic chip is totally inappropriate in an audio circuit, but bear in mind three factors:

1. Audio frequencies of 20Hz to 15KHz are slow, from the point of view of a chip that is designed to handle frequencies above 1MHz.
2. The square waves produced by a timer will never sound as mellow and melodious as a sine wave, but they are certainly audible.
3. Conventional music is already digital. Almost all the music you listen to, from CDs to MP3s, is created and processed with digital sampling.

Audible XOR

In Figure 22-1 you'll see a circuit that is somewhat similar to the one in Figure 5-1. The big difference is that the timer outputs are now linked by an XOR gate, and the output from the XOR goes through a transistor to a loudspeaker. Yes, this is a highly unusual arrangement, but all the components are operating within their specifications, and I think the results are interesting.

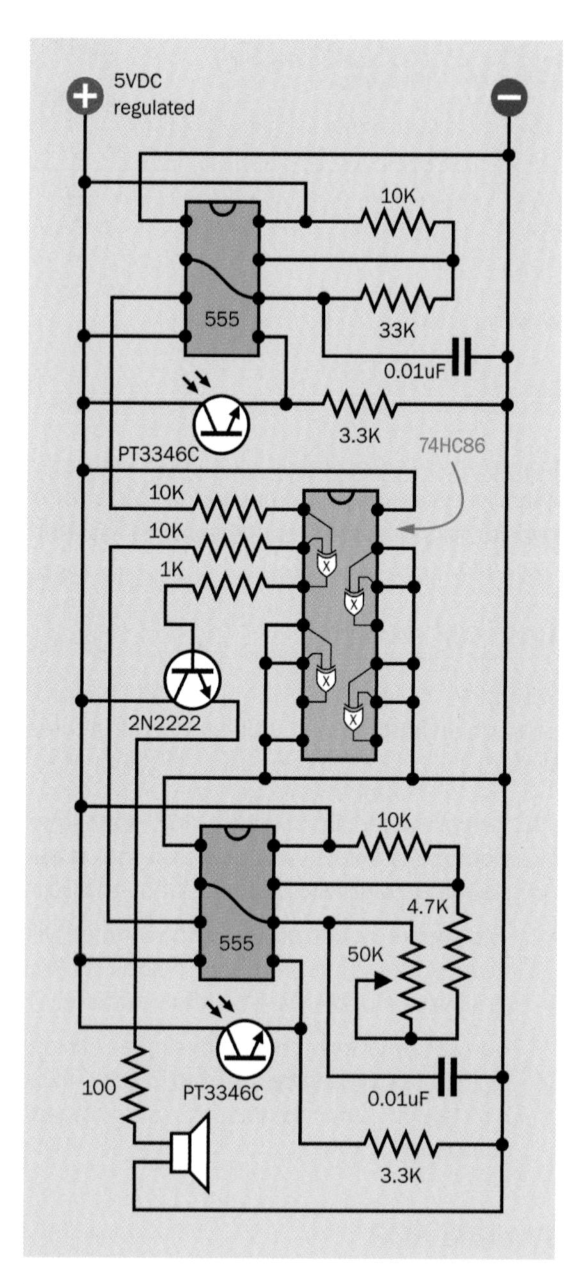

Figure 22-1. *Schematic for a circuit that XORs the audible frequencies from two timers.*

It shouldn't take you long to put these components together. Figure 22-2 shows my breadboarded version.

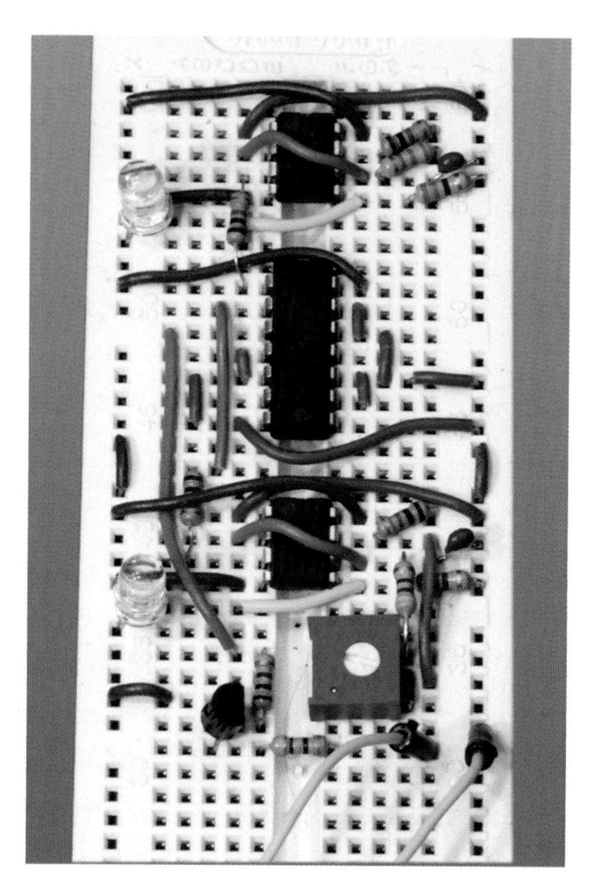

Figure 22-2. *The breadboarded version of the previous schematic.*

If you haven't made any wiring errors, you should get all kinds of sounds when you vary the light on the phototransistors while adjusting the 50K trimmer up and down. What's going on here?

All Mixed Up

Remember how an XOR gate works. Its output is only high when one of its inputs is high but not the other. Therefore, when you're feeding two audio signals into it, the output from the XOR is high when the signals are out of phase, but it transitions low when the signals are in phase.

Figure 22-3 shows this graphically, with an XOR and also with an AND and an OR gate. The pinouts of the 74HC08 quad two-input AND chip, the 74HC32 quad two-input OR chip, and the 74HC86 quad two-input XOR chip are all the same, so you don't need to change the wiring of your circuit to hear the different effects when you swap chips in and out.

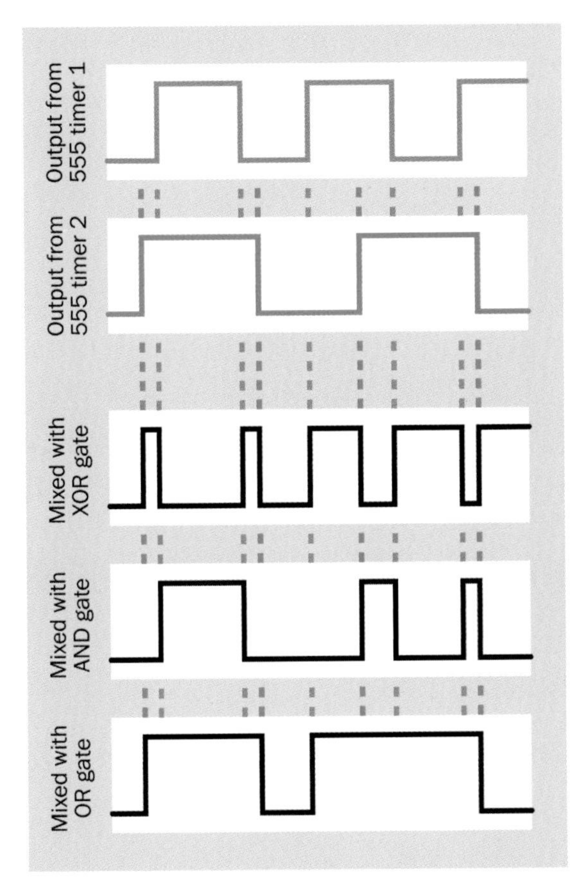

Figure 22-3. *As the highs and lows of two audio frequencies overlap, different logic gates will create contrasting audio effects.*

The OR gate gives an output when either of the timers emits a pulse, while the AND gate is fussier, only responding when both of the timers have high signals. I think the XOR gate is the most interesting, which is why I suggested you should try it first.

You can easily imagine more options. For example, you could substitute a second 50K trimmer for the 33K resistor—but put a 4.7K resistor in series with it to avoid a situation where you may end up applying zero resistance to Pin 6 of the timer when you turn the trimmer all the way down. This would be tough for even a robust 555 timer to endure.

You can increase the value of R1 relative to the value of R2 in one or both of the 555 circuits to lengthen the high cycle relative to the low cycle. This may sound good with the XOR gate, which tends to chop the cycles into smaller pieces. For instance, what happens when you substitute a 470K resistor for the 33K resistor? What happens if you substitute a 1M or even higher resistor for the 33K resistor?

When the first timer is running much more slowly than the second, you can really hear their frequencies interacting. With the 1M resistor in place, substitute a 0.1 µF capacitor for the 0.01 µF capacitor on the first timer. This creates a fluctuating two-tone effect. If you sketch the two outputs from the 555 timers, you should be able to see why.

Can you think of any applications for this circuit? Perhaps it could be the voice of a robot, changing as the robot moves around and exposes the phototransistors to different light patterns. Angle the phototransistors in different directions to heighten the effect. Or you could remove the phototransistors and simply adjust the trimmer to find the optimum sound that you want to use for an audible response in any other electronics project.

What if you created a duplicate of this entire circuit, and then XORed the two XOR outputs? I'm going to leave you to explore possibilities of that kind.

There's an ulterior motive behind this experiment. I'm going to use the concept in Experiment 26 for a totally different purpose: creating random bursts of *visible* pulses.

Experiment 23: A Puzzling Project

23

Here's another relatively simple experiment with logic, enabling a two-player game that seems deceptively easy—until you play it yourself. Like the Hot Slot game, I'm going to come back to it later in the book. In Experiment 32, I'll suggest ways to upgrade the user input when I start to explore the world of sensors. For the time being, though, the game will be playable with pushbutton switches.

Background: The British King of Puzzles

Before the advent of television—before there was even radio!—British newspapers used to entertain their readers with little games and puzzles. These were much more challenging than today's crossword puzzles.

King of the British puzzle-makers was Henry Ernest Dudeney, a master of the art of asking a question in just one paragraph that could require days or weeks for anyone to answer.

Many of his problems were geometrical or arithmetical. For instance, he challenged readers to find the two prime factors of the number 11,111,111,111,111,111—and somehow, using only pencil and paper, Dudeney was able to supply the right answer. (Even using a computer, this is not a trivial problem, because the numbers are so large.)

He had a playful streak, as in this cunning little puzzle:

- A customer wants a builder to make a window measuring two feet on each side. The window must be divided into eight panes. Each pane must measure one foot on each side. There must be no empty spaces. This sounds impossible, yet it can be done. The question is, how?

For the answer to the window problem, see "Answer to the Puzzle" on page 188—but first, try to figure it out for yourself. A hint: the secret is in the way the question is worded.

Moving Counters

Dudeney enjoyed what he called moving-counter problems. Checkers is an obvious one, because counters move around the board until one person wins. Tic-tac-toe is not a moving-counter problem, because the marks remain fixed on a piece of paper—but it can become a moving-counter problem and is much more interesting as a result. In this mode, Dudeney referred to it as "Ovid's Game," because he claimed to have found it mentioned in the works of the Roman poet Ovid. That may or may not have been true, but in any event, while the strategy is subtle, the rules are very simple:

- The game is for two players, each of whom has three counters. The counters for one player are different in color from the counters for the other player.
- The playing board consists of nine square cells, like a tic-tac-toe grid.
- Players take turns placing their counters one at a time, anywhere on the grid.
- After each player has placed three counters, the players take turns moving their counters. A move consists of shifting one counter into an empty adjacent empty cell. Diagonal moves are not allowed.
- A player wins by getting three counters in a row, horizontally, vertically, or diagonally.
- Because the center cell is important in this game, the first player is not allowed to place a counter in the center on the first turn.

The key to this game is to place your counters to obstruct your opponent. Remember, although you can shuffle your counters around, you are not allowed to jump over your opponent's counters, and you can only move one square at a time.

For example, in Figure 23-1, if it's white's turn to move, he should shift from square 6 to square 5, to obstruct black's two counters. Now black has no way to prevent white from winning on the next turn by playing from square 3 to square 6. Game over!

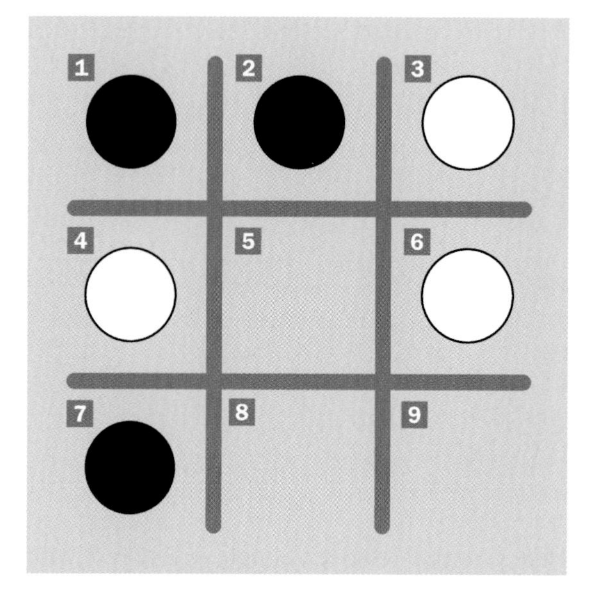

Figure 23-1. *A hypothetical situation in Ovid's Game.*

But let's suppose that it's black's move. In that case, she should move from 2 to 5 to prevent white from occupying that square. After that, the outcome of the game is hard to predict.

I don't know how to build a circuit that's smart enough to play this game, but I can show you a circuit that will detect if someone has won it. The same circuit will also respond if someone gets three in a row in a tic-tac-toe game, although I'm assuming that tic-tac-toe is too simple to interest most of my readers.

You may feel that a circuit that detects a winning combination is unnecessary, because anyone can see if there are three counters in a row. Well, yes, of course! But it's fun to have an audible signal when you win, and a circuit that can detect this will rouse some curiosity. Just how does it work?

The Logical Grid

The first question we have to decide is how to process user input. That is, how can an electronic circuit detect if someone makes a move? The simplest answer is to get rid of the two colors of counters and substitute two kinds of switches (one for each player) in each cell on the board. At the start of a game, players take turns to flip switches until each person has activated three switches. After that, a move consists of turning off one switch and turning on a switch in an adjacent empty cell. Of course you're not allowed to flip a switch in a cell that is already occupied by your opponent.

Latching pushbuttons (the type that you press once to turn on and a second time to turn off) seem ideal for this task. If there is a colored LED beside each button (perhaps red for one player and blue for the other), it will be easy to see who is occupying a square. In Figure 23-2 I'm including a rendering to show you what I have in mind.

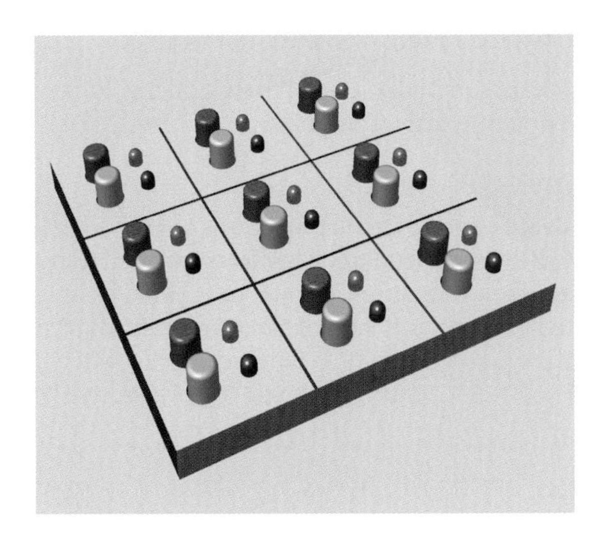

Figure 23-2. *The simplest way to provide user input for electronics in Ovid's Game is to have latching pushbuttons and LEDs for each player.*

Another option would be to use ON-OFF-ON switches—that is, a three-position switch with a center-off position. Each player occupies a cell by turning the switch toward him, and the cell is vacated when the switch is returned to its center-off position. You can buy ON-OFF-ON single-pole switches very cheaply, and because each one has two "on" positions, you'll only need nine instead of the eighteen latching pushbuttons.

Using Logic

Suppose we use logic gates to detect a winning move. In that case, as usual, I will begin by describing the logic in words. If the squares are numbered 1 through 9 as shown in Figure 23-1, I can state the problem like this:

The winning player must occupy cells 1 AND 2 AND 3, OR 4 AND 5 AND 6, OR 7 AND 8 AND 9, OR 1 AND 4 AND 7, OR 2 AND 5 AND 8, OR 3 AND 6 AND 9, OR 1 AND 5 AND 9, OR 3 AND 5 AND 7.

If you have been reading this book sequentially, you've had a lot of practice in converting this kind of sentence into an array of logic gates. You can see there are eight possible winning combinations, linked with ORs, and each combination consists of three conditions linked with ANDs. Can you draw a logic diagram? The basic concept is shown in Figure 23-3, while a complete diagram, including the switches, is shown in Figure 23-4.

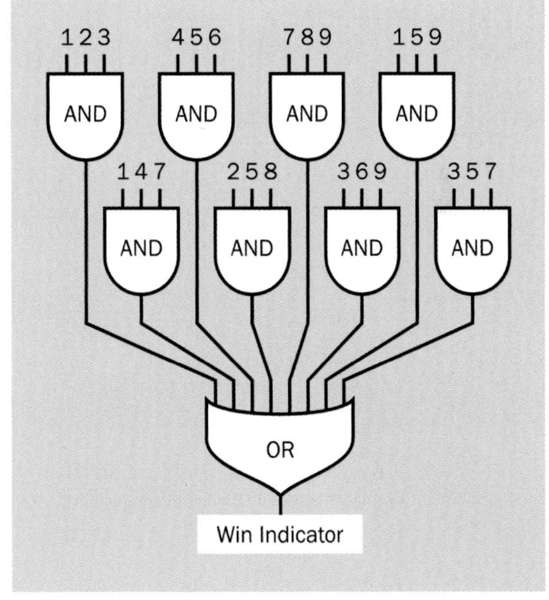

Figure 23-3. *The bare bones of a logic diagram for Ovid's Game. Each number feeding into an AND gate represents a square on a 3 x 3 board where one player's counter has been placed.*

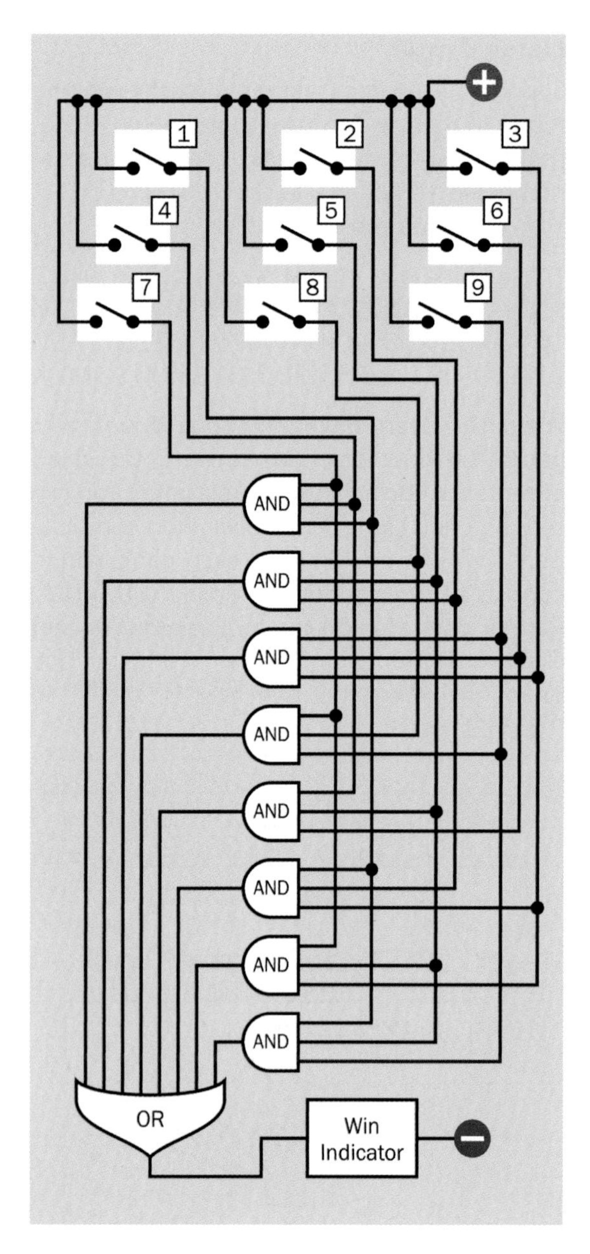

Figure 23-4. *A logic diagram for Ovid's Game showing switches that represent one player's counters placed on the board.*

Notice the eight-input OR gate at the bottom of each diagram. Does such a thing exist? Yes, it does, so this is not a problem. Likewise, three-input AND gates are easy to find. How many will be needed? There are eight different ways to win, so that will mean eight three-input AND gates. There are three of these gates on a 14-pin chip, so three chips will do it.

This will only detect one person's win-or-lose situation. To detect the other person's status, another set of switches and another set of logic chips will be needed—but if you want to give yourself less work, you can have a master switch to identify which person is taking a turn, and it will apply power to the same set of chips through one player's set of switches, or the other.

I'm not sure I like either of these options. So now let me suggest an alternative.

Switching Ovid

Ovid's Game (or tic-tac-toe) turns out to be ideal to be wired by multipole switches, without using any logic gates at all. By now you should be able to figure out this kind of circuit yourself, but I'm showing you the one that I came up with in Figure 23-5. (Can you derive a simpler version?)

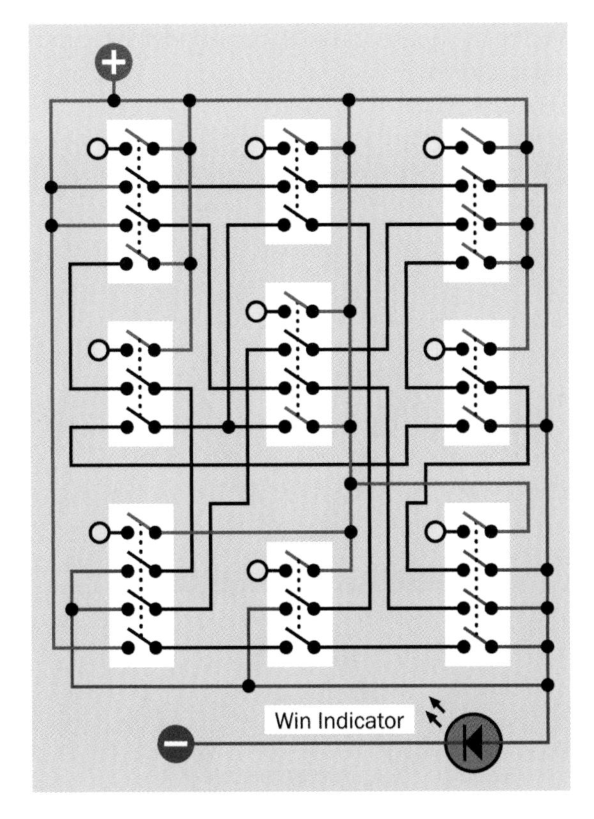

Figure 23-5. *Ovid's Game can be modelled using switches without logic chips.*

To see how this works, begin by imagining that one of the switches has been closed, and follow the positive current through other switches to the negative ground wire. It's really much simpler than it might first appear.

The little yellow circles will be small LEDs that light to confirm that a switch has been pressed. Naturally they will have to be connected with negative ground, but I omitted those connections to keep the schematic simple.

As you see, each switch will require three or four poles. Actually I was able to model the game using two-pole switches, but that required omitting the little LEDs.

You can buy 4PDT latching pushbuttons very cheaply, which will do the job. The big LED at the bottom lights up when the player has pressed three switches in a row.

A separate array of nine switches will be required for each player, making eighteen in all, as suggested in the rendering in Figure 23-2.

Do you like the logic-chip version of the game, or the switched version?

You know—moving counters around a board and using sensors to detect their position is clearly the best way to do it, so I really will come back this later, but in the meantime, I think you've had enough practice to build this game yourself. You can use wired switches or logic gates and retrofit it with sensors later.

Making Even More

I mentioned earlier that I don't know how to build a circuit that's smart enough to play this game. However, I think I could do it using software. A typical microcontroller has relatively little memory, but Ovid's Game allows so few possibilities for each move, a program might be able to handle it by using just four rules, in order of precedence. These rules would tell the microcontroller what to do (and you could test them by following them yourself, if you play the game with someone):

- If you can move a counter to make three in a row, do that. Otherwise—
- If your opponent can move a counter that will create three in a row, try to block him. Otherwise—
- If you can move a counter into the center cell, do that. Otherwise—
- Move a counter at random into an unoccupied square. If this is not possible, the game is a tie.

This program wouldn't win many games, but it would work. The problem is, it would have to represent the playing board and the counters in computer code, which really requires the use of integer arrays. On an Arduino, in C language, there is no error checking to make sure that you stay within the bounds of an array, and there are some other features that I don't like too much.

Even now, most microcontrollers are less powerful, less user friendly, and less error proof than the old IBM-PCjr on which I used to write games in Microsoft BASIC, back in the 1980s. What a sad state of affairs!

My friend Fredrik Jansson, a physicist from Finland who helped with fact-checking this book, suggested that you could implement Ovid's Game with a microcontroller if you used a desktop computer to do the initial hard work. Fredrik calculated that there are 1,680 possible positions of three black counters and three white counters on the board. Because a position is logically the same if the black and white counters are swapped, really there are only 840 logically different positions. This number is small enough to allow a computer to play out every possible game and establish the best move in every situation.

If you use the computer to compile a table of all the best moves, you can install that table in the limited memory of a microcontroller. Fredrik figured that you only need four bits for each recommended move: two bits to select a counter, and two bits to instruct it to move up, down, left, or right. Therefore, the instructions for playing 840 logical positions could fit into 420 bytes. You'd still need some extra instructions for initial placement of the first three markers of each color, but still, this scheme looks doable. Each time a human opponent made a move, the microcontroller would just look up the best response that was previously determined by its more powerful cousin, the desktop computer.

If you decide to try this programming strategy, or any other, please be sure to let me know how it works out.

Answer to the Puzzle

The answer to Dudeney's puzzle about the window is shown in Figure 23-6. The window measures two feet on each side, and each triangular pane measures one foot on each side. Therefore, the layout satisfies the requirements of the problem. Wait—you didn't assume that the window had to be square and each pane had to have four sides, did you?

Figure 23-6. *The solution to the "window puzzle" posed at the beginning of this section.*

If you like this kind of puzzle, either to amuse yourself or confound your friends, there are two compilations of Dudeney's life work: *The Canterbury Puzzles* and *Puzzles and Curious Problems*. Because they were published so long ago, they are now out of copyright and you can read them freely (and legally) online through The Gutenberg Project.

You can also get Ovid's Game as an Android app, if you want to practice against a computer-opponent. But I think building something of your own is more interesting.

Experiment 24: Adding It Up

24

Before I move on from logic gates to other topics, I wouldn't be doing a thorough job if I didn't show you the most fundamental logic application: adding numbers. You already have a calculator, which is far more powerful than anything you are likely to build yourself; but in my experience, there is something slightly magical about putting together your own little circuit that can perform basic arithmetic.

The Five Rules of Binary

Because logic chips only have two states—low and high—they are ideal for representing the 0s and 1s of binary code. Therefore, to make a circuit that adds numbers, you have to use binary addition. The good news is that there are only five rules for adding binary numbers, three of which are obvious. I'll deal with those first:

- Rule one: 0 + 0 = 0
- Rule two: 0 + 1 = 1
- Rule three: 1 + 0 = 1

I think you'll have no problem agreeing with these. Now for the fourth, which looks a bit odd:

- Rule four: 1 + 1 = 10 binary

If you're reading this as "one plus one equals ten," this is not correct. You have to let go of your ten-based thinking! The word "binary" in rule four tells you that 10 has nothing to do with "ten," because 10 is a binary number.

We all know that 1 + 1 = 2, so why don't we just write 2? Because in binary code, we only have 1s and 0s. So, we apply the fifth rule:

- Rule five: when you have an output that would be decimal 2, write 0 as the sum, and carry 1 over to the next place on the left, where it has a place value of two.

You've run into the concept of place value previously in this book, when I talked about binary inputs to chips, such as a decoder. In binary counting, a numeral 1 has a place value of 1 in the rightmost place, 2 decimal in the next place to the left, 4 decimal in the next place, 8 in the next place—and so on.

So, 1 + 1 = 10 really means that the answer has no units in the rightmost place, and a value of 2 in the next place to the left.

The four possible combinations, when adding two binary digits, are shown in Figure 24-1.

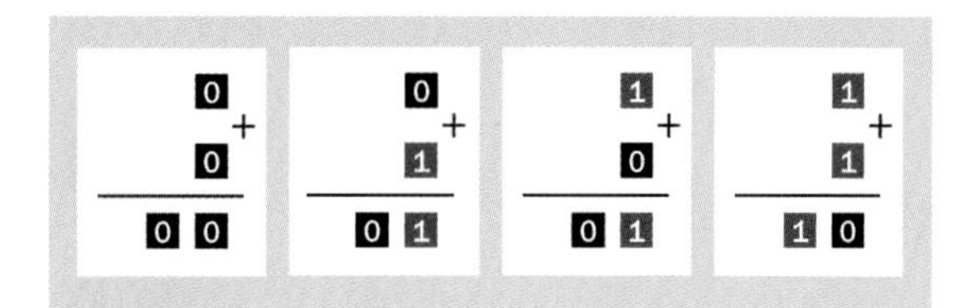

Figure 24-1. *The basic rules of binary addition, using 1-bit binary numbers.*

Now take a look at Figure 24-2 and you'll see that even when we have to handle binary numbers consisting of two digits each, the same rules still apply. This diagram shows every possible combination of two two-bit numbers.

There's only one issue that may seem unclear. What exactly is going on when we add 11 + 11 in binary?

As always, we start in the rightmost place. We add 1 + 1 and write a zero, and carry 1 to the left. But in that place, there is already another 1 + 1 waiting to be added. So now we have to deal with 1 + 1 + 1.

Well, we know that 1 + 1 is 10 binary. So when we add another 1 to that, we get 11 binary. You can see this in the very last example in Figure 24-2.

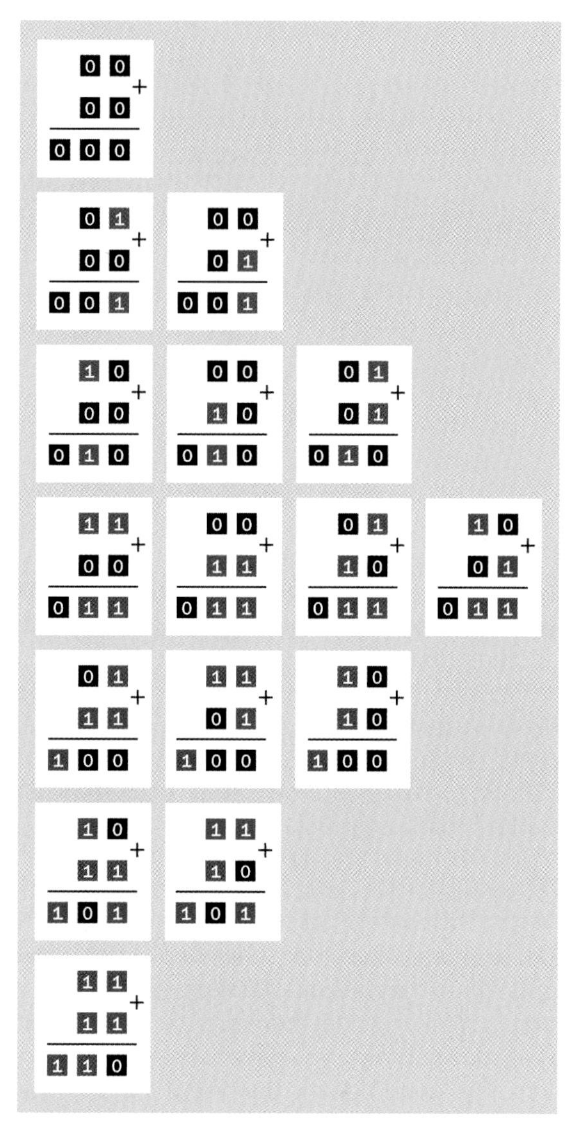

Figure 24-2. *The basic rules of binary addition, using two-bit binary numbers.*

From Bits to States

So much for the rules of binary arithmetic. You now have all you need to know to build a binary adder (and if binary numbers don't seem very user friendly, I'll show you how to convert to and from decimal numbers in the next experiment).

Let's consider how we can use logic chips to emulate the binary rules, using the high state of a logic gate to represent a 1, and the low state to represent a 0. I can rewrite the first four rules of binary addition using words instead of numbers, like this:

- Low input + low input = low output
- Low input + high input = high output
- High input + low input = high output
- High input + high input = low output (and carry high to the next place)

What does this remind you of? Well, by George, if we forget about the "carry" operation for a moment, this is a precise description of the inputs and outputs of an XOR logic gate! Just apply the two numbers to be added as low and high states to the XOR inputs, and the XOR output will be correct. We can represent it with an LED that indicates 0 when it is off and 1 when it is on.

Now, what about carrying 1 to the next place on the left? I could say:

- High input + high input = high carry
- All other input combinations = low carry

What does this remind you of? An AND gate, of course! If this is tied in to the two inputs, its output can be a second LED.

Figure 24-3 shows how the gates are combined to add any two single binary digits, which I have represented with the imaginative names "A" and "B." Figure 24-4 shows the four possible combinations of inputs, and the resulting outputs, with red indicating a 1 and black indicating a 0.

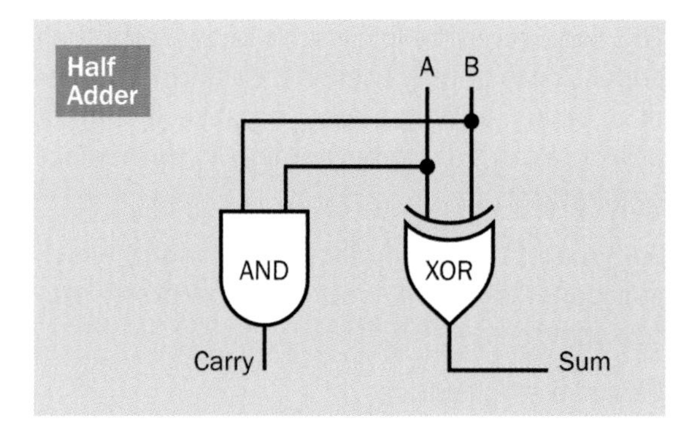

Figure 24-3. *Probably the simplest way for logic gates to add two binary digits, and generate a "carry 1" if necessary, is with an XOR gate and an AND gate.*

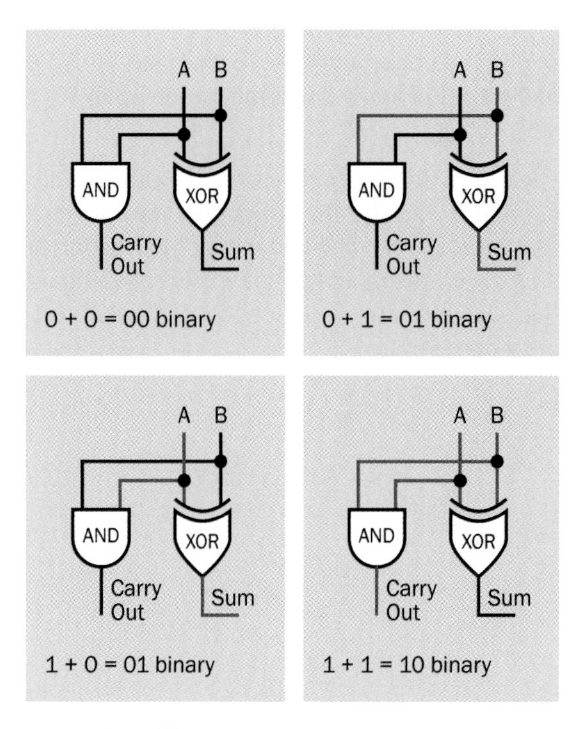

Figure 24-4. *The four possible input combinations to a half adder, and the resulting outputs. Red is interpreted to mean 1, while black means 0.*

This very, very simple circuit is known as a "half adder." It can only be used in the rightmost place of a binary addition sum, because it cannot process a carried number coming in from a place farther right.

But what if there is a place farther right, which may generate a carry-1 output? How do we process that?

We need a "full adder."

This is a bit more tricky. We can only compare two inputs at a time using an XOR logic gate, but we have three numbers to deal with: two binary digits as inputs, plus a carry digit (which may be 0 or 1) from the previous place in the calculation. We might find ourselves trying to deal with 1 + 1 + 0 or 0 + 1 + 1 or any other combination up to 1 + 1 + 1.

The way to do this is by breaking the process into two steps. Figure 24-5 shows how. You'll notice that it is really two half adders, each consisting of an AND gate and an XOR gate, plus an OR gate that says we'll carry 1 out if either of the half adders requires it.

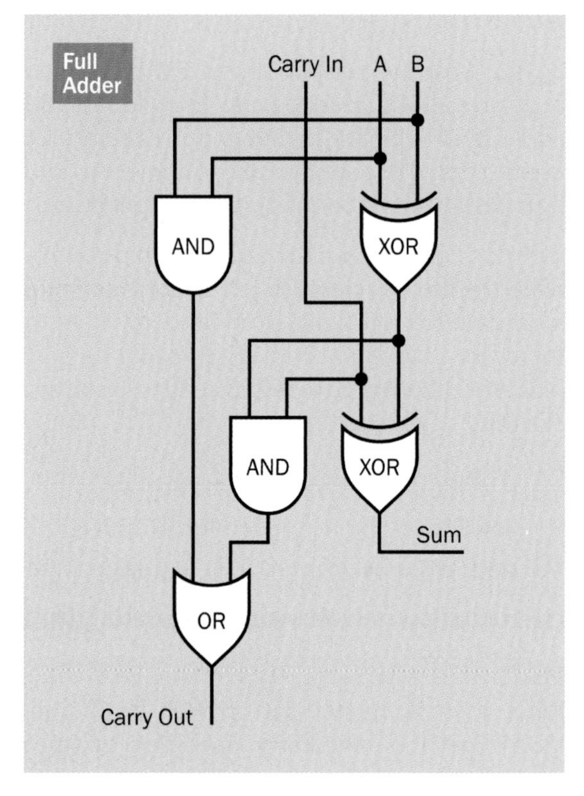

Figure 24-5. *A full adder receives a carry digit from the previous stage and adds it to two new binary digits.*

The first XOR adds two binary inputs, as before. Now we take the single-digit output from the XOR and add the incoming carry to it. This means placing another XOR gate under the first.

There will be two circumstances under which our full adder will have a new outgoing carry:

1. If we have 1 AND 1 as our binary inputs, we will carry 1 out using the first AND gate—as usual.
2. OR if the output from the first XOR is 1, AND we also have an incoming carry of 1, we will carry 1 out.

There are eight possible combinations of two inputs and an incoming carry. I have shown four of them as examples in Figure 24-6.

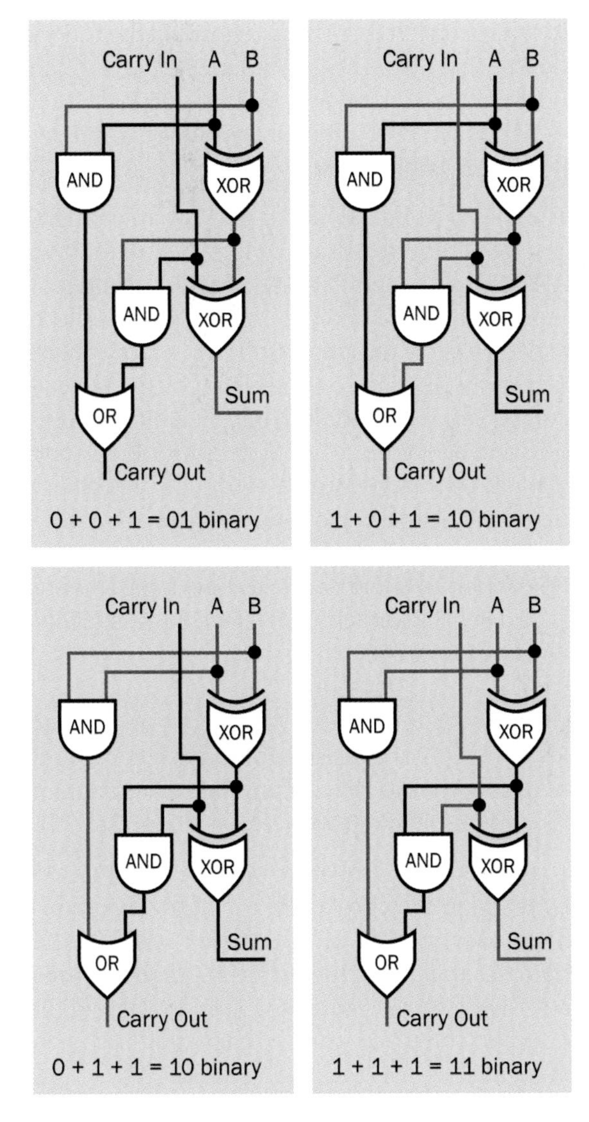

Figure 24-6. *The three inputs to a full adder can have a total of eight different combinations. Here are four of them, illustrating how the adder operates.*

You are now almost ready to build your own adder. But first, a small digression.

Background: The NAND Alternative

Because logic gates can emulate each other, depending on how they are wired together, the XOR/AND combination is not the only way to build a half adder. I used it because I think it's the easiest to understand, but NAND gates are commonly used. This alternative requires more gates (five instead of two, to make a half adder, and nine instead of five, to make a full adder) but the great advantage is that NANDs can do the job all on their own, with no gates of any other types. An all-NAND computer is easier to construct in some ways, so NANDs are often thought of as the most fundamental gate in computing. They were assigned part number 7400 in the long series of 74xx logic chips.

Figure 24-7 shows how five NANDs can form a half adder, while Figure 24-8 shows how nine can make a full adder. Once again, the full adder is composed of two half adders stacked up.

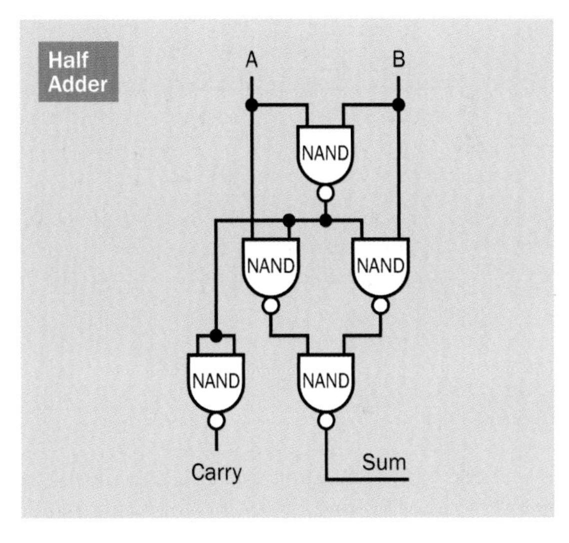

Figure 24-7. *A half adder can be constructed entirely from NAND gates.*

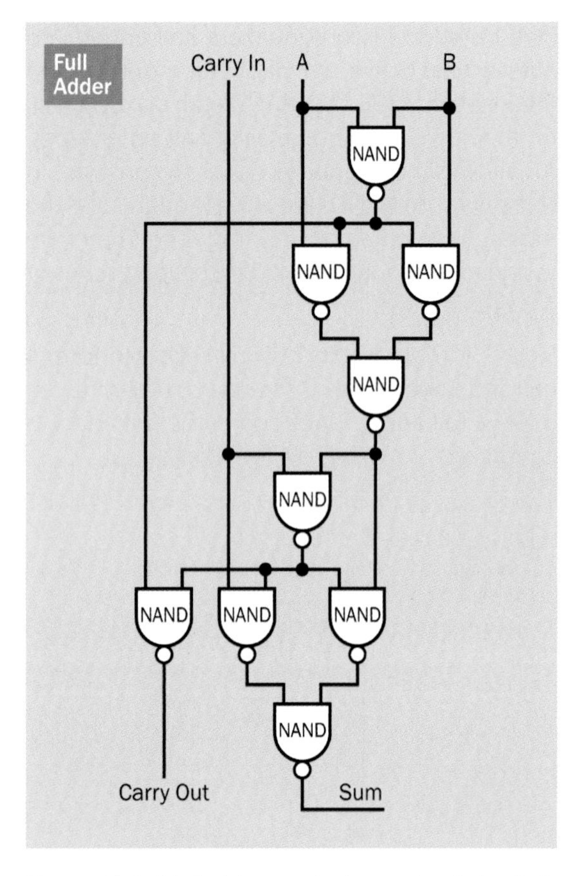

Figure 24-8. *A full adder can also be constructed entirely from NAND gates.*

If you are methodical and patient, you can print several copies of these diagrams and use red ink to trace logic paths through them to verify that the output is correct for each of the eight possible combinations of inputs. Personally, I'm not going to do this. I'm going to stay with the XOR/AND combination, because NANDs give me a headache.

Your Own Little Adder

So where's the hands-on activity, here? Normally I suggest putting components together first, and then figuring out why they do what they do. That's the basis of what I like to call "Learning by Discovery."

In this case, I felt that if I didn't explain the theory first, nothing would make any sense. But you can go right ahead now and build a calculator that will add two three-digit binary numbers and deliver a four-digit output.

This will use one half adder and two full adders. They will require a total of five XOR gates, five AND gates, and two OR gates. There are four two-input gates on a quad two-input chip, so we'll be leaving quite a few gates unused. In fact, we will have enough gates on the chips to create a calculator capable of adding four-digit binary numbers and delivering a five-digit output. But I don't want to do that, because an output of five binary digits will be too big for the next little adventure that I have in mind. Maybe you remember what I promised to show you in the next experiment —and it will be much easier if we only have four binary digits to be processed.

Breadboard Addition

In Figure 24-9 I have assembled a half adder and two full adders to make a complete logic diagram for a three-bit adder with a four-bit output.

The two binary numbers to be added are entered by pressing switches at the top of the diagram. I have used darker blue areas to identify the switches that enter the first three-digit number, and switches that enter the second three-digit number. The place value of each digit is shown in green.

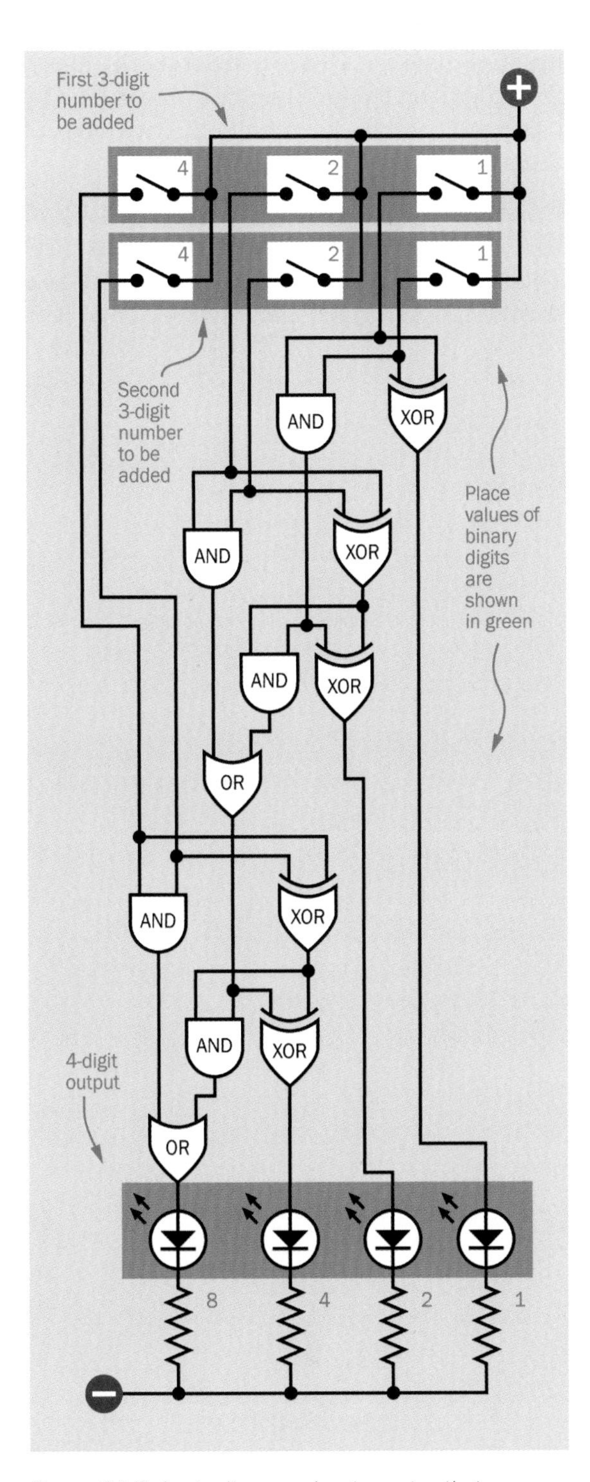

Figure 24-9. *Logic diagram showing gates that can sum two three-bit binary numbers, with a four-bit binary output.*

Logic gates have been shifted around just a little, compared with the previous diagram showing a full adder. But the connections are still the same.

To take a random example, if your little arithmetical operation is 5 + 7 decimal, you would close switches 4 and 1 in the upper row (to make 5), and switches 4, 2, and 1 in the lower row (to make 7). Assuming that the adder works properly, you should immediately see LEDs 8 and 4 lighting up to give you the answer of 12 decimal, while LEDs 2 and 1 would remain dark.

Figure 24-10 shows the pulldown resistors that you will have to include when you build this circuit so that none of the inputs to logic gates will float when switches are open.

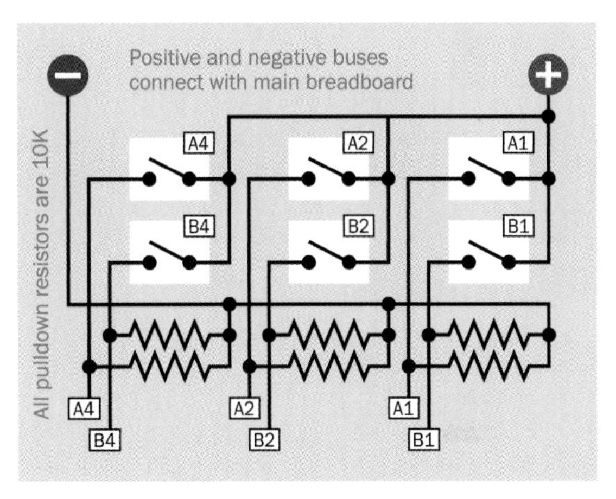

Figure 24-10. *The input switches for the three-bit adder, with pulldown resistors.*

The switches and their outputs are labelled to match the inputs to the main part of the circuit, which is shown in Figure 24-11. Little yellow circles have been used as a space-saving alternative to the usual LED symbols. Remember to add series resistors between your LEDs and negative ground, if you are not using LEDs that contain their own resistors.

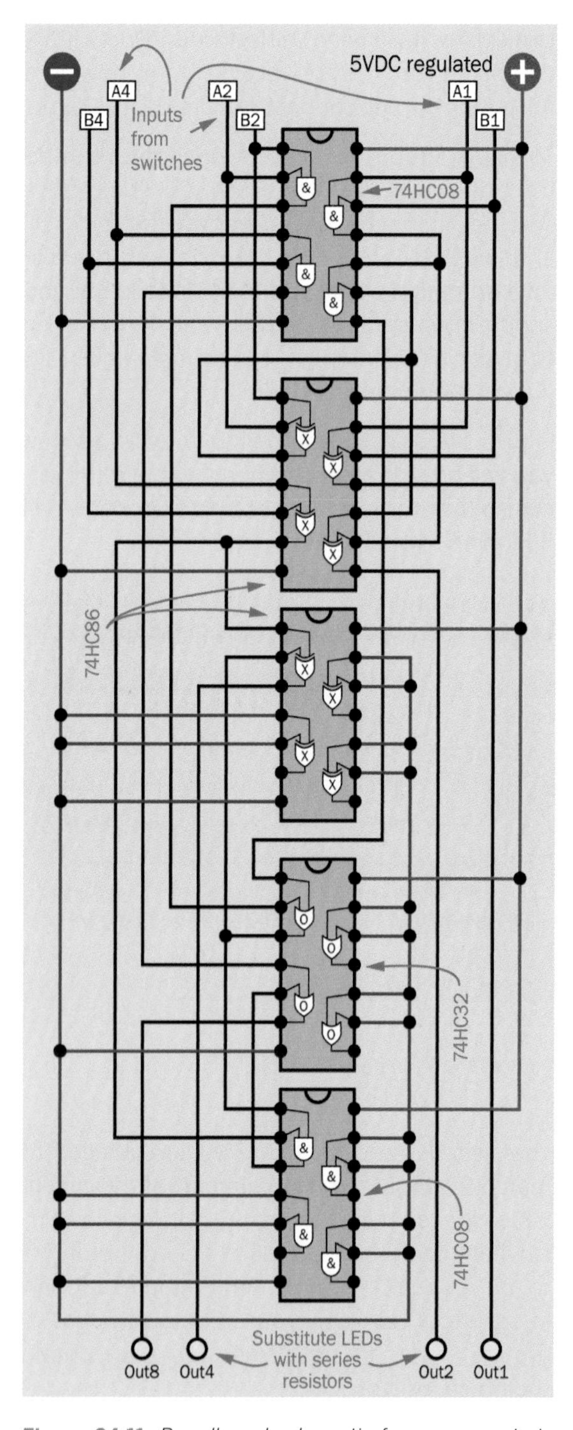

Figure 24-11. *Breadboard schematic for components in the three-bit adder with a four-bit output. Labels at the top are for connection with input switches, shown separately. Numbers beside input letters indicate their binary place values.*

Because there are a lot of parallel conductors in this schematic, I tried to reduce the confusion by using red and blue to identify the two sides of the power supply.

A photograph of the completed project is shown in Figure 24-12. Dual-inline switches (DIP switches) were used for inputs, because they take so little space on the board.

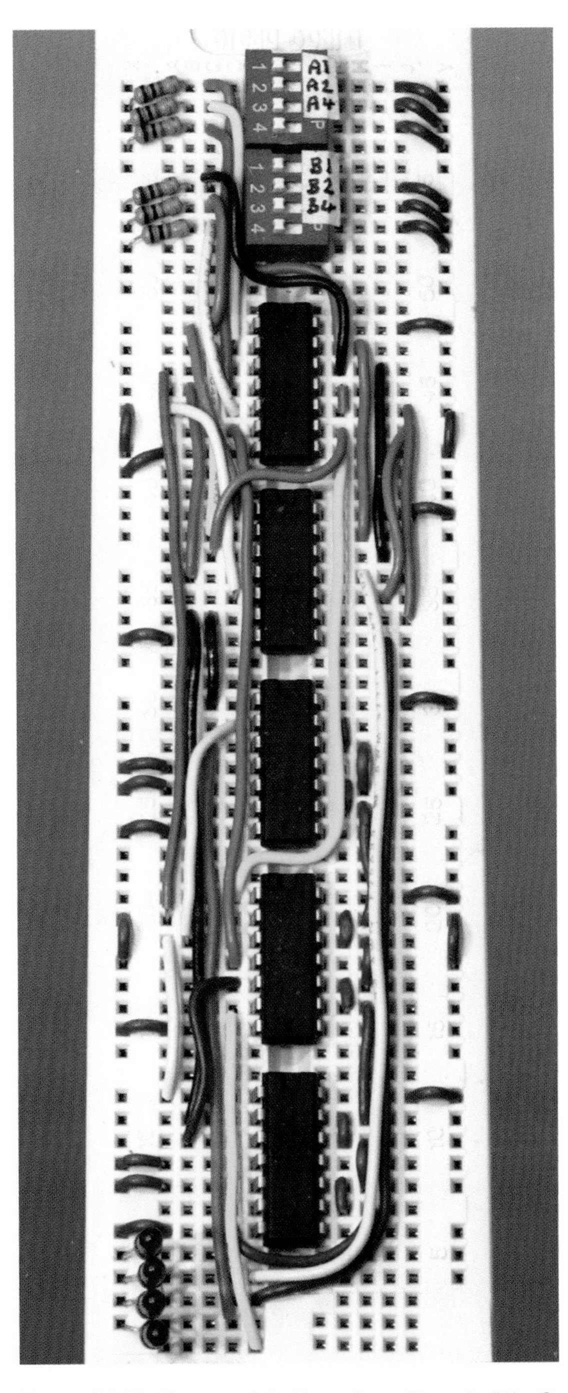

Figure 24-12. *The completed breadboard layout of the 3-bit adder. The DIP switches have place values 1, 2, and 4 for each of the binary numbers (A and B) to be added. The LEDs at the bottom have place values 1, 2, 4, and 8, reading from top to bottom.*

So, now you have a real-life, solid-state adding machine—something which would have seemed utterly magical as recently as the 1930s. The only problem is, quite apart from its limited capability, it's not very easy to understand, because you have to convert decimal numbers into binary to use them as inputs, and then you have to convert the binary outputs back into decimal. Moreover, you don't even have a numeric keypad, let alone a keyboard.

Actually, the world's first personal computer (the Altair 8800, marketed in kit form in 1975) required you to enter all data by flipping switches. This was known as "toggling in" a program. The computer didn't have a keyboard because that would have required more chips to interface it with the computer, and chips were expensive!

I realize that it's not 1975 anymore, so maybe we can simplify the data entry ritual and enhance the display for your 3-bit adder. That's the next experiment.

Experiment 25: Enhancing Your Adder

25

Adding a decimal output to your adder is an easier task than adding a decimal input, so I'll deal with the output first.

Return of the Decoder

It would be nice to have a couple of seven-segment numeric displays for the output, but they require drivers that use binary-coded decimal (BCD) inputs. In other words, the input to each driver must range from 0000 through 1001 binary (0 through 9 decimal). Your adder has an output ranging from 0000 through 1110 binary, and converting that to run a pair of seven-segment numerals is not so easy.

Therefore, I'll assume you can be happy with individual LEDs, each of which will represent a decimal value.

To take care of this, we can use the 4-to-16 decoder that you first encountered in Experiment 19, although it will have to be the HC version, 74HC4514, not the old 4514 CMOS version, because each of its output pins must be capable of driving an LED. The binary inputs to the chip will be supplied by the four outputs from your adder, weighted 1, 2, 4, and 8. Output pins from the chip will drive LEDs that are assigned values 0 through 14 decimal (bearing in mind that the largest addition sum your adder will handle is 111 + 111 = 1110 binary, which is 14 decimal).

These additional components won't fit on the same breadboard as the 3-bit adder, but you can buy breadboards cheaply, especially if they are the type that I recommend, with just one bus on each long edge. I'm assuming you may be willing to use a pair of breadboards to complete this experiment.

The schematic for the output is shown in Figure 25-1. You will want the LEDs to be in numerical order, but the output pin values of the 74HC4514 are scrambled, as indicated by the black numerals inside the chip.

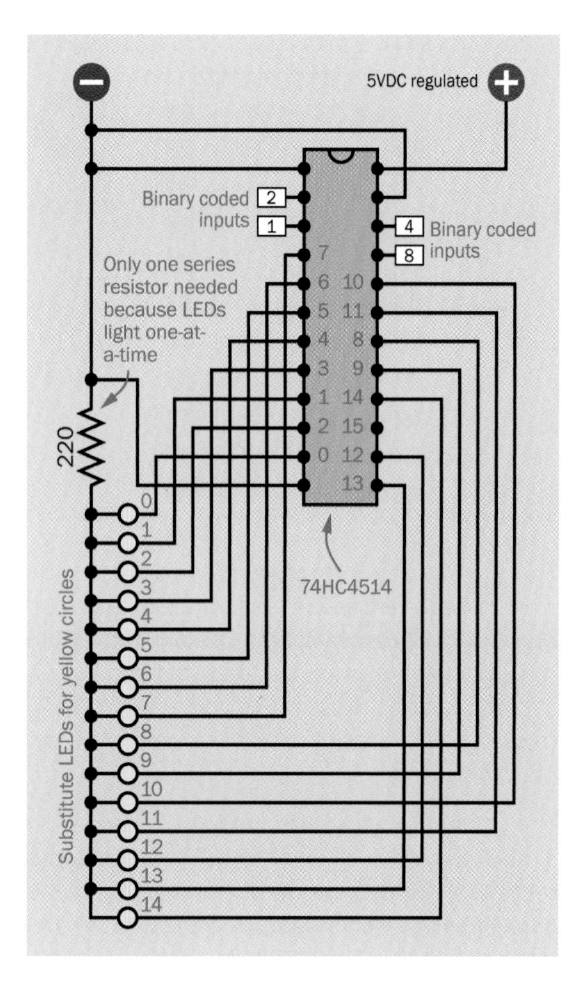

Figure 25-1. *Binary-coded outputs from the adder in Experiment 24 become inputs to the 74HC4514 decoder chip. Each of its decoded outputs, from 0 through 14, is connected to an LED. Because only one output is ever in a high state, all the LEDs can share one series resistor.*

The easiest way to deal with this is to mount the LEDs farther down the breadboard, as suggested in the figure, where yellow circles represent the LEDs. If your LEDs do not contain series resistors, you only need one resistor to serve all of them, because only one LED lights at a time. Assuming the bus on your breadboard is divided into sections, all the LEDs can share the center section, which would be linked to the upper section of the bus by the series resistor.

So much for that! Now for the harder part.

DIPping In

For decimal data entry, you can use simple SPST on-off switches. There can be two rows of eight switches, with values 0 through 7. You'll turn one switch in the first row and one switch in the second row to enter the two values that are to be added.

DIP switches are the simplest, cheapest, smallest option for this purpose. Normally, in a set of eight DIP switches, they will be numbered 1 through 8. You will have to renumber them from 0 through 7.

The eight wires from each bank of switches will go to the eight inputs of an 8-to-3 encoder chip. As you might imagine, this is opposite in function to a decoder chip. It takes an input on one of its eight input pins (with values 0 through 7) and converts it into a binary value on its output pins (with place values 1, 2, and 4). The outputs from one encoder can feed straight in to the adder circuit data lines labelled A1, A2, and A4, while the outputs from the other encoder will supply B1, B2, and B4. (See Figure 24-11.)

Figure 25-2 shows a schematic for wiring the two encoders. If you follow it carefully, it should work. But precisely why it works requires some explanation—especially if you notice that the switches are attached to negative ground instead of the positive side of the power supply. Why is this?

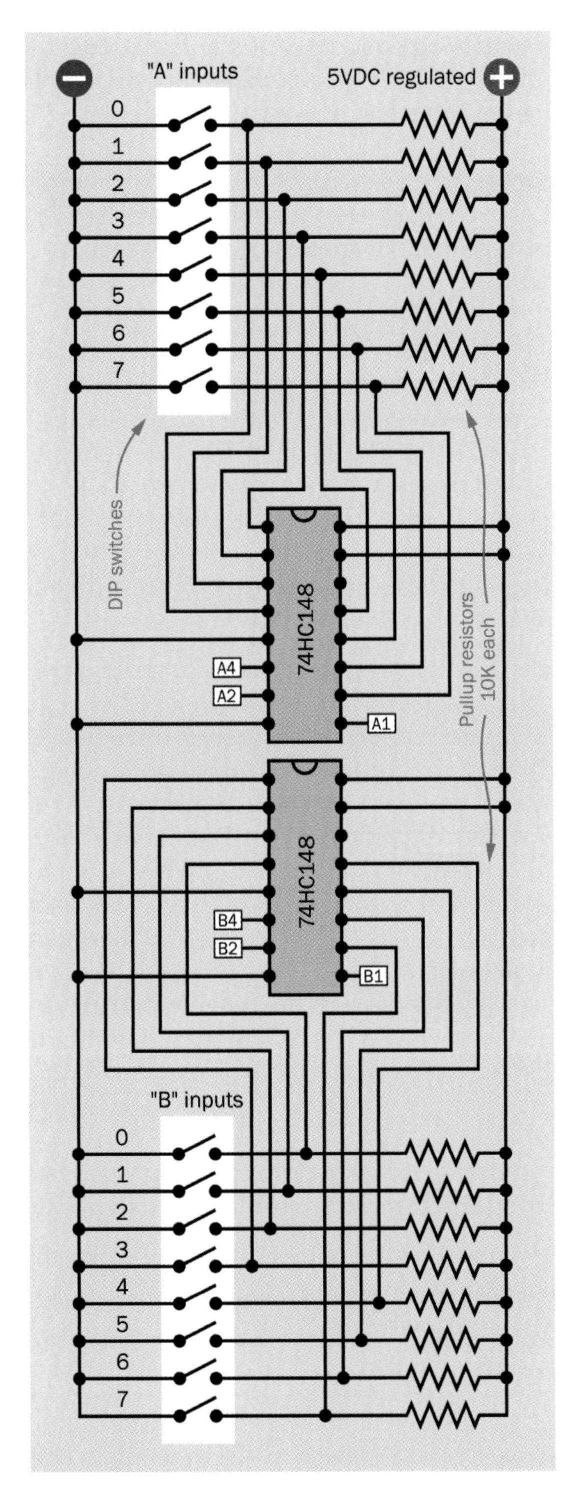

Figure 25-2. Outputs from two sets of DIP switches are connected with inputs of 74HC148 encoder chips. Note that inputs are logic low and resistors are for pullup, not pulldown. This is opposite to most logic chips.

Introducing an Encoder

Encoders have become relatively rare, and the ones still listed by suppliers use active-low logic. In other words, a low input or output represents a 1, and a high input or output represents a 0. This is irritating, because the chip has to work with the adder circuit, which uses conventional active-high logic. Without going into the historical reasons why encoders are made this way, I can still use the most common chip, the 74HC148, so long as I work around its quirks. Its pinouts are shown in Figure 25-3.

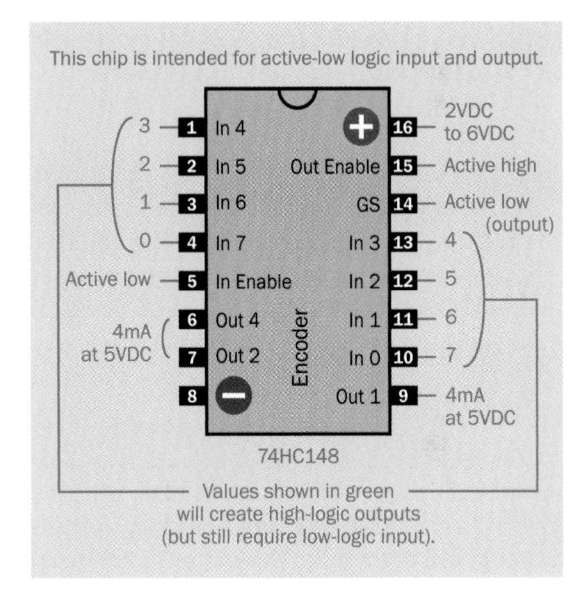

Figure 25-3. *Pinouts for the 74HC148 encoder chip. See text for details about input and output.*

There is a trick to get active-high outputs from the encoder. You reverse the sequence of the inputs. This is shown by the green numbers in the pinout diagram.

Are you confused, yet? Let me summarize. To get our usual, preferred, active-high logic output from the 74HC148:

- Assign value 0 to input pin that has value 7.
- Assign value 1 to input pin that has value 6.

(And so on, for intermediate values.)

- Assign value 7 to input pin that has value 0.

Each input will still be active low. There's no way around this. The reverse-numbering trick only fixes the outputs so that they are active high (which is what the rest of our circuit will understand). Therefore, as shown in Figure 25-2, pullup (not pulldown) resistors are applied to all the input pins, to keep them normally high (representing 0 values). The DIP switches are connected with negative ground so that when you turn a switch "on," it provides the necessary active-low signal to the chip (representing a 1 value).

Other Encoder Features

Just to make things more interesting, the 74HC148 has some additional confusing features. There is a "Group Select" pin, identified as "GS" on most datasheets. It is an output pin that goes active low when the chip is processing an input. Because it is an output pin, you don't have to worry about it, and you can leave it unconnected.

As for the two Enable pins, the input enable (often referred to as EI on datasheets) is active low, so I have tied it to negative ground. The output enable is active high (why, I don't know), so I have tied it to the positive side of the power supply.

At this point you may be wondering if there are still some practical applications for encoder chips. Indeed, they have a possible use in industrial control processes where, for example, an item is moving past a series of sensors that are either "on" or "off." If you are using a microcontroller that has a limited number of inputs, you could connect the sensing devices through an 8-to-3 encoder chip, and reduce the inputs to three. The power of this technique increases radically if you have more inputs, because you can chain encoder chips together, and if you double the number of inputs, the number of binary output lines only increases by one. This is because the place value of each digit in a binary number is double the place value of the digit that precedes it. Because the power of place values isn't readily appreciated, I'm going to go into that for a moment.

Background: The Power of Binary

So far in this book we have only dealt with binary inputs and outputs that have place values of 1, 2, 4, and 8. The doubling of place value from each digit to the next may not seem particularly significant, because our 10-based human counting system is so much more powerful, with place values in multiples of 10. Still, the process of doubling can have unexpected consequences.

In an 8-bit binary number, the leftmost digit has a place value of 128, and the whole set of digits can represent any value from 0 through 255. These values are very common in computing. A byte of computer memory usually consists of eight bits. A typical JPEG consists of three colors (red, green, and blue), each of which is represented with a value ranging from 0, meaning "completely dark," to 255, meaning "maximum brightness." Maybe 256 color shades don't sound like a lot, but because the three colors can vary independently, the total number of combinations is:

```
256 * 256 * 256 = 16,777,216
```

If you have ever heard that a computer video card can create "16 million colors," now you know why.

If we increase the number of binary digits, the situation becomes more interesting.

A 16-bit binary number can represent any value from 0 through 65,535.

A 32-bit binary number can have a value slightly above 4 billion—which is why a computer with a 32-bit operating system is usually unable to address more than 4 gigabytes of RAM.

Regarding the term "gigabyte": the international definition of "giga" is 1 billion, but in computer memory, "giga" means a binary number consisting of 1 followed by thirty zeroes. This translates into 1,073,741,824 in decimal notation. The situation originated when a kilobyte was defined as 10000000000 binary, which is 1,024 decimal. A megabyte became 1,024 kilobytes, and a gigabyte became 1,024 megabytes. But in hard drives, a gigabyte is simply 1 billion decimal. Just in case you were wondering.

But getting back to the power of binary numbers—

I own a little $10 calculator that can represent decimal numbers as high as 999,999,999,999, or almost 1 trillion. How many binary digits must the calculator use internally to deal with this? I think just 41 bits are sufficient, assuming that one of those bits is allocated to show whether a number is positive or negative.

Even though binary code is restricted to 1s and 0s, it is capable of handling very large numbers.

Background: Encode Your Own

If you prefer not to use the 74HC4514 chip, you can create your own encoder, because the logic is very simple. It is shown in Figure 25-4. Follow the connections and you'll see that pressing (for example) the switch with value 6 will activate binary outputs with values 4 and 2. Note that if you actually create this circuit, you will need to add a pulldown resistor on the output from each switch.

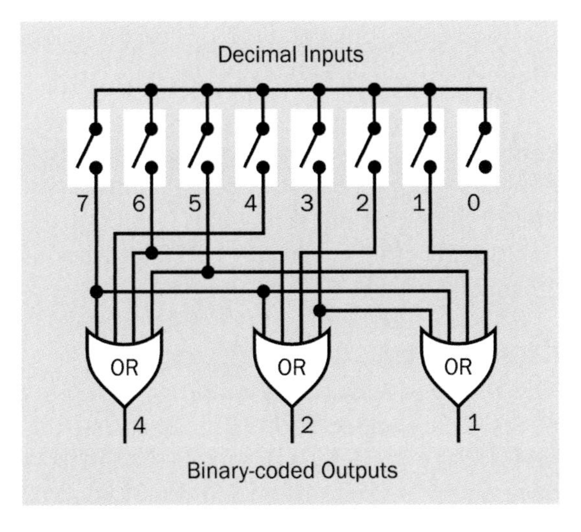

Figure 25-4. *Three four-input OR gates are all you need to emulate the behavior of an 8-to-3 bit encoder chip. Pulldown resistors (not shown here) would be added to the output of each switch.*

Maybe you remember when I wanted to use a four-input OR gate, but it didn't exist in the HC family. (See Chapter 20.) Would that be a problem here? No, because in this application, the OR outputs are not driving LEDs. They will be connected with the inputs of other chips in the adder circuit in Figure 24-11. Therefore, you can use chips from the old CMOS 4000B series. Two 4072B dual-four-input OR chips will do the job.

I like to minimize the chip count, so I specified the 74HC148 encoder instead of using two chips containing OR gates.

Make Even More: Other Input Options

For data entry, instead of using DIP switches, you could substitute eight-position rotary switches. They would prevent the selection of more than one number at a time—but they wouldn't fit on your breadboard and would be more expensive.

You could also consider keypads, especially if you can find a couple that are not matrix encoded. A matrix-encoded keypad is designed to be scanned by a microcontroller. An unencoded keypad gives you a separate output line for each digit that you enter, but this type of keypad is less common. (I used one for an experiment in *Make: Electronics*).

Lastly, you could use the type of input device that provides a thumbwheel to select a number, typically 0 through 9 but sometimes 0 through 15. Confusingly, these are often referred to as "encoders."

Can We Switch It?

Now that I've decimalized the inputs and outputs of the binary adder, you might think that this project has gone as far as it can go—but in fact I have one more suggestion. No doubt you will remember that I described versions of the Telepathy Tester, the Rock, Paper, Scissors game, and Ovid's Game that used switches instead of logic chips. Surely, I'm not thinking of using primitive electromechanical switches to build a binary calculator?

Actually, I can't resist the idea. I won't include this as a separate experiment, because it's really a Make-Even-More digression that I am including just for fun—at least, my idea of fun. I like the idea of a binary adder that contains no electronics whatsoever (unless you count the LEDs).

Also, while this kind of thing has no practical application, it helps you to think as a computer thinks, which is a prerequisite for designing good logic circuits (or good software). Therefore, I think it has some value.

Make Even More: Switched Binary Adder

I showed you the switch equivalents of logic gates in Figure 18-1. Since a half adder can be built from an XOR gate and an AND gate, you can represent it with only two switches, as shown in Figure 25-5.

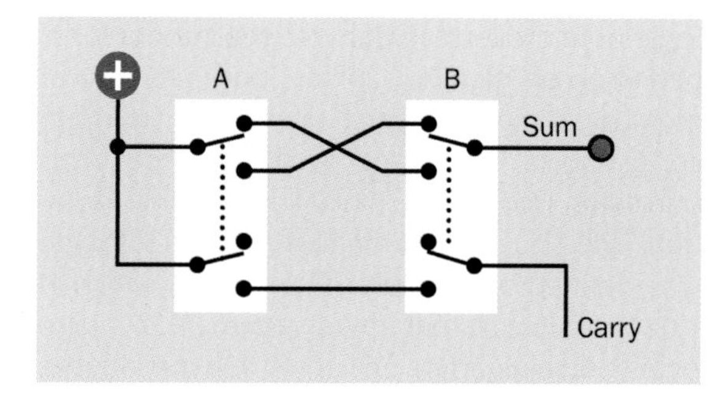

Figure 25-5. *The function of a half adder can be emulated with a pair of DPDT switches. A grounded LED, with series resistor as required, should be substituted for the "Sum" indicator.*

When switches are used as a substitute for logic gates, an input consists of pressing a switch. So, to enter the binary addition sum 1 + 0, you would close the lefthand switch while leaving the righthand switch open. This correctly gives an answer of 1 by illuminating the LED, with no carry output. And if you entered 1 + 1 by pressing both switches, you'd get the correct answer 10, represented by a carry output while the LED stays dark.

This seems easy enough, but a full adder is more problematic. Refresh your memory of the logic diagram for a full adder by taking a look at Figure 24-5. The top part is exactly the same as the half adder, which I just dealt with. No problem there. But the output from the first XOR gate goes to another XOR, where it is compared with the carry from the previous place.

The problem is, we are now using fingers pressing switches as inputs. But the second XOR only has electrical inputs—nothing that can be pressed by a finger.

Here's another problem. A low output can be interpreted by a logic gate just as easily as if it were a high output. But when switches are used, a low output becomes an open switch, which cannot be interpreted, because its state is indeterminate.

My answer to both these problems is to have an active "no carry" output from the half adder, as well as a "carry" output. This will provide voltage that each new stage of the adder can either block or pass along.

A revised half-adder schematic is shown in Figure 25-6. The "no carry" output is active if neither switch is pressed, or if either switch is pressed, but not both. The "carry" output is only active if both switches are pressed, so the "carry" and "no carry" outputs are always the inverse of each other.

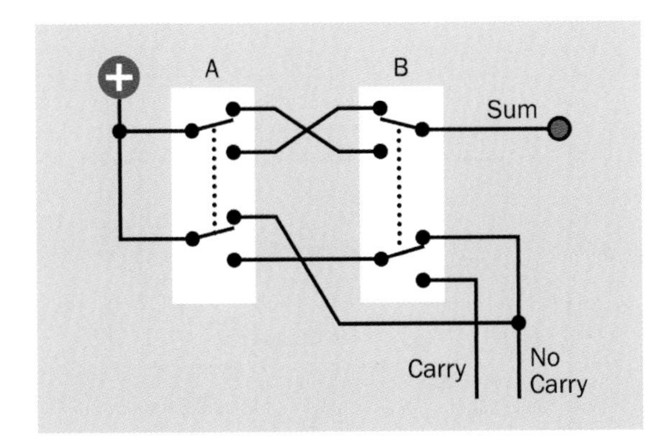

Figure 25-6. *The previous schematic for a switched half adder has been revised to include a "No Carry" output to provide voltage for the next stage.*

Making a Table

The next step is to nail down exactly what we want from a full adder. To do that, I made a little table, shown in Figure 25-7. This shows the three possible outputs that the adder can create and all the input combinations that can create each output.

Switch A	Switch B	Carry In	No Carry In	Output
open	open	1	0	Sum=1
open	closed	0	1	
closed	open	0	1	
closed	closed	1	0	
open	closed	1	0	Carry Out = 1
closed	open	1	0	
closed	closed	1	0	
closed	closed	0	1	
open	open	0	1	No Carry Out = 1
open	open	1	0	
open	closed	0	1	
closed	open	0	1	

Figure 25-7. *The outcomes from a full adder, for all possible input combinations.*

For instance, the second line of the table shows that when Switch A is open (representing a 0 input), and Switch B is closed (representing a 1 input), and there is no carry coming in from the previous place, then Sum = 1. The eleventh line of the table shows that the same combination of inputs results in a "No Carry" voltage going out. This makes sense because if you have one switch being pressed, and no carry coming in, there should be no carry going out.

Switch Specification

I was able to figure out (with some difficulty) a pattern of switches that would match the requirements shown in the table. The schematic is shown in Figure 25-8.

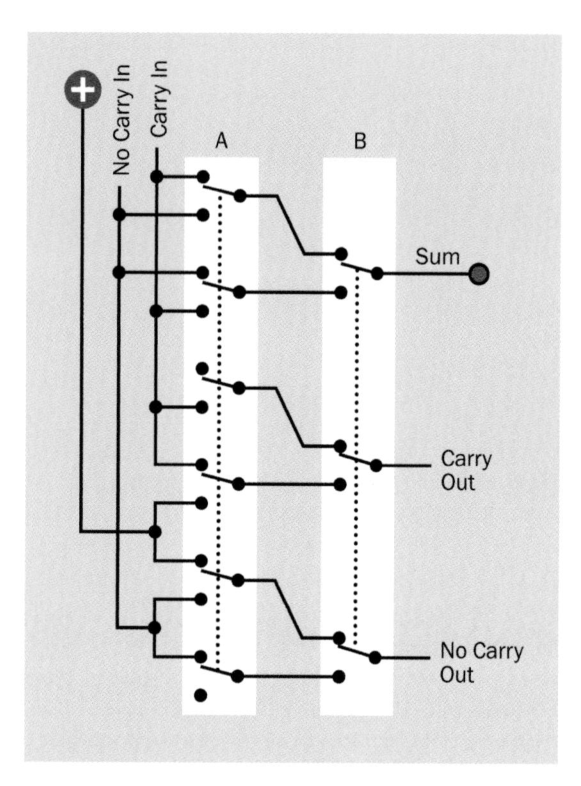

Figure 25-8. *A full adder built from a six-pole switch and a three-pole switch, creating the outputs specified in the preceding table.*

As you'll see, this requires a six-pole switch and a three-pole switch. Such switches are readily available in the latching pushbutton format, but I haven't finished, yet. I want to add an LED to each switch, to show when it is pressed.

This is easy enough in Switch B, because right now it only has three sets of contacts, allowing me to add another set to activate the LED. But adding extra contacts to Switch A would be more problematic, as it already has six poles. Some eight-pole switches do exist, but they tend to be expensive unless you can dig some up from a Chinese source on eBay.

After pondering this problem for a while, I recalled that a computer programmer named Graham Rogers, in England, has always shared my interest in puzzles—and because Graham is almost as obsessive as I am, he created his own switched version of a binary adder a few years ago. I couldn't find a schematic of it, so I checked with him and found that he and I had come up with the same basic circuit independently. However, he swapped the switch functions that create the sum, which freed up a pole on the lefthand switch to drive an LED.

This revised circuit is in Figure 25-9. It's not quite as simple as the previous version, but it only requires six-pole switches.

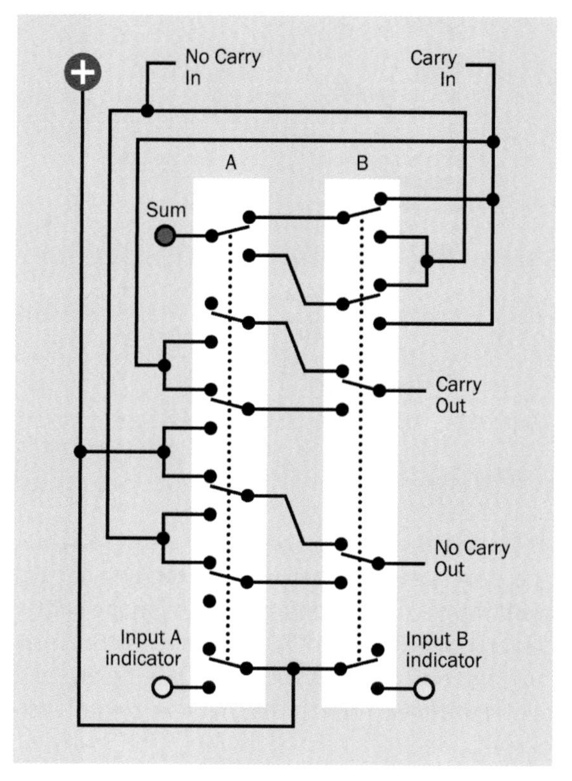

Figure 25-9. *Revised schematic for a switched full adder, allowing allocation of an extra pole in each switch for an LED indicator.*

You can string together as many of these switched adders as you like, to handle binary numbers of any size. The version that Graham built is capable of adding two eight-bit numbers (two bytes) and can also do subtraction. I can't remember the rules for subtraction in binary code, so I'll leave that problem to you.

Make Even More: Other Possibilities

Can the wiring for a switched binary adder be simplified any further? Can the number of switch contacts be reduced? I don't think so, but I'd like to be proved wrong. If you come up with a plan, let me know.

Now, I can't resist asking—could we get rid of the OR gates in Figure 25-4 and build an encoder just from switches, to process decimal input and turn it into binary? It should be possible. Any time you see a simple set of gates like those in Figure 25-4, you can feel fairly sure that they can be emulated with switches.

Indeed, this is quite easy. The circuit is shown in Figure 25-10.

The output from this circuit could be used for input to the electronic version of the adder in Figure 24-11, if you don't mind getting rid of the DIP switches and substituting an assortment of SPST, DPDT, and 3PDT switches.

However, I see no way to interface it with the switched adder. The problem is that in a switched adder, the input switches aren't just providing input—they're doing dual duty, performing the arithmetical operations. I'd be very surprised if a decimal-input switched adder were possible, using switches with a reasonable number of poles. On the other hand, I cannot prove that it's impossible, and human ingenuity sometimes produces amazing results.

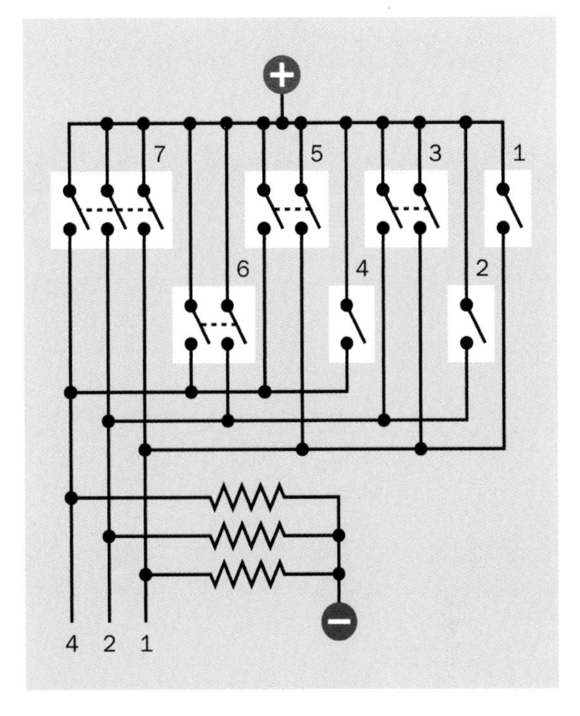

Figure 25-10. *The switched circuit shown here would emulate the function of an 8-to-3 encoder chip.*

Remember that before the modern era of electronic calculators, adding machines (which could also do multiplication sums) were entirely mechanical. When you looked inside, you found an insanely complicated assembly of gears and levers, and of course the designs were all done with pencil and paper, long before the era of computer-aided design.

Going further back, Charles Babbage's "Difference Engine" (the first part of which was built in 1832) was a calculating machine, leading Babbage to plan his "Analytical Engine," a general-purpose, programmable mechanical calculating machine. Vision combined with persistence can achieve remarkable results, if someone has enough time and money to work on a problem.

But I'm going to move on, now, to ring counters and shift registers. These will lead to an in-depth exploration of randomicity, ultimately involving various types of sensors, which will be the closing topic of the book.

Experiment 26: Running Rings

26

A ring counter is a type of counter that has "decoded outputs." This means that it activates one pin at a time, beginning with the pin that represents 0 and continuing up to a value that is limited by the number of pins. After that, the cycle automatically repeats.

By comparison, a binary counter has "coded outputs." In other words, the high and low states on its output pins represent a number in binary code.

The ring counter that I will be using here is the 74HC4017, which can also be described as a decade counter, meaning that it can count up to 10 (actually, from 0 through 9). I plan to apply it in two different reflex-testing games: one that flashes LEDs sequentially, and another that flashes them randomly.

Ring Demo

Demonstrating a ring counter is very easy. You just attach ten LEDs to its ten outputs and drive it with a slow-running timer. In my test, I found that the classic bipolar version of a 555 timer created such a noisy signal, the ring counter often interpreted it as two pulses instead of one. Probably I could have resolved the issue by adding a smoothing capacitor, but I felt more confident about using a less-noisy component: the 7555 timer. This has the same pinouts as the old 555 and the same timing characteristics but has a clean output that is properly compatible with logic chips.

Warning: Timer Incompatibilities

Be careful to store your 7555 timers separately from your old-style 555 timers. They look the same and they seem to function identically, but the output from some versions of a 7555 has a more limited current-sourcing capability. You can use it to drive other chips, or a single LED, but don't try to use it with a relay.

Also, while the 7555 gives you the benefit of a cleaner output, it can be fussier about its input. When driving its input pin through a coupling capacitor, for instance, the value of the capacitor can be important. A spike in the voltage can cause the timer to end its current cycle prematurely. You won't find this mentioned in the datasheet, but I have seen it happen.

Annoying Pin Sequence

The demo schematic is shown in Figure 26-1. As in the previous experiment, I am using yellow circles to represent LEDs, because there isn't space for the proper symbols. Add a series resistor if your LEDs do not have resistors built in. You will only need one series resistor, because only one LED will be illuminated at any time.

Note that Pin 8 of the ring counter does not connect through an LED, but is tied directly to negative ground, because it's a power pin on the chip.

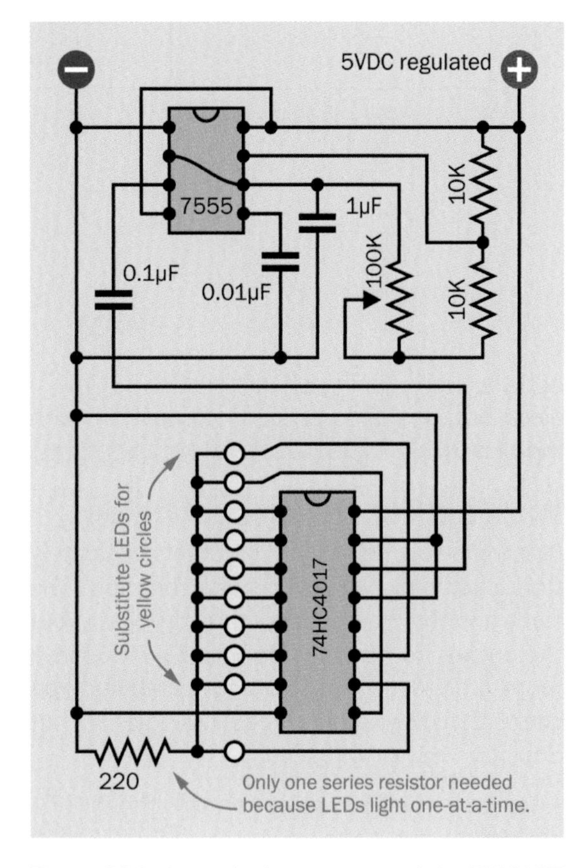

Figure 26-1. *A simple demonstration of the 74HC4017 decade counter with a slow sequence of pulses applied to its clock input pin while the enable and reset pins are held low.*

If you put this circuit together and apply power, you'll see the LEDs flashing one at a time as the counter responds to pulses from the timer. Unfortunately, the pin values of the 74HC4017 are scrambled, and their sequence is even more erratic than the pin values of the 74HC4514 decoder chip in Experiment 25. Figure 26-2 shows how the values are assigned to the chip.

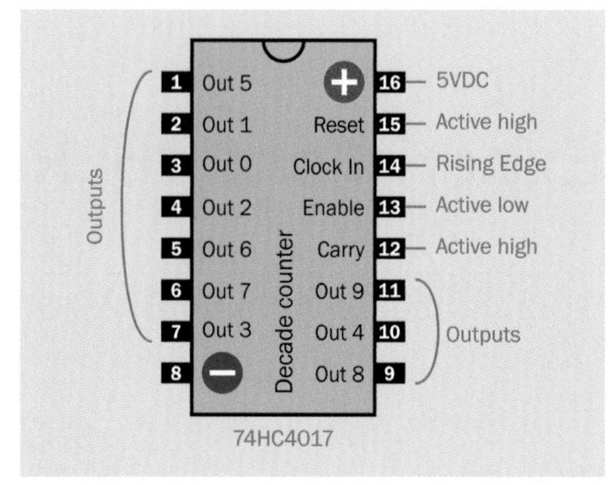

Figure 26-2. *Pinouts of a 74HC4017 counter. This is a ring counter, meaning that its outputs are decoded—they are powered one at a time instead of in a binary-coded combination. Because it has ten outputs, this is also classified as a decade counter.*

You could shift the LEDs farther down the breadboard and unscramble the value sequence with jumper wires, as I suggested with the decoder chip in Experiment 25. However, for the application that I have in mind, I'll need all available breadboard space for more chips. Really a second breadboard is necessary to mount the LEDs. You can then run wires across to them from the counter, as shown in Figure 26-3. The connections are shown in color so that you can follow them more easily.

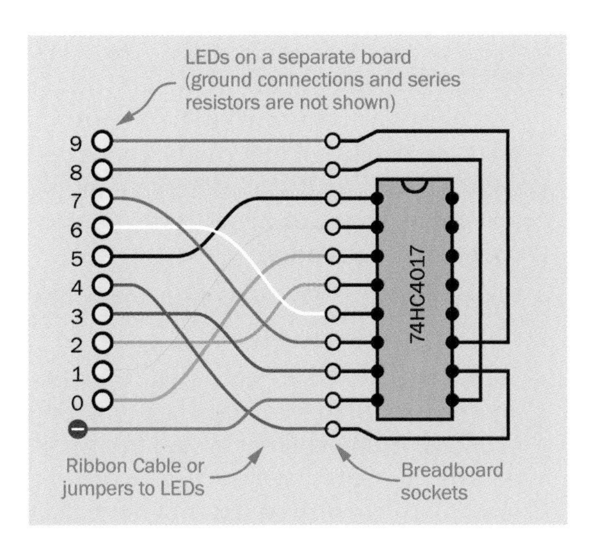

Figure 26-3. *Unscrambling the output pin sequence of a 74HC4017 ring counter to drive LEDs that will light up in numerical order. The colored connections can be jumper wires or a ribbon cable.*

You can use flexible jumper wires, as in the test circuit shown in Figure 26-4. Use as many colors as possible to make it easier to trace connections from one board to the other.

Figure 26-4. *Jumpers can be used to unscramble the counter outputs and power a line of LEDs that will light up in numerical order. The components at the top of the breadboard in this photograph will run a demo of the counter, described below.*

However, in my experience, the little plugs at the ends of jumper wires don't always make a firm and reliable connection. A better solution is to use a ribbon cable soldered to headers. This is shown in Figure 26-5.

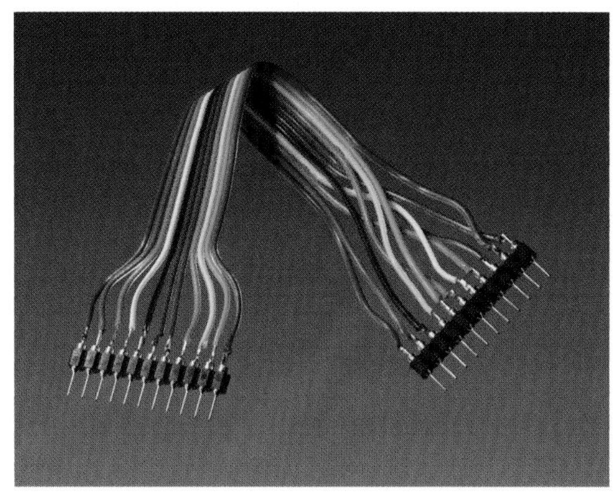

Figure 26-5. *A piece of ribbon cable with header pins soldered to each end. The conductors are swapped around to create connections shown in the previous figure.*

Maybe you have never encountered the term "headers" before. Okay, I'll pause for a moment to explain this, because the usage is confusing.

Quick Facts About Headers

The term "header" is ambiguous. It often refers to a series of "header pins" embedded in a strip of thin plastic, with shorter terminals sticking out at the other side. This is what you see in Figure 26-5. The plastic can be snapped off to give you the number of pins that you need, and they can push securely into a breadboard (provided you are careful to buy header pins with 0.1" spacing). You can solder the wires of a ribbon cable onto the terminals, resequencing them at the same time.

For permanent installation using a perforated board, you must mount "header sockets" on the board. The header pins will then plug into the sockets. Confusingly, both the pins and the sockets can be referred to as "headers" in parts catalogues.

A strip of header pins and two strips of header sockets are shown in Figure 26-6.

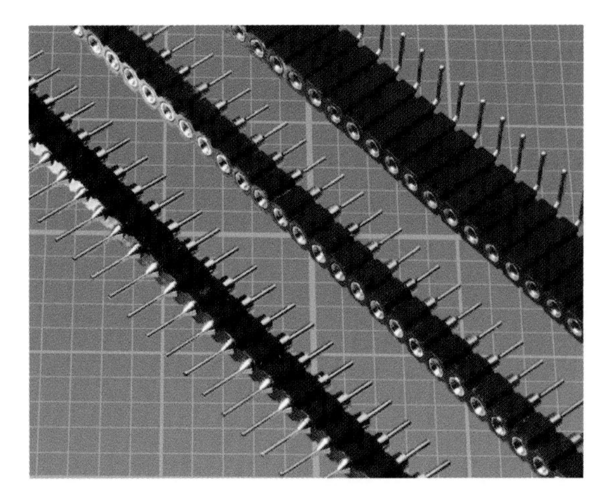

Figure 26-6. *Header pins and header sockets are sold in long strips that can be snapped off to provide you with the number of pins and sockets that you require. Both the pins and the sockets may be referred to as "headers." Various configurations are available, such as the angled pins in the strip at top-right.*

Quick Facts About Ring Counters

Before I go any further with the test circuit, I'll just sum up the most important facts about ring counters:

- The 74HC4017 is a decade counter with ten outputs that are activated in sequence, one at a time.
- This is also known as a ring counter.
- Ring counters often have eight, ten, or sixteen outputs.
- A counter with binary-coded outputs will represent 0 by holding all outputs low. A ring counter includes an output line to represent value 0, and this output will be high when the counter is at 0.
- A ring counter repeats its output sequence indefinitely, so long as the clock input keeps running.
- The 74HC4017 is normally used with its reset pin held low and its enable pin held low. The carry pin is an output, which can be left unconnected. The clock pin responds to a low-to-high transition; this is known as the rising edge of the clock signal.
- Because the carry output changes from low to high when the counter transitions from a 9-value output back to a 0-value output, the carry pin of one counter can be connected with the clock input of a second counter, enabling the display of values up to 99. More counters can be added.
- When the reset pin is high, it forces the counter back to a 0 output.
- Some ring counters are not decade counters, and some decade counters are not ring counters. A ring counter always has single decoded outputs that are activated one at a time, but there may be fewer or more than ten of them. A decade counter always counts from 0 through 9, but it may have binary outputs or single decoded outputs. The binary outputs of a decade counter are referred to as "binary-coded decimal," or "BCD," in datasheets.

Making a Game of It

In Las Vegas, you can play a game where LEDs flash in sequence around a giant circle, and you have to press a button at exactly the moment when the LED nearest to you lights up. You are allowed a limited number of button presses before you have to pay to try again.

We can make a simple version of this game, but so long as it's on a breadboard, a circle of LEDs will be difficult to create. I'll compromise by using a vertical column of LEDs, where the player has to press a button at the exact moment when the LED at the bottom is illuminated. You can rebuild the circuit with a ring-shaped output if you have the time to do so.

To prevent a player from winning simply by holding the button down continuously, I can add another timer in one-shot mode. A timer will normally retrigger itself while its input is held low, but if the pushbutton connects with the timer input through a coupling capacitor, the capacitor will only pass the initial transition from high to low (just like the one-shot timer in the chronophotonic lamp switcher, back in Chapter 7). Also, if the capacitor has the right value, it will ignore any little transient spikes caused by "contact bounce" when the button is pressed. (I went into some detail about "contact bounce" in *Make: Electronics*).

However, the pushbutton must have two poles (DPST or DPDT type) so that the capacitor is connected to the positive side of the power supply when the pushbutton is not being pressed. Otherwise, the capacitor will have no charge to discharge when the button is pressed.

Figure 26-7 shows the game concept in block-diagram mode.

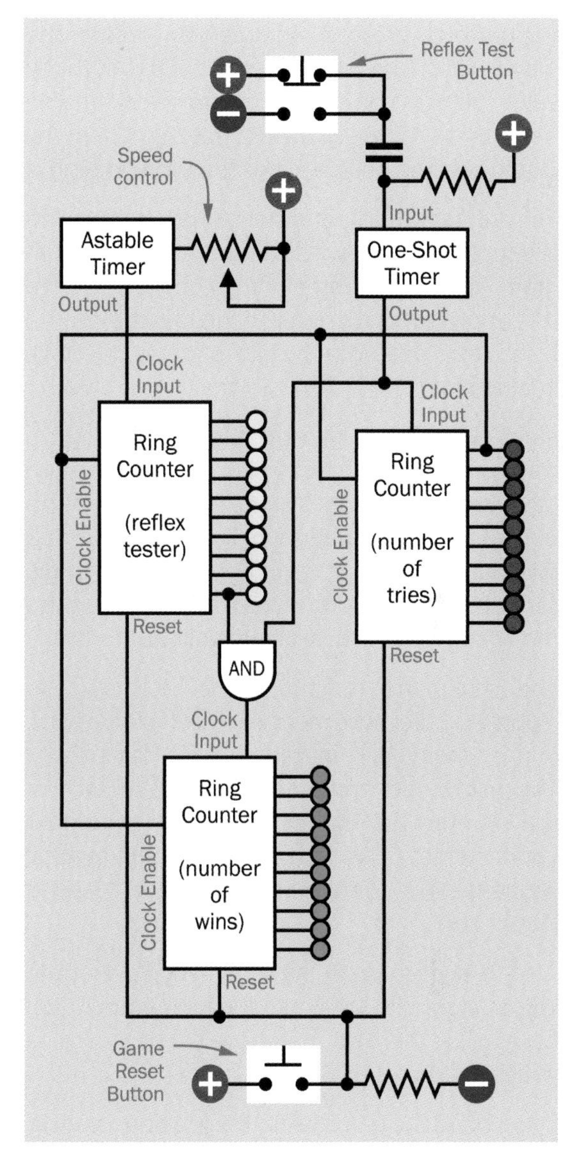

Figure 26-7. *This block diagram shows the simple logic of the ring-counter game.*

Additional Features

You can see in the block diagram that when a player is successful, an AND gate detects that the button press has synchronized with the ring counter. The AND gate triggers (guess what) another ring counter, which keeps track of the score using its own set of ten green LEDs.

To end the game, I've used a third ring counter, using ten red LEDs. This one advances every time the button is pressed, regardless of whether the player wins. When it has gone from 0 to 9, the 9-value output connects with the clock-enable pins on all three counters, freezing them by changing their enable states from low to high.

The only way to play another game is by pressing a reset button, shown at the bottom of the diagram. This sets all three of the counters back to 0 by applying a high signal to their reset pins. The reset signal overrides the high state on the clock enable pin.

As you can see, you'll need thirty LEDs to set up this game. Fortunately, LEDs are cheap these days (around two cents apiece from China), and because a ring counter only lights one LED at a time, each column of ten LEDs can share a single series resistor. You don't need to spend the extra money for LEDs with resistors built in.

The schematic for the game is shown in Figure 26-8. Because there was insufficient space, in this schematic I omitted the LEDs completely. The yellow, red, and green numbers in each of the counters tell you that these pins should be connected to LEDs of these colors, in numerical sequence. The colors match those that I used in Figure 26-7.

Also to save space in the schematic, I offset the chips diagonally. On a breadboard, naturally you have to line them up, one above the other. I think you'll find that they fit.

If you build the game using the previous test circuit as a starting point, remember to disconnect pins 13 and 15 of the first ring counter from negative ground. I tied them to ground originally to disable the reset function and to keep the chip in its enabled mode. In the full circuit, those functions are now controlled to start a new game and to freeze the game after ten turns, respectively.

Photographs of the dual-breadboard circuit are in Figure 26-9 and Figure 26-10.

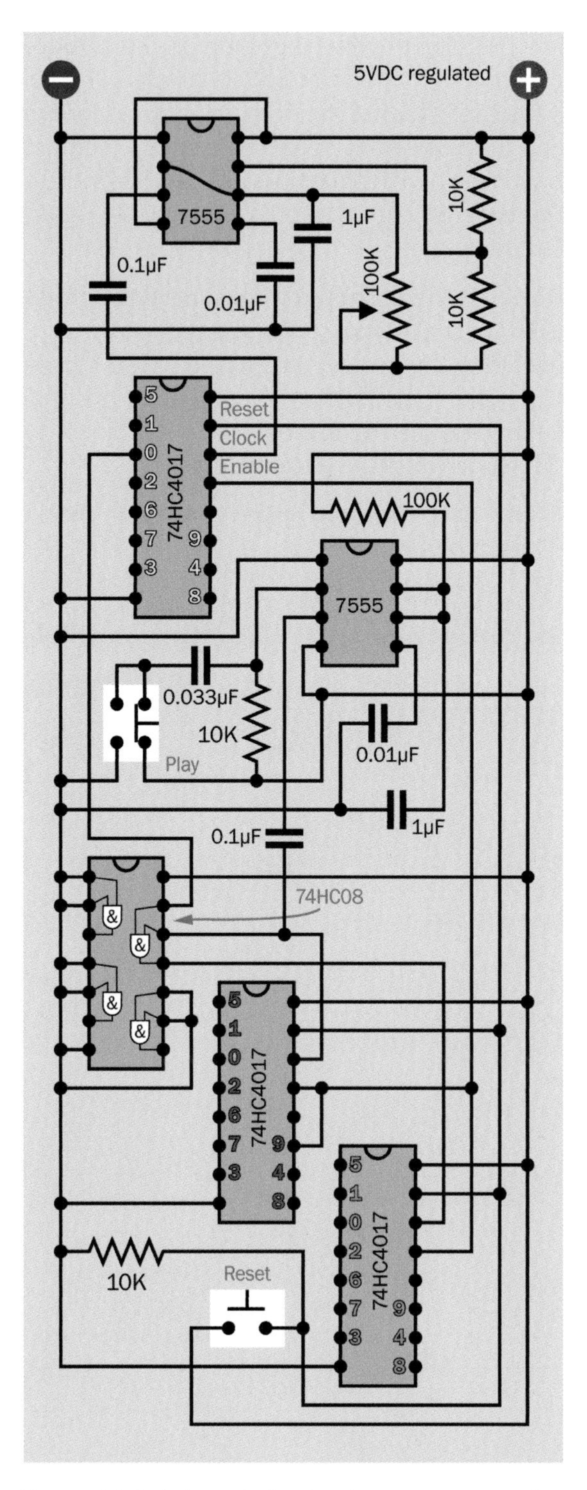

Figure 26-8. *Schematic for the ring-counter game. Numerals inside the three counter chips indicate that those pins should be connected with LEDs of the same color, arrayed in numerical order.*

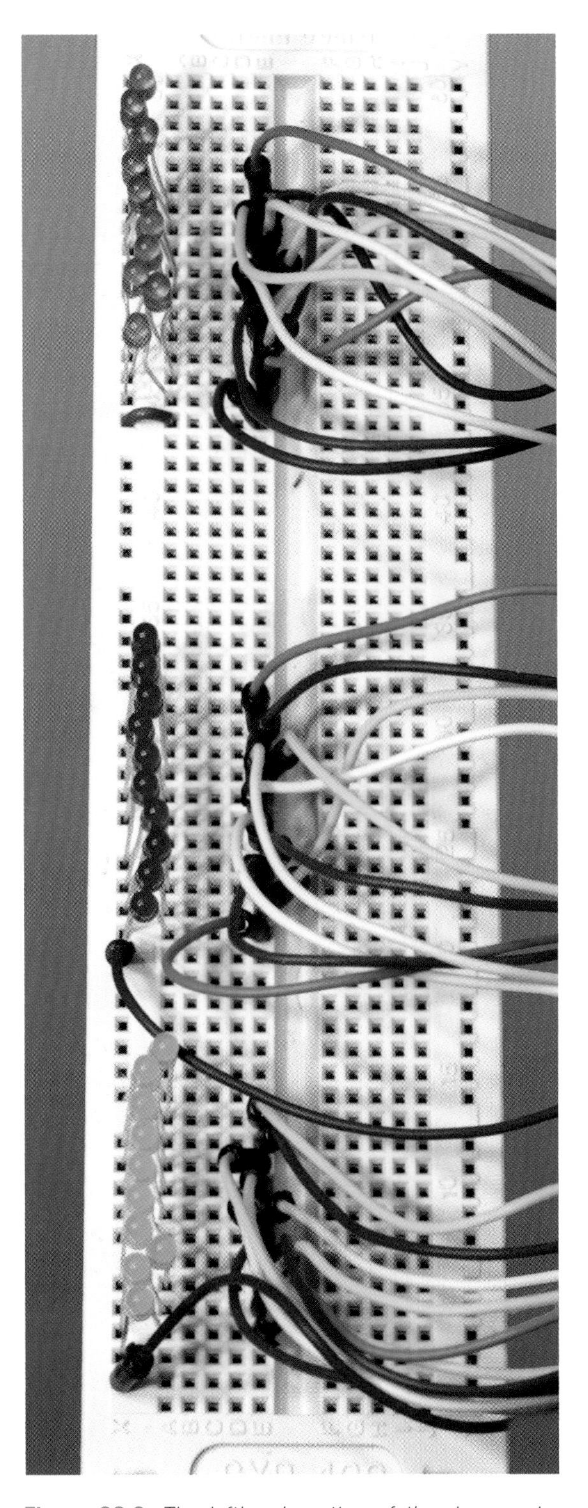

Figure 26-9. *The lefthand section of the ring-counter game.*

Figure 26-10. *The righthand section of the game, with jumper wires that run across to the other section.*

Playability

You can use the trimmer potentiometer to control the game speed. After you play the game for a while, you may wish to replace the trimmer with two or three fixed resistors with DIP switches, for preset levels of difficulty.

I tested the game with three different pushbutton switches, and found that it worked well with a 0.033µF capacitor on the input to the one-shot timer. Remember, I mentioned that the 7555 can be a little fussy about this. If you find that the timer is sometimes unresponsible, or abbreviates its output pulse, try a higher or lower value capacitor, or try a different pushbutton switch.

The duration of the pulse from the one-shot timer is critical. If the pulse is too long, the player will be able to win by pressing the button a moment before the bottom LED lights up, because the duration of the pulse will overlap with the "on" state of the LED. The 100K resistor and 1µF capacitor will create a pulse lasting about 1/10 of a second. To make the game more difficult, substitute a 47K or 22K resistor, which will create a pulse lasting about 1/20 or 1/40 of a second, respectively.

Make Even More

I'm thinking that the game would be more interesting if the string of LEDs varies its flashing speed unpredictably. That sounds difficult to do, but really, it isn't. In Experiment 22, I already showed how you can connect the outputs from a pair of timers through an XOR gate so that their signals move in and out of phase, creating unpredictable audio effects.

All you have to do is slow the timers in that experiment by a factor of about 1,000 and take the output from the XOR gate as the clock input for the reflex-testing ring counter, instead of using the astable timer that I included originally.

There won't be room to do this all on one breadboard, but you can put your pseudorandom cycle generator on a separate board and run a signal wire between them (with additional power and ground connections). Adjust the two trimmers until the lights flash in a stop-and-start way that pleases you. Are they really random? No, but they can seem to be random, which is all we need. Sooner or later they will tend to repeat the same sequence, but if you get the timers running just slightly out of phase, this can take a long time to happen.

The breadboarded version of this circuit is shown in Figure 26-11. The circuit is derived from Figure 22-1, except that the XORed timers are now connected with a decade counter.

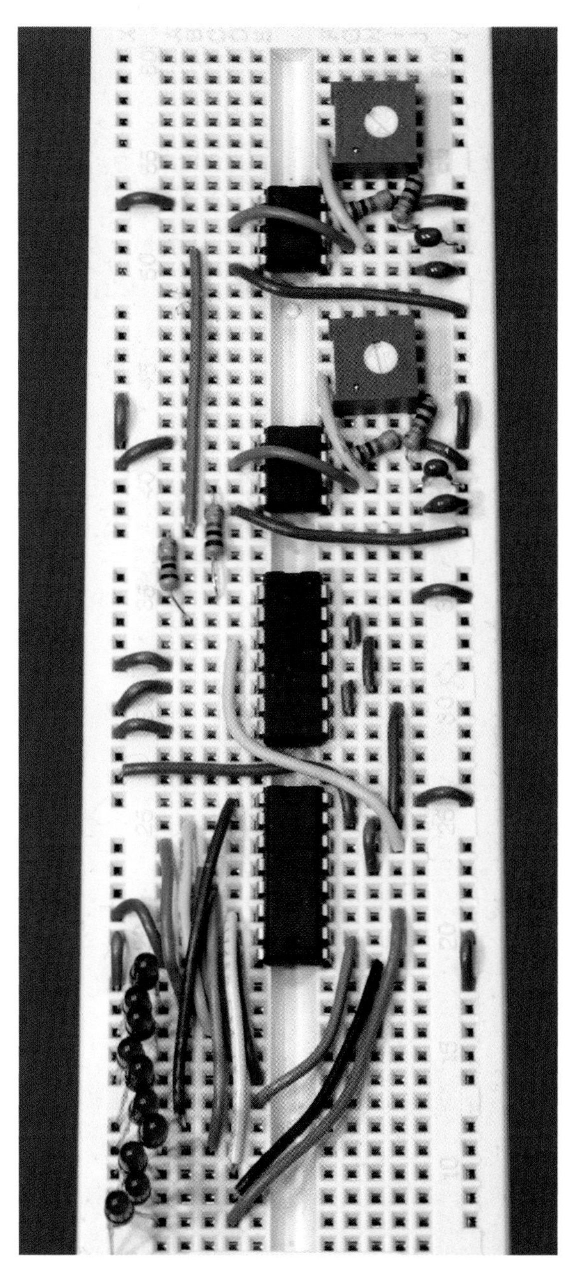

Figure 26-11. *The idea first mentioned in Experiment 22, of XORing the outputs of two timers that are slightly out of phase, is applied here to create a seemingly random fluctuation of the counter controlling ten LEDs.*

And a Microcontroller?

The main challenge for replicating this game with a microcontroller is that you need thirty outputs to drive all those LEDs. Or do you? Really, of course, you only need the ten yellow LEDs to make the game playable; the score and the number of tries can be displayed on an LCD screen. And if even the ten LEDs exceed your available number of output pins, you can flash the LEDs by driving them through a binary-to-decimal decoder chip, which only requires four binary inputs. Of course, that means you have to send numbers to the decoder in binary code.

Checking the user-input button with a microcontroller can be done with an interrupt. But you may want a routine that tells the microcontroller to ignore the button if it is being pressed before the 0-value LED lights up. A premature button press would tend to indicate cheating.

Most microcontrollers that run a high-level language such as C or a version of BASIC will have a pseudorandom number generator built in. That eliminates any need for XOR'd timers. And you can adjust the speed of the game by attaching a trimmer potentiometer to an input pin that has an analog-digital converter.

So, yes, it can be done, and you'll certainly reduce the chip count. Oddly enough, though, I think that making the game work would be difficult with a microcontroller, because writing and debugging the program would be nontrivial. The reason is that you'll be compressing all the functions of various chips into one chunk of code. It has to flash the LEDs in a pseudorandom way, and while each LED is lit, it has to go check the user-input button and the reset button—and it has to stop the game if the number of tries reaches the maximum—and it also has to update the LCD display and the game-speed trimmer input, which must be decoded to change the value of a variable that is used in conjunction with an internal clock. Some of these tasks can be handled by interrupts, but then you have to write code to handle the interrupts when they occur.

Sometimes, it's easier to wire chips together. And in any case, I like having all of those thirty colorful LEDs as the game display.

Experiment 27: Shifting Bits

27

A counter with decoded outputs can be used for entertainment in games and flashing-light displays, but perhaps you don't always want single LEDs to flash in sequence. Perhaps you want to create a sequence of your own.

The component that will do this for you is known as a shift register. Very interesting, but—why is this interesting? What would you use it for?

I can provide some practical answers to these questions, but first I'd like you to set up a circuit that you'll need to trigger the shift register and control other experiments in the future. It provides a clean pulse of a fixed length, in much the same way as the one-shot 7555 timer in the previous experiment (see Figure 26-8).

No Bouncing!

I talked about switch bounce, also known as contact bounce, in *Make: Electronics*. This is the nasty habit of mechanical switch contacts that vibrate for a very brief moment when they open or close. Because digital chips are so sensitive and can respond so quickly, they misinterpret vibrating contacts as multiple presses of the switch.

Switch bounce hasn't been a problem in most of our experiments because I haven't been using a switch or a pushbutton to send pulses to a chip that counts them. In Figure 26-1, for instance, a timer running in astable mode controlled the counters.

To test a shift register, which is going to be the subject of this experiment, you really need to advance it manually, and a bounceless switch is the only sensible way to do this.

Specifics

In my previous book, I showed that when two NOR gates or two NAND gates are wired as a flip-flop, they can debounce an input. But I prefer to use a timer for this task, as the timer can provide a pulse that has a fixed duration. This in itself can also be useful.

Figure 27-1 shows the circuit. A double-throw pushbutton must be used, because in its normal "up" position, it maintains a positive charge on the 0.033μF coupling capacitor. Meanwhile, Pin 2, the input pin of the timer, is also maintained in a high state through a 10K pullup resistor. So long as the input pin is high, nothing happens.

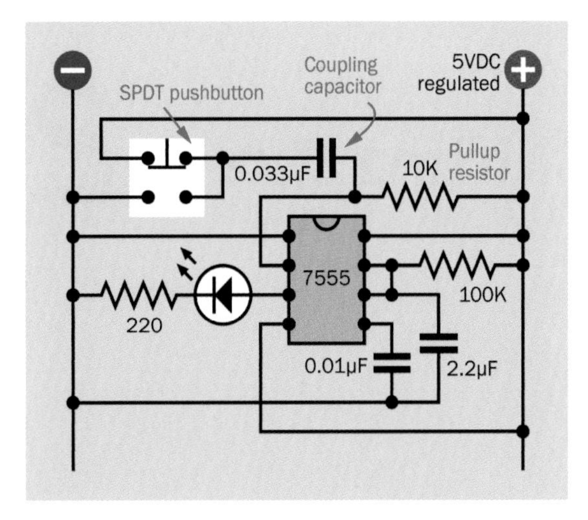

Figure 27-1. *A basic circuit to deliver a clean pulse of fixed length, suppressing contact bounce that tends to be created by a pushbutton.*

When the pushbutton is pressed, it grounds the coupling capacitor. This event is transmitted to the input pin of the timer, pulling it down for long enough to trigger the timer in its one-shot mode. The 100K timing resistor and 2.2µF timing capacitor deliver a pulse lasting about a quarter of a second. While this is going on, any vibration from the switch contacts is ignored.

- The output from the timer must last longer than the contact bounce. However, the contacts stop vibrating within a couple of milliseconds.

When the timer finishes its pulse, it would normally be retriggered if its input pin was still low. In this circuit, however, even if the button is still being pressed, the coupling capacitor now blocks the DC connection, while the pullup resistor holds the timer input pin high.

- If the pushbutton is still being held down when the output from the timer ends, the timer ignores the pushbutton, completes its cycle, and its output pulse ends.

Now suppose the pushbutton is released before the end of the timing interval. The capacitor is immediately recharged, and the pullup resistor holds the timer input high.

- If the pushbutton is released before the output from the timer ends, the timer still completes its cycle, and its output pulse ends.

The only possibility for error is if the pushbutton is released at virtually the same moment that the timer pulse ends. Under these circumstances, if the switch contacts vibrate as they open, it is possible for retriggering to occur.

- The timer should deliver a short pulse (a quarter second or less) or a relatively long one (one second or more) to avoid conflict if the switch contacts open and vibrate at the same moment as the timer pulse ends.

The timer is shown in the schematic with an LED attached to its output for demonstration purposes. Depending on the application, another coupling capacitor may be placed on the output from the timer to transmit a brief pulse to the next stage in the circuit, while blocking DC.

A Bit-Shifting Demo

Now we're ready for the shift register. The setup in Figure 27-2 bears some resemblance to the ring-counter test schematic in Figure 26-1, except that it is now manually controlled with the debounced circuit that I just described. Note that it still uses a 7555 timer instead of an old-style 555.

A photograph of the breadboarded version is in Figure 27-3.

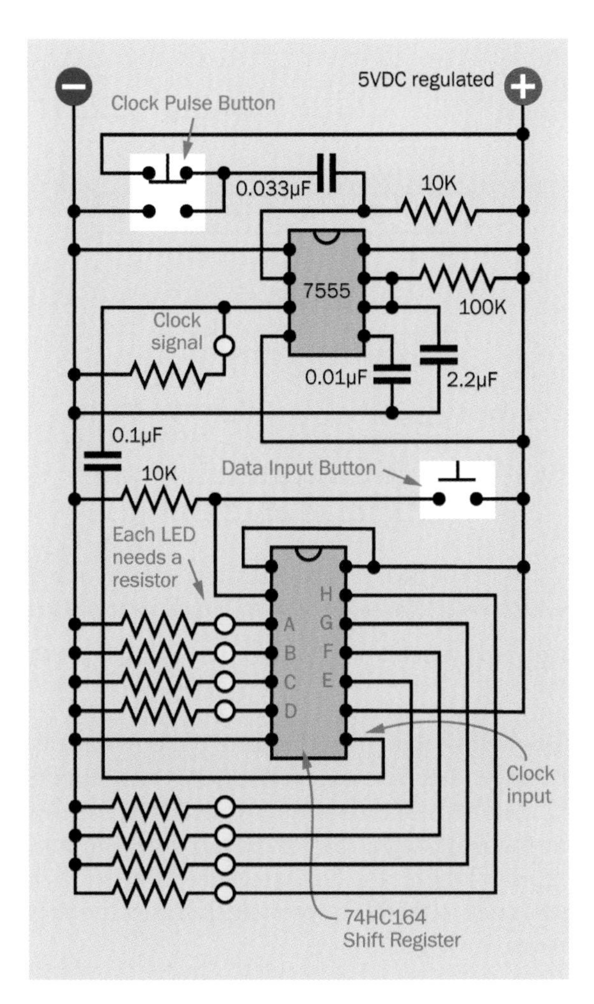

Figure 27-2. *A test circuit that shows how a shift register moves the contents of its memory locations in response to a clock signal. The pushbutton loads data into the shift register.*

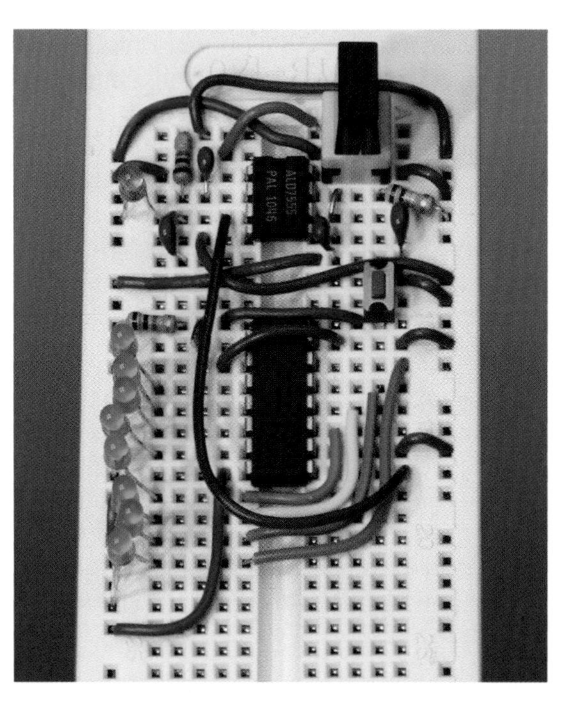

Figure 27-3. *The breadboarded version of the shift-register test circuit.*

While the pin values of the ring counter that you used previously (and the decoder before that) were not in numerical sequence, the outputs from the 74HC164 shift register are much more convenient, proceeding sequentially from pin 3 around the chip, counterclockwise. This makes it easy to attach a vertical line of LEDs that will light up in sequence, although the sequence will be from the top down instead of bottom-up.

As before, the yellow circles in Figure 27-2 represent LEDs. If your LEDs do not contain their own resistors, you will need one resistor for each and every LED in this circuit. While the decoder and the ring counter lit only one LED at a time, the shift register can illuminate any combination, including all eight simultaneously.

I added a "clock signal" LED just to show that the output from the timer is alive and well, because when you apply power to this circuit, the eight output LEDs are likely to do nothing. This is because the memory locations in the shift register are empty until you put some data into them.

Hold down the Data Input button. While you are holding it down, press the Clock Pulse button repeatedly. The Data Input button applies a high state to the input buffer of the shift register, and each time you press the Clock Pulse button, the state of the input buffer is copied into the first memory location of the shift register (identified as "A" in the schematic) while the contents of the other memory locations are bumped along to make room.

Now let go of the Data Input button, and the 10K pulldown resistor applies a low state to the input buffer. If you continue to press the Clock Pulse button, low states are "clocked in" to the shift register, and as before, the memory contents are moved along to make room. What happens to the last one, identified as "H" in the schematic? It is discarded.

A single shift cycle of the register is represented in Figure 27-4. Each of the eight memory locations can be thought of as containing a binary digit, or bit. In this diagram, initially the bits in locations C and H happen to be high. First, the pushbutton inserts a high state into the input buffer. Second, a rising clock pulse causes the shift register to move all the bits along and copy the high state in the input buffer into location A.

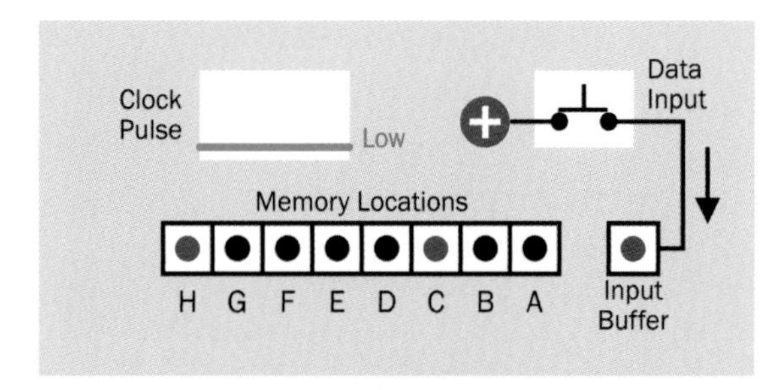

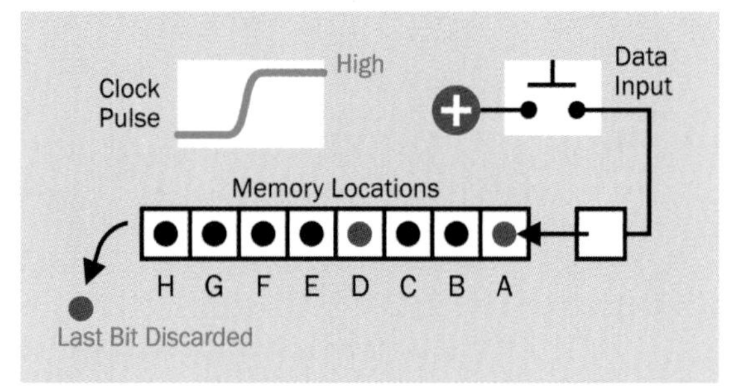

Figure 27-4. *Data moving into and through a shift register.*

The tactile switch in the schematic is not debounced because the shift register only checks its state at the moment when the clock pulse rises from low to high. The rest of the time, the shift register ignores the tactile switch—and in fact, if you press it briefly between clock pulses, the shift register will not notice.

Quick Facts About Shift Registers

- The shift register contains memory locations, each of which holds a high state or a low state. You can think of them as binary digits.
- Most shift registers are eight-bit components. However, it is possible to chain several of them together.
- A signal on the clock pin of the shift register tells it to throw away the value of the last memory location, move all the preceding bits one step along, and load a new value into the first memory location.
- The new value is determined by the high or low state of an input pin at the beginning of

a new clock cycle. Most shift registers respond to the rising edge of a clock pulse.

- The shift register ignores the state of its input pin until it is triggered by a clock transition.
- Some shift registers can convert parallel to serial data instead of, or as well as, serial to parallel data.
- There is a shift register with part number TPIC6A595 that has "power logic" outputs capable of delivering 100mA or more. This can be useful in some applications.

Pinouts

Pinouts of the 74HC164 are shown in Figure 27-5. Eight pins connect with memory locations inside the chip. I've labelled them A through H, although some datasheets may identify them as 1A through 1H, or QA through QH, or something similar.

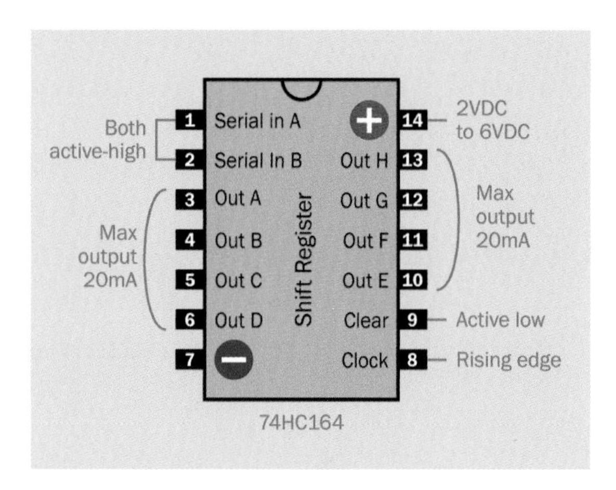

Figure 27-5. *Pinouts of the 74HC164 shift register.*

The Clear input is active low, and will zero out all the memory locations. Therefore, it is normally held in a high state. The chip has two serial data inputs, on pins 1 and 2. For our purposes, one input is kept high while the other receives data. The functions of pins 1 and 2 are interchangeable.

The 74HC164 is a relatively simple chip with only 14 pins. Other shift registers have additional features. I won't be discussing those chips here.

Background: Bit Streams

Long, long ago, communication between computing devices used to be done through a serial cable, which contained three wires. One of them was negative ground, another signalled an external device to start and stop the data stream, and the third wire carried data.

When the device on the receiving end accumulated seven binary digits, it assembled them to form a binary value from 0000000 through 1111111 (0 through 127 decimal), and each of these values was interpreted as a code that could represent a letter of the alphabet in upper- or lowercase. You also had a few control codes, such as "start a new line of text." (Subsequently the code system was extended to use eight bits, but the meaning of the additional codes was not standardized.)

This is how text was transmitted, back in the day. The character-coding system that was used for this very basic system was called ASCII, which is an acronym for American Standard Code for Information Interchange.

It was slow and primitive—yet today we still use serial data transmission, meaning that bits of data travel down a single wire, one at a time. This is true of a USB device or a SATA connection for an internal hard drive. The speeds have increased tremendously, but basic principles remain the same.

The ASCII codes are still the same, too, although they have become a subset of "Unicode," which permits up to 32 bits per character, facilitating the inclusion of foreign languages such as Japanese.

The tricky part, which I skipped over in my explanation, is how a device on the receiving end of a serial stream "assembles" the data. The early computers could process eight bits (one byte) at a time. Therefore, they had to receive eight bits serially, shift them into eight memory locations, and then spit them out simultaneously along eight parallel wires for processing.

As I'm sure you've guessed, the chip that did this was the shift register. It was a register that could shift bits along. It also functioned as a serial-to-parallel converter.

Modern Applications

Today, the capability of a shift register is built into bigger chips that do other things. Still, the old chips have their uses.

For instance, suppose you want to switch eight devices on and off from a microcontroller, but you don't have eight output pins to spare. You send eight on-or-off states over a single wire to a shift register, at high speed, using an additional wire for the clock signal that tells the shift register when each bit is coming so that it can be clocked in. The states of the eight output pins of the shift register can control the eight devices, and you can update the register so quickly, the result will seem to be instant.

Moreover, you can chain shift registers to control sixteen devices—or twenty-four, or thirty-two—and you still only need one wire to send the data. This is a very powerful concept.

Here's a thought. Suppose we use seven bits in a shift register to represent a binary number. If we shift all the digits one space to the left, and replace the rightmost digit with a 0, we just multiplied the original value by two. Why is that? Because when you move one step left through a binary number, each binary place value is twice the one that preceded it.

Hmm . . . do you think you could adapt the binary adder that you built to do multiplication sums? It's an intriguing idea—but I'm not going to mess with it. I'm going to use a shift register (three of them, actually) in a device that will tell your fortune.

Experiment 28: The Ching Thing

28

In this experiment, you can build a device to display a pair of hexagrams in an electronic version of the I Ching. I'm calling it the "Ching Thing."

If these terms mean nothing to you, there is no cause for concern, as I will be explaining them almost immediately.

A much shorter version of this project appeared in *Make* magazine. Really it was too complicated to be compressed into just a few pages, so I'm presenting a new version of it here, with more illustrations and detailed explanations.

Also, I simplified the circuit with help from my friend Fredrik Jansson, who once built his own computer entirely from 4000-series logic chips. In fact, my first contact with Fredrik was when he sent an email to me at the magazine, pointing out that I could eliminate one of the OR gates that I had included in the schematic. I mention this to make it clear that I really do read all the messages that come my way. And I take them seriously!

Hexagrams

Getting back to the Ching Thing:

The I Ching (pronounced "ee-ching") is an ancient Chinese book containing enigmatic advice about your present situation and future prospects. You can think of it as telling your fortune. This strange and remarkable compendium of advice is more than 2,000 years old, and the basis for it may be more than 3,000 years old. Some people believe it really does have predictive power. I'm not sure they're right, but on the other hand, I can't prove they're wrong.

Many English translations of the I Ching are available, and some of them are free online. Every version contains sixty-four basic descriptions of your status, and each description is identified by a graphic image called a hexagram.

To give you an example, two hexagrams are shown in Figure 28-1, with abbreviated explanations of their meaning. Anyone who takes the I Ching seriously will complain that these interpretations are much too simplistic—which is true. The thing is, I don't claim to be an I Ching expert. I'm just here to show you how to simulate it electronically.

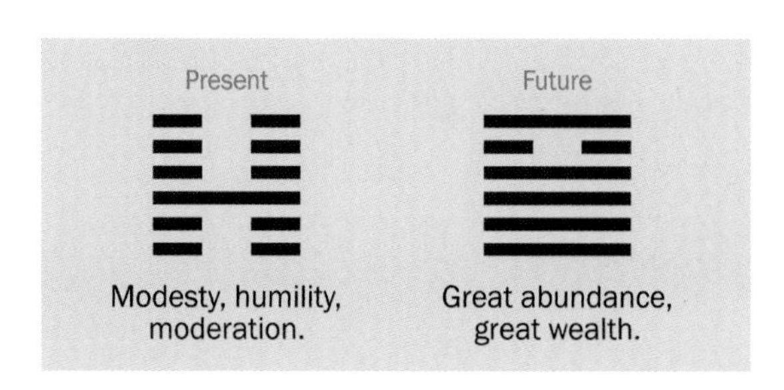

Figure 28-1. *Two sample hexagrams, and some very approximate interpretations.*

Each of the six lines in a hexagram can be solid or broken. In other words, each line can have two states. There are six lines, which is why the number of different possible hexagrams is sixty-four:

```
2 * 2 * 2 * 2 * 2 * 2 = 64
```

The important part is that the hexagram on the left relates to your situation at the present time, while the hexagram on the right tells you your future. You always need a pair of hexagrams to figure out where you are and where you're going.

The Display

When I started planning this project, I decided to display the hexagrams electronically with light bars—little rectangular components with LEDs inside. A closeup of a light bar is shown in Figure 28-2, and a rendering showing how the Ching Thing could look is in Figure 28-3.

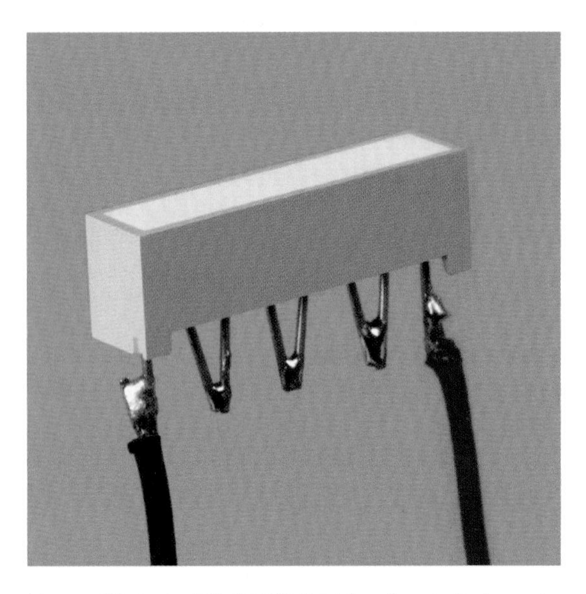

Figure 28-2. *An LTL-2450Y light bar (or similar) can be used to make a neat display.*

Figure 28-3. *A 3D rendering of the Ching Thing, displaying a couple of hexagrams.*

To use the Ching Thing, I'm imagining that you will hold down a button while two hexagrams are generated for you. Then you can look up their interpretation in the I Ching translation of your choice to acquire some insight about your fate.

The challenge is to generate the hexagrams in a way that is an accurate simulation of the traditional method. This required me to do a bit of research.

The Straight and Yarrow Path

In ancient times, you determined the solid and broken lines in a pair of hexagrams by "casting yarrow stalks." Yarrow is a weed, and a yarrow stalk is what you end up with when the weed dries up and dies. Casting the stalks involved a complicated procedure of dividing them into piles and counting them, but the underlying principle was clear: your fortune depended on the way in which the stalks fell by chance.

Casting yarrow stalks was a complicated business, and when the I Ching experienced a surge of popularity in the 1960s, most people lacked the patience to follow the correct procedure. In any case, hardly anyone knew what a yarrow stalk was, and there was nowhere to buy them, because the Web had not been invented. This may seem hard to believe, but back in the 1960s, you couldn't order stuff from Amazon or eBay.

Consequently, people started using an easier system to generate hexagrams, based on tossing coins. Unfortunately, this system created a set of probabilities for hexagrams which was different from the set associated with yarrow stalks.

When I started planning an electronic simulation, I decided that it should be as authentic as possible. I wanted to follow the original, stalk-based probability set—but how could I figure that out? No problem! I looked it up in Wikipedia, which has a pretty good entry on the I Ching.

Now, remember, you must have two hexagrams, the one on the left describing your present state and the one on the right relating to your future.

The I Ching goes into a lot of detail about situations where a broken line in the lefthand hexagram becomes a solid line in the righthand hexagram, or vice versa. These are referred to as "changes," which is why the I Ching text is often called "the book of changes." In fact, I think it was the source of that old 1960s saying, "My life is going through so many changes," which Buddy Miles recorded as a song with Jimi Hendrix—but, I digress.

So, to do this job right, we need to know the odds of a broken line changing into a solid line, or a solid line changing into a broken line, or a solid line staying the same, or a broken line staying the same. That's how the hexagrams are constructed: one pair of lines at a time, six times in succession. For convenience, I'm going to refer to a single pair of lines as a horizontal "slice" that spans the two hexagrams.

The Numbers

Figure 28-4 shows the four ways that solid and broken lines can be paired in a slice. Each of these combinations is not equally probable because of the complicated way in which the yarrow stalks were counted. The probabilities are shown on the right.

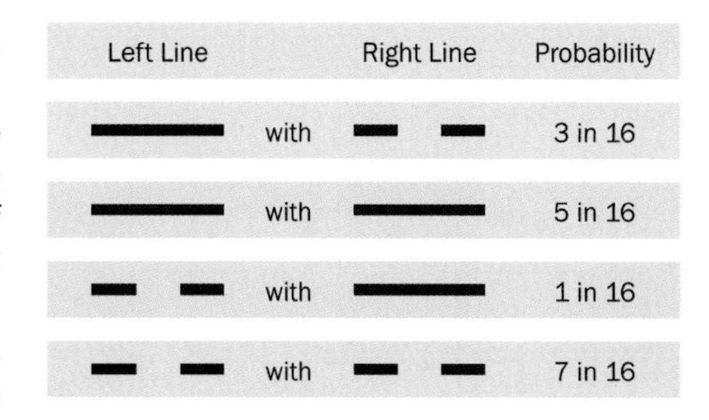

Figure 28-4. *Probabilities of each combination of hexagram lines in a "slice" composed of one line from the left hexagram and one line from the right hexagram.*

How to make this work electronically? Well, the number 16 is very convenient, because (as I'm sure you remember) a decoder has sixteen outputs. Suppose a timer is running very fast, controlling a binary counter, and the counter output is going to the decoder. Now suppose the timer stops arbitrarily so that one of the decoder outputs is selected at random. The decoder outputs can be grouped to create probabilities of 1 out of 16, 8 out of 16, or whatever we want.

Figure 28-5 shows the general idea. The decoder has sixteen outputs, numbered 0 through 15. We could decide that if any of the bottom eight outputs has a high state, the left line in a slice will be solid. That is, the light bars will all be illuminated.

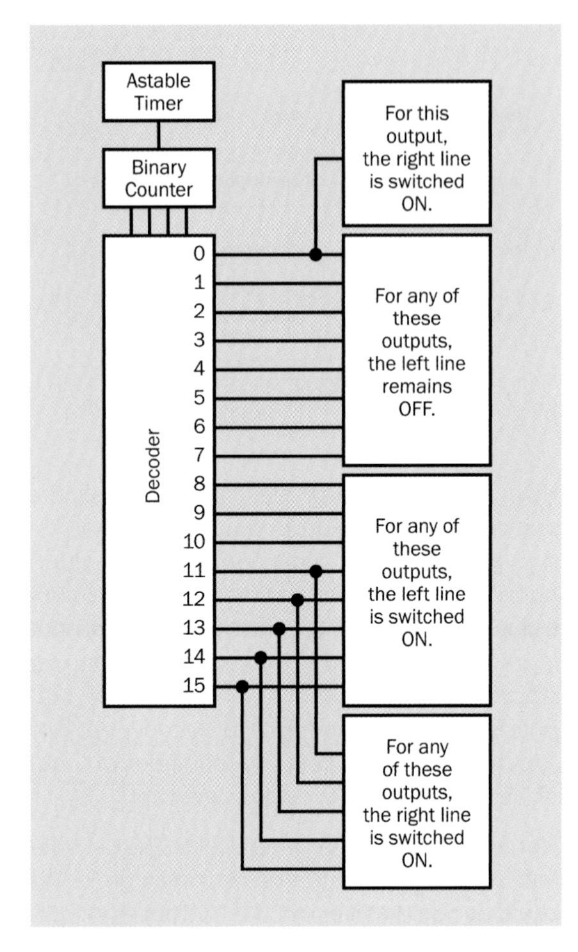

Figure 28-5. *How the outputs from a decoder can be grouped so that a random selection has the correct odds for creating a pair of hexagram lines.*

Looking back at Figure 28-4, you can see that in the 8 times out of 16 when we have a solid line on the left, 3 of those 8 times there will be a broken line on the right, and 5 of those 8 times there will be a solid line on the right. You can see that I have taken care of this in Figure 28-5.

Now, if any of the other eight outputs from the decoder has a high state, the left line will be broken—that is, we'll have two light bars illuminated at either end, but the one in the middle will be off. One time out of those eight, we will have a solid line on the right. This, again, matches the specification in Figure 28-4.

The natural state of the light bars is off, so we only have to worry about switching them on. The rules for doing this can be summarized using the decoder output numbers, like this:

- Rule 1: if output 8 OR 9 OR 10 OR 11 OR 12 OR 13 OR 14 OR 15 has a high state, the left line in a slice is switched on.
- Rule 2: if output 1 OR 11 OR 12 OR 13 OR 14 OR 15 has a high state, the right line in a slice is switched on.

Sounds like we will need a couple of big OR gates. Although—wait a minute. In rule 1 (as Fredrik Jansson pointed out to me), the left line is solid for all numbers in the range 8 through 15 (1000 through 1111 binary), and off the rest of the time (0000 through 0111 binary). What do binary numbers 1000 through 1111 all have in common? They all have a 1 in the leftmost place. So I can rewrite the rule like this:

- Rule 1 (revised): if the 8-value output from the binary counter has a high state, the left line in a slice is switched on.

We will no longer need an OR gate for rule 1.

As for rule 2, it will require an OR gate with six inputs. Does such a thing exist? No, but there is an eight-input gate that has an OR output (it also has a NOR output, which I can ignore). I'll tie two of its inputs to negative ground and use the remaining six.

This is sufficient to create one slice in a pair of hexagrams. There are six slices, so I'll have to go through the procedure six times.

Random Sampling

My method for choosing a random number will be to sample or stop the rapid counter at an arbitrary moment. Okay, but—how?

I want the process to seem automatic. I don't want the user to have to press a button six times. So how about this: I can add a timer running slowly (about one pulse per second) in asynchronous mode. If its speed varies unpredictably, I can use it to sample the fast-running counter six times.

How can I make it unpredictable? I have an idea. If you moisten your finger, the skin resistance between two points that are fairly close together will range from maybe 500K to 2M. I can use this as the resistance that controls the pulse speed of the slow timer.

Now all I need is an automatic system to generate the bottom slice of the two hexagrams, shift it up one space, generate another slice, shift it up, and repeat this process until I have six slices. The phrase "shift it up" strongly suggests to me that I need a shift register.

In fact, I need two shift registers, one to store and display the lines in the left hexagram, and the other to store and display the lines in the right hexagram. I'll call them Register 1 and Register 2. You can see them in Figure 28-6--which also shows a third shift register, but I'll deal with that in a moment.

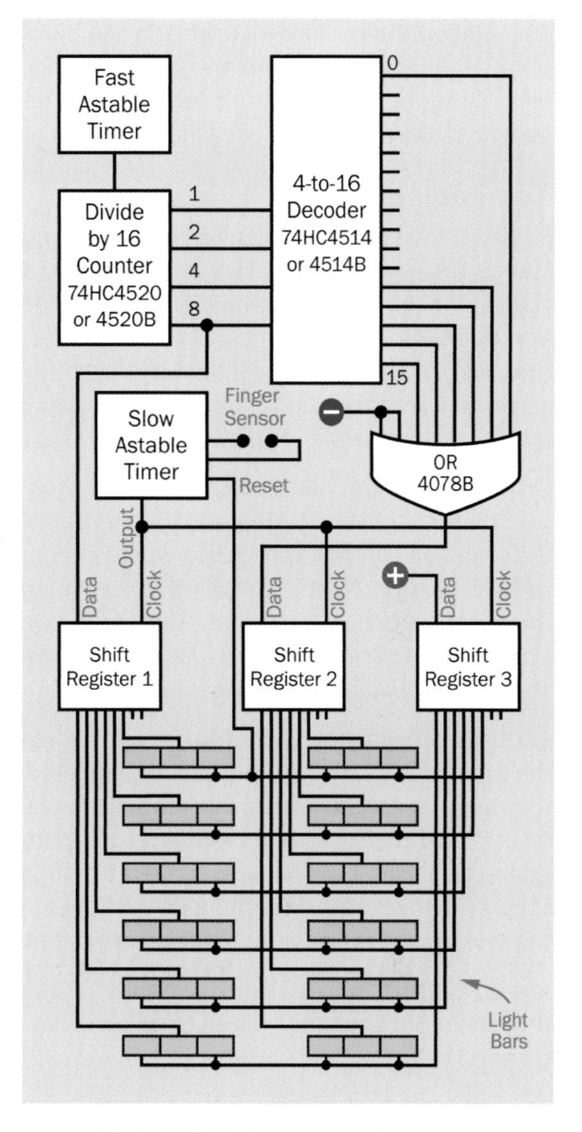

Figure 28-6. *Digital-logic components to generate two hexagrams.*

The counter runs continuously, the decoder runs continuously, and they are hard-wired into the data inputs of the shift registers. (Fortunately, chips don't wear out when we run them fast like this, on a continuous basis.) But as you may recall from Experiment 27, the shift registers don't do anything until they receive a clock signal. This tells them to shift the contents of their memory locations, "clock in" the new data, and display the result.

The slow-running timer will provide the clock pulses. It doesn't matter that the pulses are relatively long, because the shift register only responds to the rising edge of each pulse.

The Look and Feel

So here's the scenario. You switch on the Ching Thing. You moisten your finger and press it against the terminals. Depending how moist the skin is, and how hard you push, the slow timer runs at a variable speed. It samples the fast timer at random intervals, and two hexagrams scroll slowly up the light-bar display.

I especially like this plan because the slow timer won't run at all until you press your finger against the terminals. The resistance between the terminals will be almost infinite, preventing the timing capacitor in the slow timer from charging. So we don't need a "start" button. You plug in the Ching Thing and it waits for your finger press.

Ideally, it should stop itself, too. Perhaps when the topmost slice of the hexagrams is generated, I can take some voltage from there to activate the reset pin of the slow timer so that it ceases to generate pulses. I will have to convert the high state of the topmost slice into a low voltage on the reset pin, but a transistor can take care of that.

In case this isn't entirely clear, I have summarized the "chain of command" in the circuit in a flow diagram in Figure 28-7.

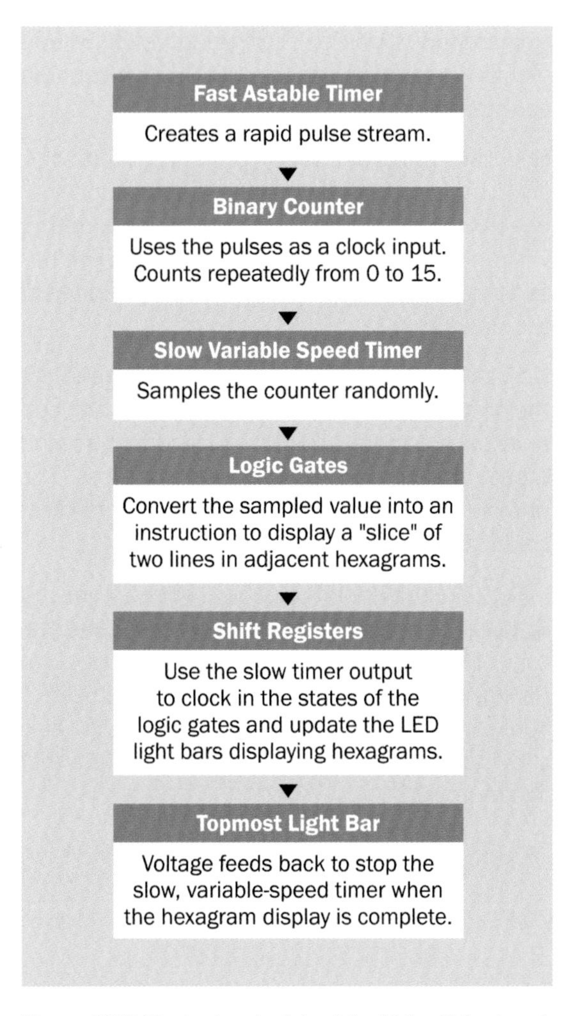

Figure 28-7. *The basic principle of the Ching Thing is outlined in this flow diagram.*

One question remains. Why is there a third shift register in Figure 28-6? Notice that its outputs drive the two outermost light bars in each hexagram. Those light bars will be on all the time, regardless of the state of the center light bar. I could just connect all of the bars permanently to the positive side of the power supply—but the hexagram display will look nicer if each slice is illuminated in succession, with all the light bars coming on and scrolling upward together. So, the third shift register is used just to make the process look good. It has its data input hard-wired to the positive side of the power supply, and it jogs the positive states upward with each clock pulse, in sync with the other shift registers.

The Details

I omitted a few details in this circuit. First, you could add a reset button that would apply a pulse to the "clear" pins of all three shift registers. Since the 74HC164 shift register requires a low input to reset it, the "clear" pins should be held high with pullup resistors, and a pushbutton would connect them momentarily with negative ground.

Second, the 74HC4078 single eight-input OR chip is listed by some suppliers as being obsolete. I still see it advertised online for less than 50 cents apiece, but in the future it will be less readily available. Fortunately the old 4078B CMOS version is still in plentiful supply and can be substituted, as we won't be taking any significant current from its output. You can use either the 74HC4078 or the 4078B in this circuit.

Both of these chips actually contain a NOR gate and an inverter, because an inverted NOR is the same as an OR. The OR output and the NOR output are available from separate pins, as shown in Figure 28-8.

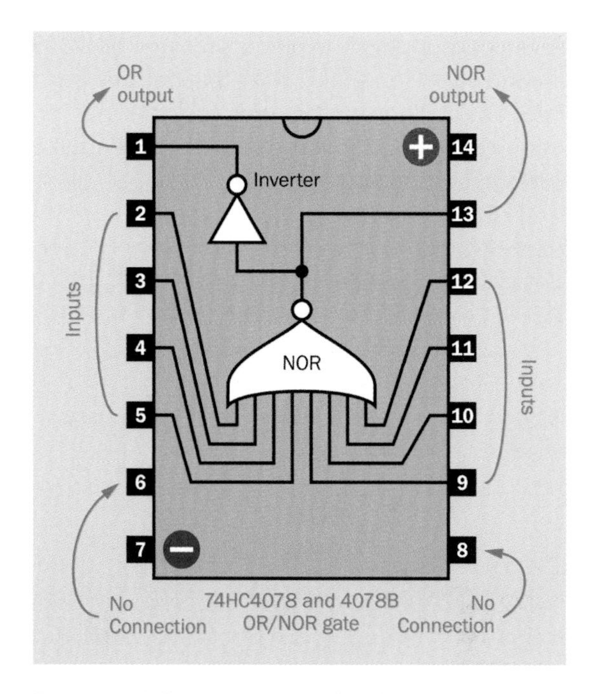

Figure 28-8. *Pinouts of the 74HC4078 and the 4078B, either of which may be used in the Ching Thing circuit. These chips provide a choice of an OR output or a NOR output.*

As for the binary counter, it can be the same 4520B chip that was used in Experiment 21. Its pinouts are shown in Figure 21-6.

Bars or LEDs

The light bars will draw too much current to be illuminated directly by shift-register outputs. The righthand register in Figure 28-6 has to illuminate four light bars in each row and will be powering a total of 24 light bars when the hexagrams are fully lit. At 20mA per light bar, that's a total of almost 500mA.

You could deal with this by using a TPIC6C596 "power logic" shift register, which can drive 100mA from each output when all outputs are high. But its mode of operation is slightly different from that of the 74HC164 shift register, and I don't want to get into those variations. You can check its datasheet if you decide to try using it.

I prefer to drive the light bars with a Darlington array, such as the ULN2003. This contains transistors that can handle up to 500mA on each of seven output pins. The pinouts for the ULN2003 are shown in Figure 28-9. Note that because each internal Darlington pair of transistors has an open-collector output, the chip does not source current, but sinks current.

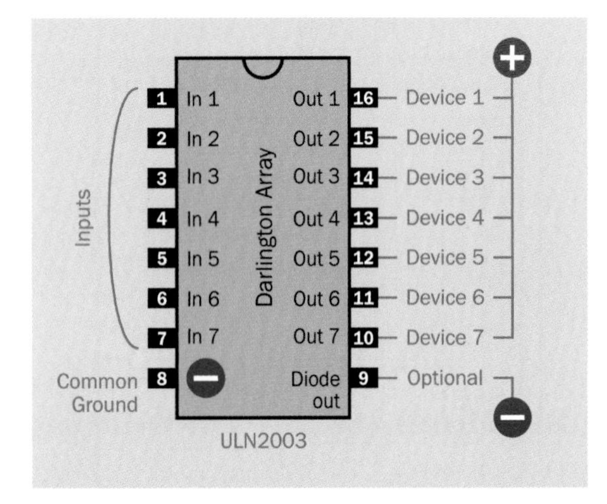

Figure 28-9. *Pinouts of the ULN2003 Darlington array, which contains seven transistor pairs, each capable of sinking up to 500mA.*

A negative ground connection must be made on pin 8 of the ULN2003 to sink the current. The optional ground on pin 9 is needed only where inductive loads may create "back EMF." Diodes included in the ULN2003 are intended to divert such transients if the optional ground is included. For a circuit that is only driving light bars, the optional ground is not required.

You can reduce the power consumption of the light bars slightly, while eliminating the chore of soldering series resistors. Most light bars contain an array of LEDs, with each LED being individually accessible via external leads. Using the Lite-On LTL-2450Y light bar, I found that if I connected the LEDs in series by soldering pairs of leads together, as shown in Figure 28-2, I could run 9 volts through them, and they would take about 16mA, which is less than their rated 20mA. This does require a 9VDC supply. You can run the whole circuit from 9VDC, which will be fine for the Darlington arrays. But remember to pass the voltage through an LM7805 voltage regulator to create 5VDC for the logic chips.

Some sample schematics illustrating three different ways to wire LEDs with a Darlington array are shown in Figure 28-10.

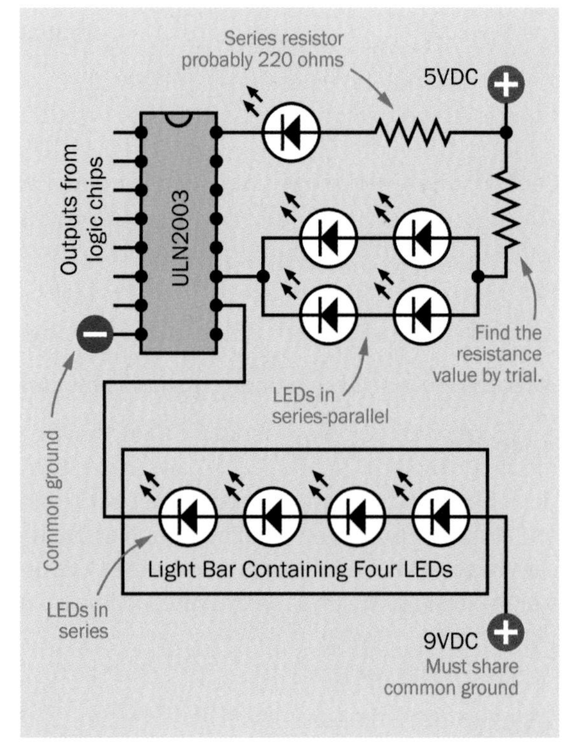

Figure 28-10. *Three different options to wire LEDs with a Darlington array.*

These various options can be boiled down into two alternatives.

1. For a fully finished circuit:
   - Use 74HC164 shift registers, which you have already tested earlier in this experiment.
   - Add a ULN2003 Darlington array to amplify the outputs of each shift register.
   - Use Lite-On LTL-2450Y or similar light bars, with the four internal LEDs connected in series.

- Power the light bars with an unregulated 9VDC supply. The rest of the circuit must have 5VDC, which can be provided by passing the 9V through an LM7805 five-volt regulator.

2. For a demonstration circuit:

    - Substitute low-current individual LEDs instead of light bars. Each LED must draw no more than 8mA.
    - Connect the four outside LEDs in each slice of the hexagrams in series-parallel, which is the middle configuration shown in Figure 28-10.
    - Drive all the LEDs directly from 74HC164 shift registers after verifying that the load on each output from a shift register does not exceed 8mA. (The total current consumed by a whole shift register chip should not exceed 50mA.) No Darlingtons are required.
    - Power the whole circuit with 5VDC regulated.

Either way, note that the circuit uses too much current to be battery powered.

Boarding the Ching Thing

The entire circuit is too large to fit on one breadboard, so I have broken it into part 1 and part 2. The schematic for part 1 is shown in Figure 28-11. This section of the circuit will be the same for both the demonstration and the fully finished versions. I had to offset the chips in the schematic to make everything fit.

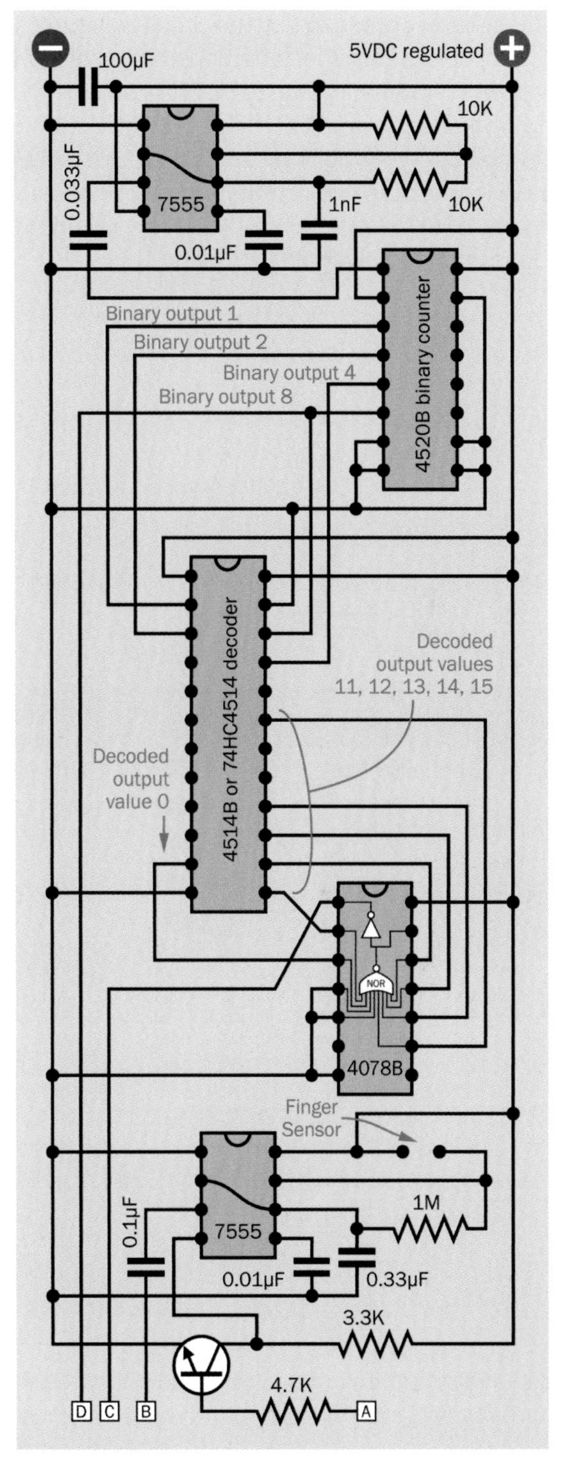

Figure 28-11. *Part 1 of the Ching Thing schematic is the same for either a demonstration version or a fully finished version of the circuit.*

The schematic for part 2 of the circuit is shown in Figure 28-12. This is for a demonstration version where the three shift registers drive LEDs directly. To upgrade this to a fully finished version, Darlington arrays would be added to the shift-register outputs, and light bars would be substituted for LEDs. A separate 9VDC line would be brought in to power the light bars.

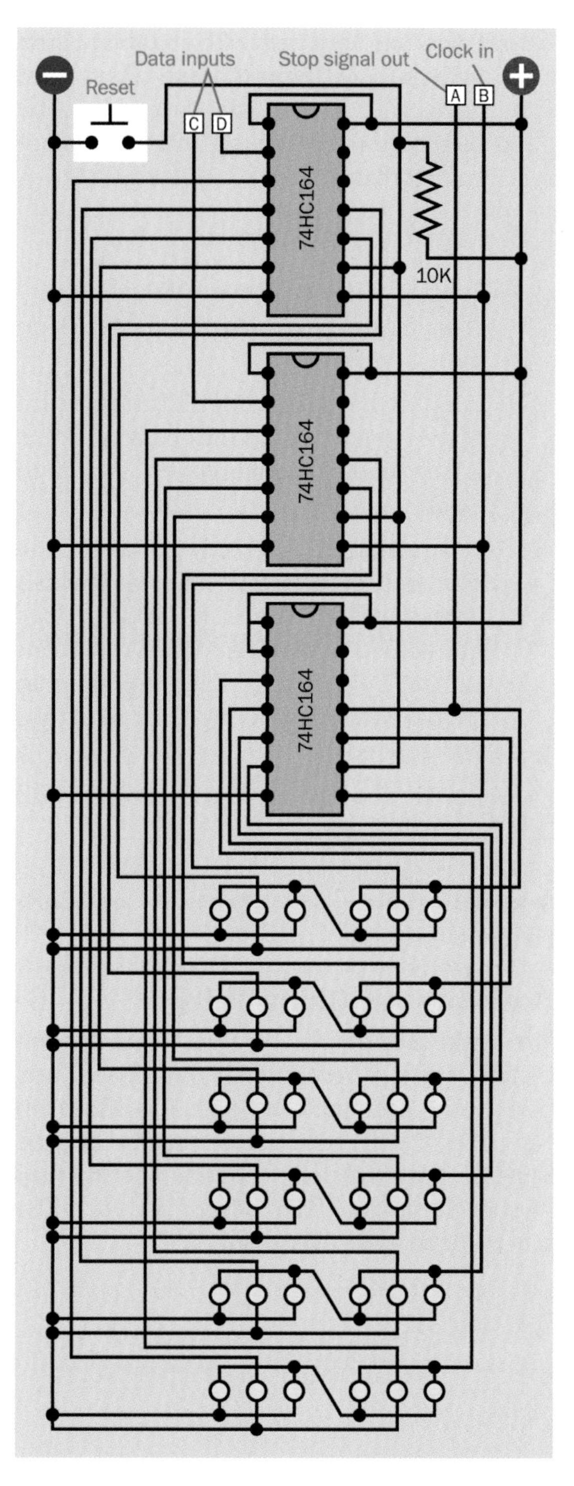

Figure 28-12. *Part 2 of the Ching Thing schematic (demonstration version, with LEDs instead of light bars).*

I regret that space limitations compelled me to draw many conductors very close together in Figure 28-12. If you place a ruler alongside each conductor in turn, you can see how they power the LEDs. The arrangement repeats itself for each row of LEDs.

Also because of space limitations, I used yellow circles to represent LEDs, and omitted series resistors that may be required.

- All LEDs have their positive (anode) leads at the top and the negative (cathode) leads at the bottom.

The always-on LEDs at the left and right side of each hexagram are wired in series-parallel, and if they don't contain their own resistors, they will need a series resistor of a value that is different from the value of the resistors you use with individual LEDs. In both cases, you must try a variety of series resistors while checking the current that they are passing. Each LED that is powered singly, and each set of four LEDs wired in series-parallel, must not take more than 8mA.

Figure 28-13 shows how LEDs that require series resistors can be wired on a breadboard to occupy minimal space. The six LEDs here represent one slice of a pair of hexagrams. The LEDs are numbered to indicate which of the shift registers drives them. The thick gray lines represent the conductors inside the breadboard.

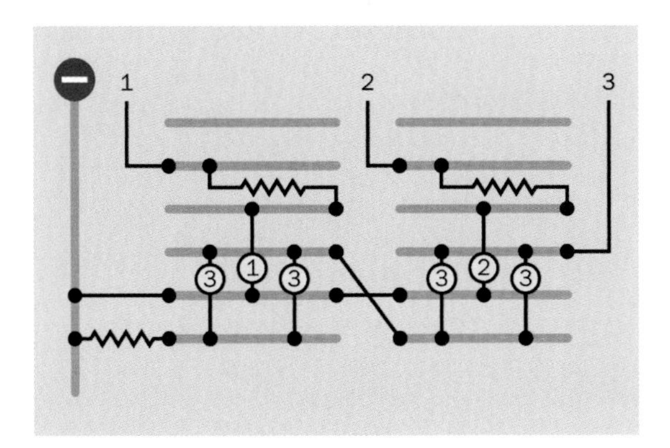

Figure 28-13. *How to wire two LEDs that are driven individually by shift registers 1 and 2, and four LEDs wired in series-parallel, driven by shift register 3. If your LEDs do not contain their own resistors, adjust the series resistors so that each draws no more than 8mA.*

This layout requires only five rows of holes so that six sets of LEDs will occupy thirty rows on the breadboard. This allows sufficient room for the shift registers in the upper half of the board.

The complete demonstration circuit is shown in Figure 28-14 and Figure 28-15.

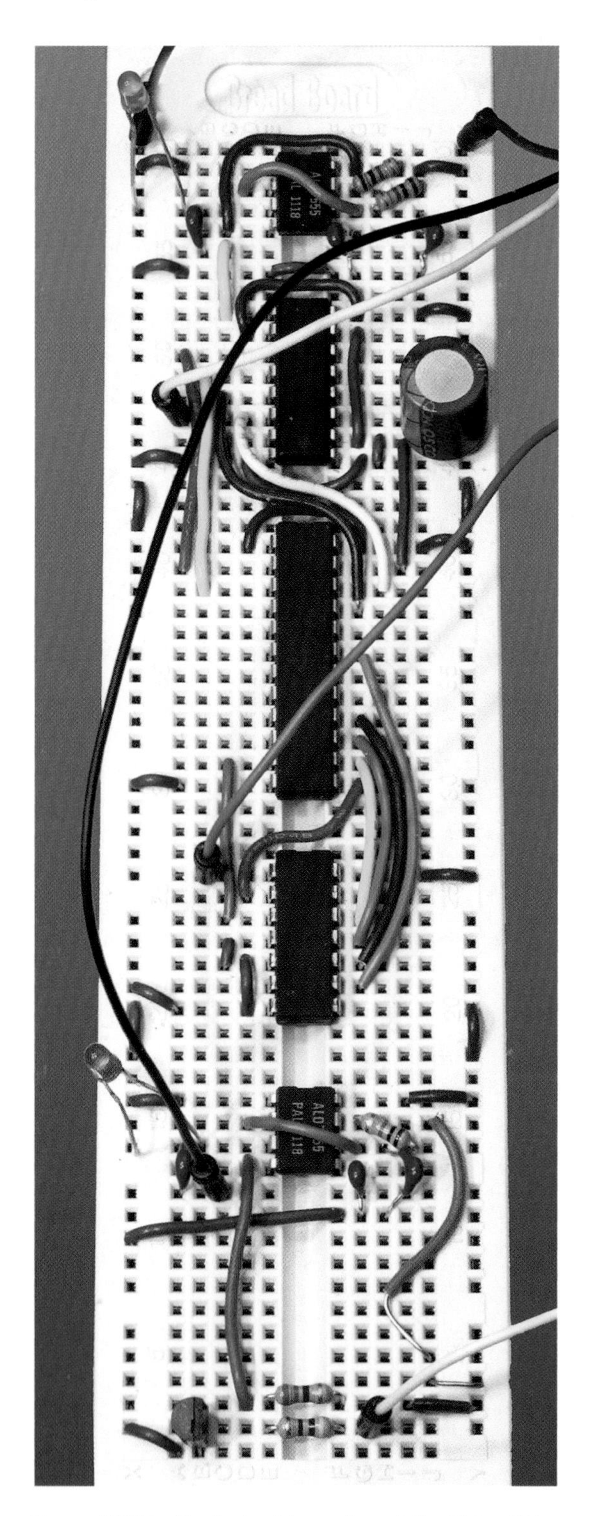

Figure 28-14. *The logic-chip section of the Ching Thing circuit.*

Figure 28-15. *The second section of the Ching Thing circuit.*

Assembly and Testing

Because this is a relatively large project, you should verify sections of it as you build them. When I was building my proof-of-concept version, I followed this sequence of steps, beginning with the LEDs and moving backward:

1. On the second breadboard, place all the jumpers and resistors associated with the LEDs. Then install the LEDs and apply voltage to each of them to make sure there are no bad connections.
2. Install the three shift registers and wire them to the LEDs. Trigger the shift registers manually using inputs C and D (in the schematic) to load them and input B to clock them. You'll have difficulty getting a clean clock signal into input B. Try grounding it through a 10K resistor and then touching a positive wire to it very, very briefly.
3. Set aside the second breadboard. Install the fast 7555 timer at the top of the first board. Use a 33μF capacitor instead of the 1nF capacitor to slow the timer for testing purposes. Leave this capacitor in place until step 10. Add an LED to the timer's output pin.
4. Add the binary counter. Test it with LEDs on its output pins, using high-value series resistors so that you don't try to draw more than 2mA maximum.
5. Add the decoder and use LEDs with high-value series resistors to check its outputs.
6. Add the OR/NOR gate and use high-value series resistors to check its output. The OR output should be high for decoder output values 1011, 1100, 1101, 1110, 1111, and 0000 binary.
7. Test the slow-running 7555 timer with an LED. The LED should come on as soon as you apply power. This is okay. Moisten your finger and press it against the sensor contacts (which can be just a pair of stripped wires). The contacts should be no more than 0.1" apart. After a second or two, the LED should blink off and then on again.
8. Add the transistor and connections between the two breadboards.
9. Remove all LEDs from the counter, decoder, and OR/NOR logic chips. This is important! If these chips are driving LEDs at high speed, they will not communicate properly with each other.
10. Substitute a 0.001μF capacitor for the 33μF capacitor on the fast 7555 timer.
11. Don't forget to link the two positive buses and two negative buses of your two breadboards so that both boards will be powered when you are ready for testing. In your eagerness to see if it works, be careful not to connect the power the wrong way around!

Ching Usage

The 100μF capacitor (shown at top-left in Figure 28-11) should suppress transient voltage spikes that occur when you first apply power. If the capacitor is doing its job, all the LEDs should remain dark. If some light up, use the reset button on the second breadboard.

Press your finger against the sensor contacts. For a faster response, moisten it first. Be patient; it may take a second or two to charge the capacitor. The output from the timer will go low, and then high again, at which point the first slice of the display should be displayed. This process should repeat in six steps, and then stop. If the display continues to scroll upward, use your meter to check the voltage on the reset pin of the slow timer. It should be above 4.5VDC while the display is being generated, and should fall below 0.5VDC when the display is complete.

If you disconnect and then reconnect power, probably there will be enough voltage left on the 100μF capacitor to regenerate the display. You will need to let the circuit sit without power for a minute or two to allow this capacitor to discharge.

Long jumpers are unavoidable to connect the two boards. Jumpers with plugs at each end may not make good connections. If your circuit behaves erratically, the jumpers will be the first thing to check.

Packaging

Figure 28-3 shows how the finished Ching Thing could look while displaying a couple of hexagrams, and Figure 28-16 shows the cutouts that you would need for the top panel of the box.

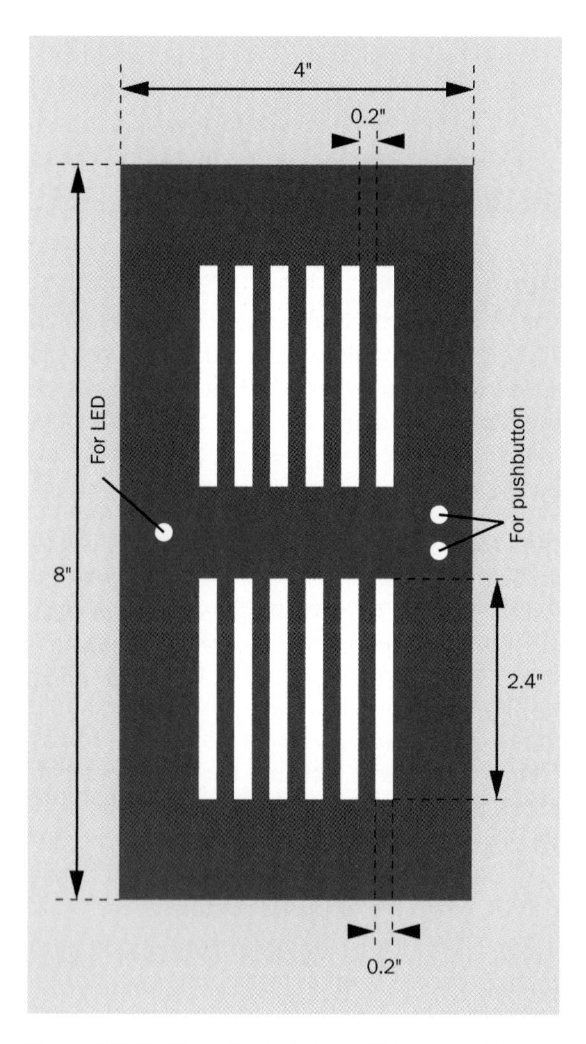

Figure 28-16. *The cutouts that would be needed for the top of a box that could be used for the Ching Thing.*

This project is not quite as ambitious as it looks because the chip connections are relatively simple. Most of the wiring relates to the hexagram displays. Of course, this becomes a little more complicated if you use Darlington arrays. There is also the issue of cost: if you use light bars, they are likely to be about $40 for the whole project.

Still, so far as I know the Ching Thing is unique as an electronic version of the world's oldest system for predicting the future.

Can it really work comparably to the yarrow-stalk version? There's certainly something compelling about the process of casting the stalks and drawing hexagrams by hand. But if you believe that fate is controlling the positions of the stalks, it seems to me that fate could also control the behavior of electrons inside silicon chips.

This leads me to my conclusion: may your fortunes all be electrically positive!

Experiment 29: Common Sensors

29

In this and the next five experiments, I'm going to be discussing sensors. This is an exciting field, because it's still developing rapidly. While basic designs and capabilities of the logic chips that I've been using were established long ago, the evolution of affordable sensors is proceeding all around us.

The key word, here, is "affordable." To take one example, in the year 2000, perhaps I could have bought an accelerometer, if I had shopped around—although I might have had difficulty finding a distributor that would sell me just one, and it would have been expensive. Moreover, after I bought it, I could have had some problems figuring out exactly how to use it.

Today, in the United States, I can buy a three-axis accelerometer from Hong Kong for $3, with free international shipping, and it will plug right in to an Arduino microcontroller.

The evolution of handheld devices has encouraged the mass production of sensors that are small, reliable, easy to use, and cheap. A modern smart phone may contain as many as ten different sensors, including a microphone, touch screen, wireless antenna, GPS, ambient light sensor (to adjust screen brightness), accelerometer (so that it knows which way up you are holding it), thermometer, barometer, hygrometer—and a proximity sensor so that when you raise the phone to your ear, it senses the nearness of your head and turns off its display to ignore touch inputs and save power.

The field of sensors is now so extensive, I don't have space to describe more than a few. Use the search term "sensor" on sites such as *http://www.jameco.com* and *http://www.sparkfun.com*, and you'll be surprised by what you find.

The Little Magnetic Switch

Perhaps the oldest sensor of all is the humble reed switch. I mentioned it briefly in *Make: Electronics,* but only regarding its application in alarm systems, where it is packaged in a small white plastic module to sense if a window or a door has been opened. I'll go into more detail here.

Let's get acquainted with some samples of this helpful little device. Two reed switches are shown in Figure 29-1.

Figure 29-1. *Reed switches. Note that each square in the graph-paper background is 0.1" x 0.1".*

The glass capsules contain an inert gas to protect the contacts from oxidation. This is a nice feature, but the glass is so thin and fragile, it can break if you merely bend the leads too sharply. You really have to handle reed switches with care. If you need a more rugged package, some switches are sealed in plastic, as shown in Figure 29-2.

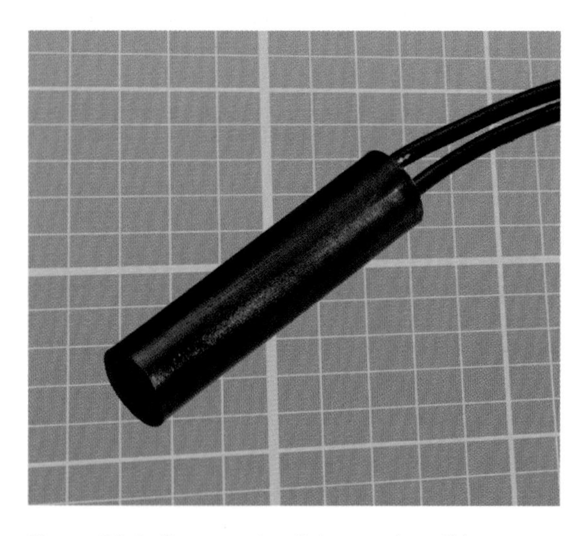

Figure 29-2. *Some reed switches, such as this one, are encapsulated in plastic to provide some protection.*

Reed Test

The following experiment is going to be the simplest in the book—even simpler than the glue-operated transistor in Experiment 1.

While there is no standardized schematic symbol for a reed switch, it is often depicted as in Figure 29-3. Connect it as shown, and bring a small bar magnet near it. The magenta and cyan color bands indicate the poles of the magnet.

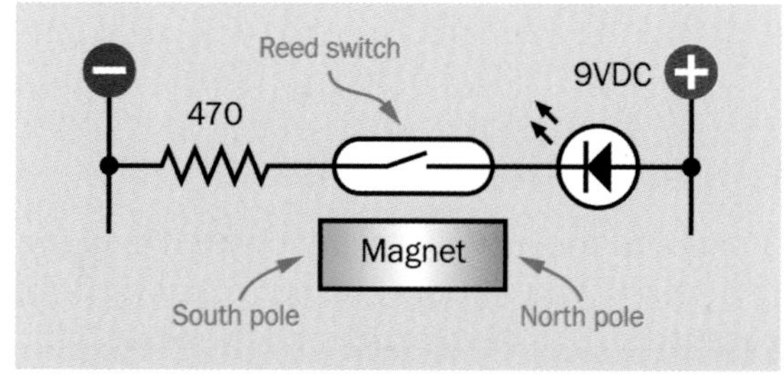

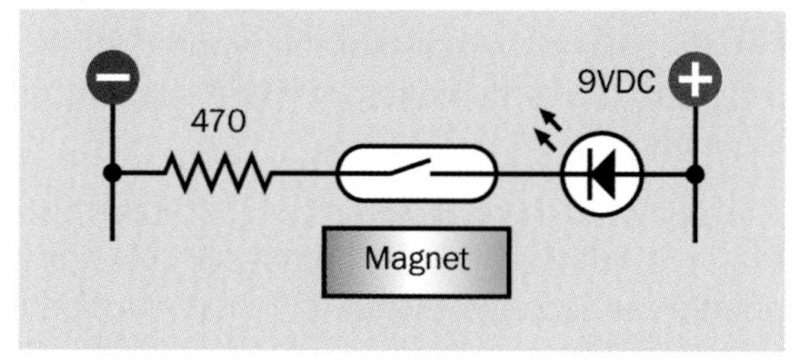

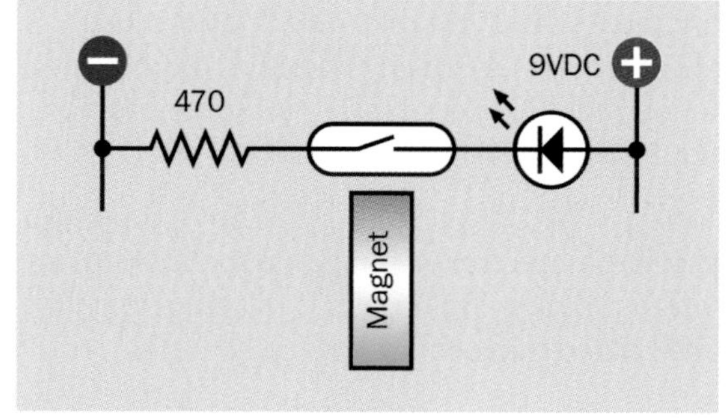

Figure 29-3. *Testing a reed switch.*

You should find that the switch will be about equally sensitive in the top and middle versions of this figure. It will be least sensitive when one pole of the magnet is much closer to it than the other, as in the bottom section. This is important information if you want a reed switch to give a reliable response.

- Even though it's called a reed "switch," it behaves like a pushbutton. It only stays on while it is exposed to a magnetic field. When the magnet moves away, it turns off.

How It Works

The contacts inside the switch are mounted on flexible metal strips (the "reeds"), which are not magnetized but are magnetically conductive. Bear in mind that a magnet creates a field that induces magnetism in other objects. In fact, when a magnet pulls a piece of iron toward it, the nearest pole of the magnet creates a temporary, opposite pole in the iron.

If a magnet is aligned parallel with the switch, the reeds become temporarily magnetized with opposing polarities. This creates a force of mutual attraction, causing the reeds to bend toward each other and create a connection. When the magnet is withdrawn, the reeds spring apart.

The reason the switch is almost equally sensitive with the magnet oriented in the top and center sections of Figure 29-3 is that the reeds become magnetized relative to each other in each case. Even though the field is reversed, they still attract each other.

Level Sensor

A few years ago, when I was building a prototype of a rapid-cooling device for a research lab, I needed a liquid-level sensor. The most basic type is an on-or-off device called a float switch that detects when liquid has reached a maximum or minimum level. A cylindrical float made of closed-cell plastic foam rests in the surface of the liquid. The float has a hole in the middle, like a donut, allowing it to slide freely up and down a plastic rod containing a reed switch. The switch is closed by a magnet mounted inside the float. Figure 29-4 shows a low-cost sensor of this type.

Figure 29-4. *A simple, low-cost level sensor in which a magnet inside the float activates a reed switch in the central plastic rod.*

Perhaps you can think of some applications. For instance, despite many decades of refinement, the shutoff valves inside toilet tanks are still not entirely reliable. Where I live, in the high-desert wilderness, water is a valuable commodity, and a malfunctioning toilet tank is not a trivial matter. Maybe I should mount a float switch inside the toilet tank, with wires leading out to an LED and a 9V battery. Then if the shutoff valve starts to fail and the water level rises up to the overflow, the LED will warn me. Alternatively, I could use a small beeper.

Here's another possibility. Some people live in houses where the basement is vulnerable to flooding. That's another situation where you could locate a float switch. What other uses could you imagine?

Fuel Gauge

In some applications it's useful to have a level sensor that varies its output in proportion with the volume of liquid, instead of just giving an on-or-off output. This would be commonly used as a fuel gauge.

The traditional way of doing this is with a float mounted on an arm, which rotates the wiper of a potentiometer. For decades, cars used this system in their gas tanks—but it's bulky, not very accurate, and vulnerable to dirt and moisture if you use it outside of a sealed environment.

While looking for a better way to measure a liquid level, I ran across a fuel-level sensor on eBay that consisted of just an eight-inch metal rod with a float that would slide up and down it. I couldn't figure out how it worked, so I ordered one. When I received it, I sawed the end off the rod and discovered that it was actually a hollow tube. Inside was a very narrow circuit board, about a quarter inch wide. Mounted at intervals along the board were seven reed switches and six resistors.

At first, I was puzzled. How could the reed switches be magnetically triggered, when they were enclosed in a steel tube? Then I realized that the tube was made of stainless steel, which is nonmagnetic. The field from a magnet mounted in the movable float had no trouble penetrating to the switches.

As the float moved up and down, its field closed one switch after another, and since the resistors were wired in series, they functioned as a voltage divider tapped at six locations. This enabled a terminal at the top of the assembly to detect a varying total resistance. A simplified version of this fuel-level sensor is shown in Figure 29-5, with just four resistors and five reed switches.

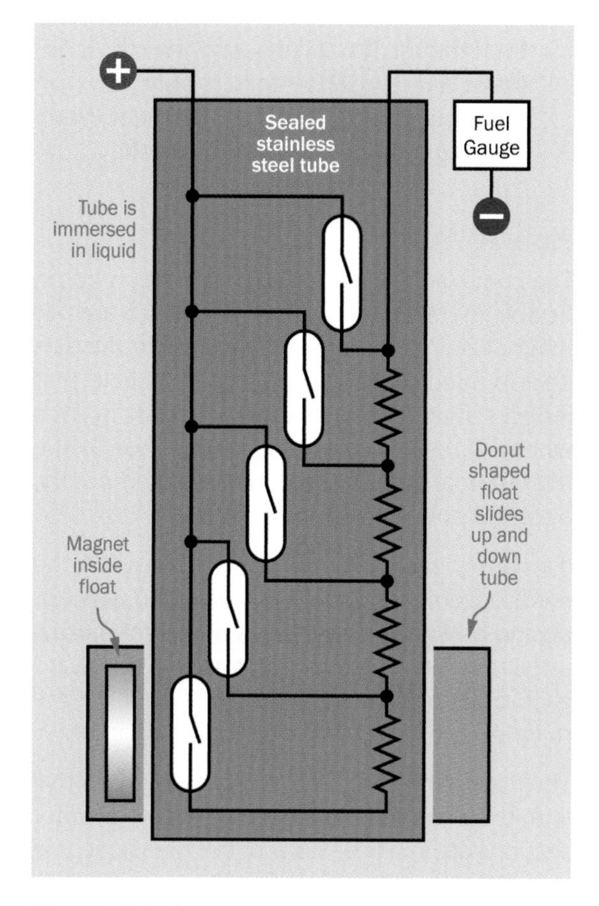

Figure 29-5. *A liquid-level sensor using reed switches and resistors that form a multistage voltage divider.*

Of course, the resistance in this type of sensor does not vary gradually, but apparently seven steps are considered acceptable for a fuel gauge in a car.

I mention this to give you an idea of the versatile way in which a sensor as simple as a reed switch can be used. Incidentally, I believe that many automotive fuel sensors still rely on a similar system, except that it now uses Hall-effect sensors instead of reed switches. I'm going to get to Hall sensors in Experiment 30.

Quick Facts About Reed Switches

- Most reed switches are SPST with contacts that are normally open. A few have normally-closed contacts. Either way, a reed switch behaves like a pushbutton, with a magnet providing the "push."

- SPDT and DPDT reed switches are available. Figure 29-6 shows a single-pole, double-throw reed switch. The pole of the switch is connected to the single wire at one end. The longer of the two wires at the other end is connected with the contact that is normally closed.

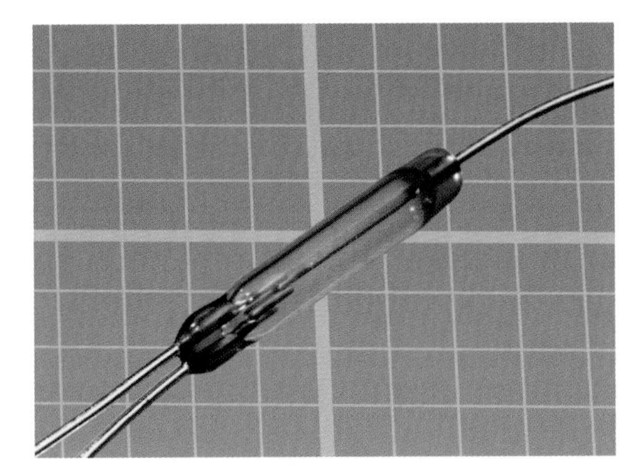

Figure 29-6. *A single-pole, double-throw reed switch.*

- A reed switch works regardless of the polarity of the magnetic field, because the field induces opposing magnetic poles in the reeds.
- Some types of very small relay contain a reed switch with a coil wrapped around it. When the coil is energized, it functions as an electromagnet and closes the switch.
- Reed switches naturally display some hysteresis. The force required to begin closing the contacts is greater than the force required to keep the contacts closed. Therefore the magnet that closes the switch can be withdrawn slightly before the switch will reopen.

Reed switches have obvious limitations:

- A reed switch is typically enclosed in a thin-walled glass capsule, which cracks or shatters easily.
- The tiny contacts will quickly deteriorate if you try to switch too much current.
- Contacts may break if they are subjected to excessive vibration. Always try to avoid dropping a reed switch, especially onto a concrete floor.
- Although reed switches are designed to be durable, they can never be quite as reliable as solid-state switches (no matter what the manufacturers may claim).
- The switch can be accidentally triggered by stray magnetic fields.
- It may fail to be triggered if the magnet is not oriented correctly.
- When the contacts of a reed switch close, the impact causes "reed vibration," also known as contact bounce. This is a greater problem in larger switches where contacts have greater momentum. Contact bounce can be misinterpreted by digital components as multiple closures of the switch.

However, reed switches have some advantages compared with solid-state sensors:

- No external power supply needed.
- No power consumption, either when the switch is opened or when it is closed.
- No interfacing components, amplifier, or similar circuitry needed.
- Negligible current leakage when the contacts are open.
- Negligible resistance when the contacts are closed.
- Higher-voltage versions are available.
- Higher-amperage versions are available.
- Can switch AC or DC.
- Less vulnerable to electrostatic discharge.
- Not significantly affected by ambient temperature.

Easy Substitution

Generally, you can use a reed switch anywhere in a circuit where you have used a SPST switch. For instance, reed switches are commonly substituted for snap-action limit switches to stop a motor when it reaches the limit of its rotation.

You could substitute sixteen reed switches for the sixteen pairs of coin contacts that I described for the Hot Slot game in Experiment 21. The rest of the circuit wouldn't have to be modified at all. The only problem, of course, is that you wouldn't be able to play that game with coins anymore. You would need sixteen disc-shaped magnets.

I think there's a better way to add sensors to that game, allowing the continuing use of coins. I'll get to it in Experiment 31, when I deal with optical sensors (see Chapter 31).

Installing a Reed Switch

Reed switches are available with differing sensitivity. The samples I have used can be activated by a tiny neodymium magnet measuring about 1/8" × 1/4" × 1/16". So long as the magnetic axis is parallel with the switch, the switch will work when the magnet is still half an inch away.

Miniature magnets and small reed switches provide a lot of flexibility in the way that you use them. For instance, a switch can be glued in place, and the magnet can be hidden behind a thin layer of plastic.

For most of the projects in this book, you could build a small box that seems to switch itself on when you open the lid. Note that the switch would not draw any power, either while the lid was open or closed.

Reed switches can also be used to provide security in a car or home. You could mount a small magnet on your key ring (so long as it remains safely separate from your credit cards) and use it to activate a hidden reed switch connected with a latching relay, to switch off an alarm system before you enter.

Another possibility is to mount a magnet in a small pointer or stylus, as the input device in a game.

Background: Magnetic Polarity

I'm going to describe some aspects of magnets, because magnets are required to trigger switches and Hall-effect sensors (coming up shortly).

A permanent magnet always has two poles, which we refer to traditionally as "north" and "south." Perhaps you are thinking that these must be similar to the north and south poles of planet Earth—in which case, let me just mention that the north pole of a magnet was originally known as the "north-seeking" pole.

Didn't you learn, somewhere or other, that opposite poles attract? Indeed, if you have two magnets, the north pole of one will be attracted to the south pole of the other. So how can the north-seeking pole of magnet point toward the North Pole of the Earth? The answer is that the Earth's so-called north pole actually has south-pole polarity.

It would be much too confusing to rename the North Pole as the South Pole at this time, so we're stuck with a North Pole that isn't really a north pole at all.

Magnetic Types and Sources

Neodymium magnets were developed in the 1980s and are far more powerful than the old iron magnets that preceded them. Inside any small, modern DC motor, most likely you'll find neodymium. In fact, neodymium has enabled the miniaturization of devices from cameras to lightweight power tools.

It's fun to play with powerful magnets, but most reed switches are so sensitive, cheaper iron magnets are sufficient. You can find many sources online. There are always some magnets for sale in assortments on eBay.

Bear in mind that it is actually disadvantageous to use a magnet that is more powerful than necessary, as it can affect other adjacent switches or components.

The most familiar style of magnet is perhaps the bar type, with a square or rectangular cross-section. Usually, but not always, the poles are at oppposite ends of the bar.

Some suppliers (including my favorite source, K&J Magnetics) list the magnetized dimension last. Thus, if a magnet is listed as measuring 1/4" x 3/4" x 1", the poles are probably at each end of the 1" dimension. But if the magnet is listed as measuring 3/4" x 1" x 1/4", each flat face of the magnet can have a polarity opposite to the other. Check carefully before ordering!

Magnetic Shapes

A classic horse-shoe magnet is a bar magnet that has been bent into a U-shape, like a horse shoe, so that the two poles are located side by side. This increases the lifting power of the magnet, because the power is greatest where the lines of force between the poles are shortest.

Some sample magnets are shown in Figure 29-7. The dull gray one is iron, while the others are neodymium. All the magnets are holding onto each other, because there was no easy way to keep them apart for the photograph. The little disc-shaped magnet is in two pieces because it broke when it was grabbed by a larger magnet. Bear this in mind: neodymium magnets are brittle.

Figure 29-7. *A selection of magnets (one of which broke when it was grabbed by another). The dull gray magnet is iron; the others are neodymium.*

Discs, cylinders, and ring-shaped magnets are likely to be "axially magnetized." The axis of a cylindrical magnet is an imaginary line running through the center, as shown in Figure 29-8. You can think of the cylinder as rotating symmetrically around that line. If it is axially magnetized, the opposite ends of its axis have opposite polarity. One flat face of the cylinder contains the north pole, while the other has the south pole—as in the rendering in Figure 29-9, where red and blue indicate opposite poles.

Figure 29-8. *The imaginary line running through the center of the cylinder is its axis.*

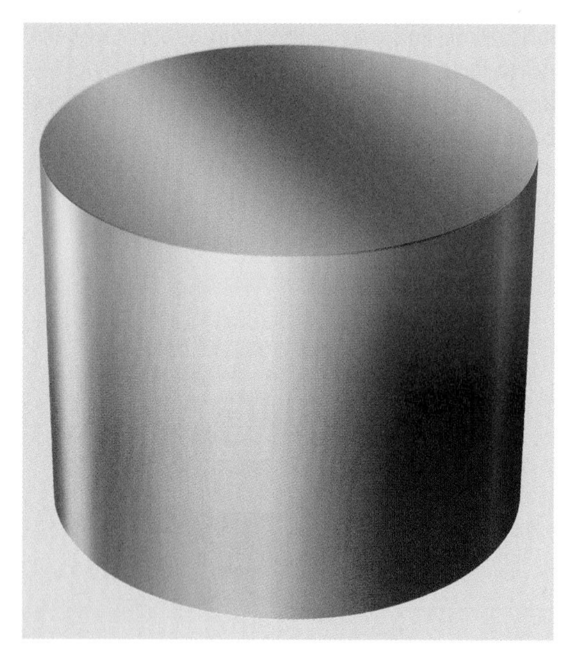

Figure 29-9. *An axially magnetized cylindrical magnet has its poles at opposite ends of its axis.*

Most round magnets are axially magnetized, including ring magnets. Figure 29-10 shows a classic demonstration in which alternate ring magnets are turned over so that north poles are opposite north poles, and south poles are opposite south poles, creating opposing forces. The magnets in this photograph are not stuck to the central rod in any way; they can slide freely.

Figure 29-10. *Most ring magnets are axially magnetized. You can stack them on a nonmagnetic rod (stainless steel, in this photograph) with their similar poles facing each other, so that the rings force each other apart.*

The cumulative weight of the rings pushes them closer together at the bottom end of the rod. If you exert additional force with your finger, you can push the rings together, but they jump apart as soon as you let go. In fact, in this little demo, the topmost magnet will jump right off the top of the rod.

I never get tired of this elementary demonstration of magnetic force. Where does the energy come from to make the magnets behave this way? The answer, of course, is that it comes from you, when you push them together. Magnets do not create energy; they merely store it.

Some circular magnets are radially magnetized, but they are a small minority. Figure 29-11 suggests this concept. One curved side of the cylinder has a polarity opposite to the other side.

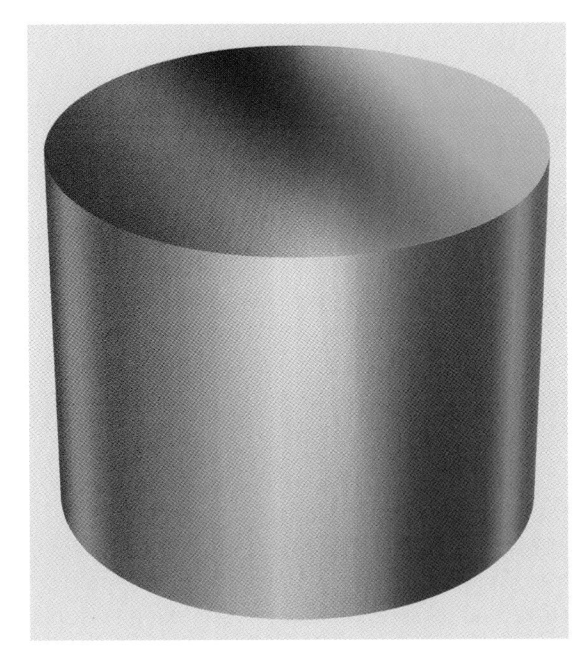

Figure 29-11. *A radially magnetized cylinder has its poles on its opposite curved sides.*

To determine the polarity of a magnet, it helps to have two of them. Observe their orientation as they attract or repel each other. If two circular magnets are axially magnetized, the force between them will not change when they are placed face to face and one magnet is rotated around its axis relative to another.

Make Even More: Eddy Currents

One of the most surprising attributes of a magnet is that it can interact with nonmagnetic metals such as aluminum. If you take a neodymium ball magnet and drop it into a vertical aluminum tube that is just slightly larger than the ball, the magnet will drift slowly down as if it is falling through molasses. The thicker the aluminum is, the more slowly the ball will fall. Thus, a tube with walls that are 1/8" thick will be much more effective than one with walls that are 1/16" thick. The same effect can be demonstrated with a copper tube.

Figure 29-12 is of a ball magnet in a tube where a slot has been cut to reveal the ball. It takes about a second to drift down from one end of the 12" tube to the other. This is a graphic demonstration of the interaction between magnets and metals.

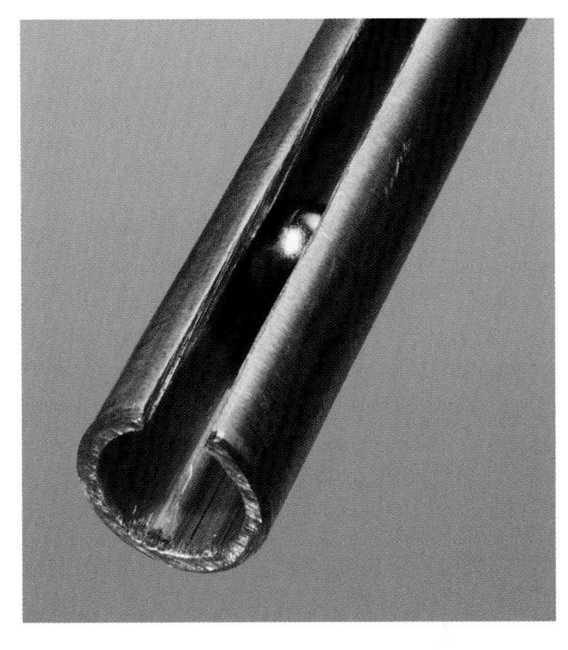

Figure 29-12. *A neodymium ball magnet is powerful enough to generate electrical eddy currents when it moves close to a nonmagnetic but electrically conductive metal, such as aluminum or copper. Generating the eddy currents requires energy, causing the ball to fall very slowly.*

The reason for this strange effect is that a moving magnet induces electric "eddy currents" around it in a nearby electrical conductor, such as aluminum or copper. In fact, this is how most electricity is generated in the world: by moving copper windings through magnetic fields. (Solar cells are the exception to this rule.)

If you're wondering what happens to the eddy currents that are generated, they create a tiny amount of heat—as is always the case when electricity flows through a conductor. In this way, the experiment demonstrates the principle of conservation of energy.

The demo could be enhanced by mounting a series of reed switches at the back side of the tube, in conjunction with resistors in series, as in the gas-tank gauge that I mentioned previously. If the resistors are put in a circuit between the discharge and threshold pins of a 555 timer running in astable mode at an audible frequency, the falling ball will create an ascending series of tones as it passes the switches.

This demonstration will also work if you use aluminum with an angle-shaped or channel-shaped cross section, although the ball won't fall quite so slowly.

Warning: Magnetic Hazards

I would be irresponsible if I didn't include a note of caution. You may find it hard to believe that you could be injured by a magnet, but neodymium magnets are quite capable of hurting you, as I have learned myself the hard way.

An N52-grade cylindrical neodymium magnet measuring only 3/4" in diameter and 3/4" tall can be rated to lift a weight of around 40 lbs. That's the weight of five gallons of water. If you have two of these magnets with opposing poles facing each other, the force between them will be doubled, and if they slam together with one of your fingers in the way, at the very least, you're going to end up with a blood blister.

You will run a different kind of risk when you try to separate the magnets afterward. Expect some torn fingernails and a lot of exasperation. You'll find some good YouTube videos online, showing how to deal with powerful magnets.

Neodymium is brittle, and although the magnets are nickel plated, they can still chip unexpectedly as a result of hard impacts caused by magnetic attraction. The metal chips are very sharp and can fly at high speed. You should wear eye protection if you start playing with heavy-duty magnets.

It should be obvious (but I'll mention it anyway) that you must keep strong magnets away from hard drives and other magnetic storage media, including credit cards. In fact, magnets should be stored at a safe distance from all electronic devices.

Lastly, a pacemaker can be affected by a strong magnetic field. Use powerful magnets with appropriate caution.

Experiment 30: Hidden Detectors

30

Hall-effect sensors are all around you. When you close the lid of your laptop, probably a Hall sensor under the plastic skin of the case detects the action and puts the computer into sleep mode. When you switch on your pocket camera, a Hall sensor detects that the lens is fully extended. Hall sensors are inside your hard drive, detecting the rotation of its motor and controlling its speed. They're in the electronic ignition of your car—and in the car's door lock, which switches on the interior light when your key turns. Your modern washing machine may use a Hall sensor to determine if the door is closed. Similarly, your microwave oven.

Each sensor generates a tiny electric current in response to a magnetic field. It can be used like a reed switch, but is totally solid state. The principle of operation was discovered in 1879, but the effect was so small, widespread application was impractical until the development of integrated circuit chips that could contain amplifiers in addition to the sensors.

The great advantages of Hall sensors are cheapness, reliability, and miniaturization—unlike reed switches, they can be tiny surface-mount devices. They have a very fast response time, they're difficult to damage, and they are made in four types that have differing behavior to match your particular application. Integrating them into a circuit requires a little extra trouble, but they are a versatile alternative to reed switches.

A selection of Hall-effect sensors is shown in Figure 30-1.

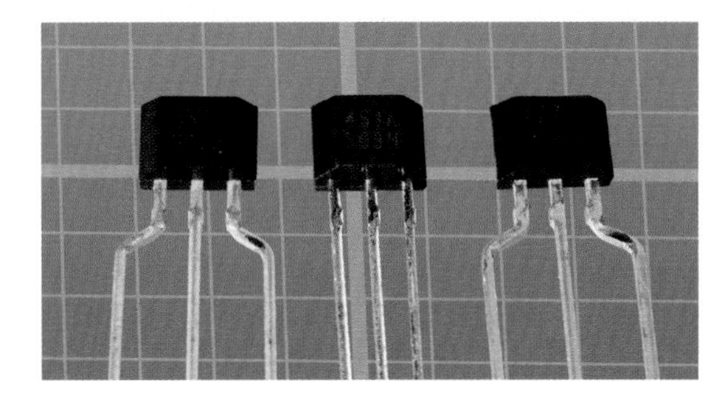

Figure 30-1. *A variety of Hall-effect sensors.*

Hall Test

This test will use the most common and cheapest type of Hall sensor, which is "bipolar," meaning that one magnetic pole will switch it on while the opposite magnetic pole is required to switch it off. (In this usage, the term "bipolar" has nothing to do with bipolar transistors.) A bipolar sensor is also referred to as "latching," since it stays in one state until you force it back to the other state.

Figure 30-2 shows a simplified view of an ATS177 sensor made by Diodes Incorporated. Most Hall sensors have the same pinouts. The code below the part number usually relates to the date of manufacture and is not relevant to us here.

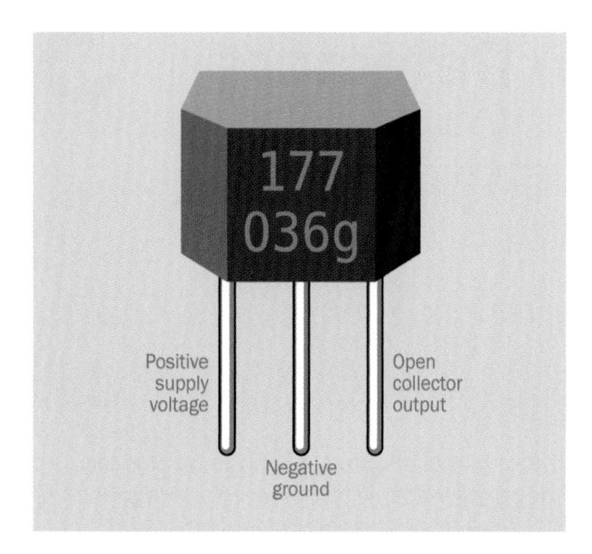

Figure 30-2. *Pin functions for an ATS177 bipolar Hall-effect sensor made by Diodes Incorporated. Other Hall-effect sensors usually have the same pin functions.*

In a datasheet, the positive power-supply pin will be identified as Vcc or Vdd (which mean the same thing, in this case). Negative ground is almost always identified as Gnd. The output is often identified as Out, but in the case of sensors that are intended for connection with logic chips, DO may be used instead, meaning Digital Output.

A sensor is most easily tested with the same kind of rectangular bar magnet that you used to investigate a reed switch in Experiment 28. However, unlike a reed switch, a Hall sensor is designed to be activated when one pole of a magnet is much closer to it than the other pole.

Assemble the schematic shown in Figure 30-3. Note that the sensor has an open-collector output, just like the phototransistor, the electret microphone, and the comparator that you have encountered previously. If you need to remember how this works, see "The Output" on page 36.

I'm suggesting a 9VDC power supply so that you can use a 9V battery. The supply doesn't need to be regulated, because this sensor is not a digital device. Most Hall sensors can use up to 20VDC without any trouble, but a few are designed for lower voltages. Always check the datasheet to be sure.

Apply the magnet as shown in Figure 30-4. Because the poles of a bar magnet are not usually marked, you'll have to use trial and error to find which end of the magnet switches the LED on. Once the LED is on, use the opposite end of the magnet to switch it off.

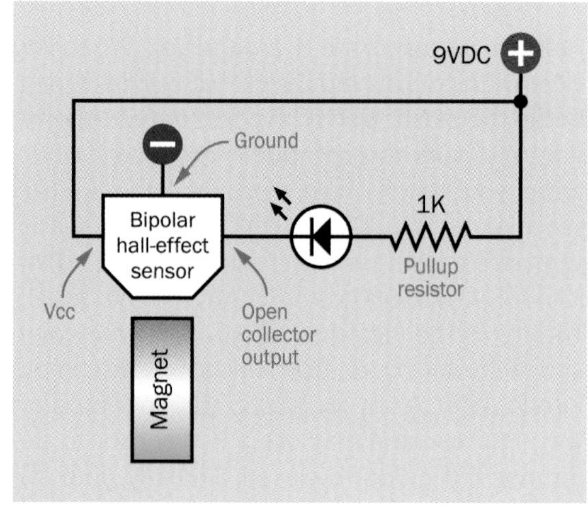

Figure 30-3. *A very simple schematic for investigating Hall-effect sensors.*

Figure 30-4. *Apply one pole of a magnet to the bevelled face of the bipolar sensor, keeping the other pole farther away. Reverse the magnet to flip the sensor into its opposite state.*

The LED should switch on and off cleanly, without any dimming or flickering. This is enabled by circuitry included with the sensor known as a Schmitt trigger.

Typically a Hall sensor can sink up to 20mA, but that will be an absolute maximum value. Using a 1K pullup resistor with a 9VDC supply, you should find that the sensor is sinking about half of that. It's a good idea to check this with your meter. A typical LED rated for 20mA will not glow brightly when it is only passing 10mA, but certainly should be bright enough for you to see that it makes clean on-and-off transitions.

Applications

You may feel that a bipolar Hall sensor is less convenient than a reed switch, not just because it requires a power supply, but because you have to use one pole of a magnet, and then the other, to switch the sensor on and off. However, the sensor is designed to work conveniently in a situation where a series of poles passes it—or, conversely, where it passes a series of poles.

Pulses from the sensor can be used as feedback to provide precise control of the speed of a motor. The simplified rendering in Figure 30-5 shows this concept, where a rotating object has magnetized teeth with alternating polarity.

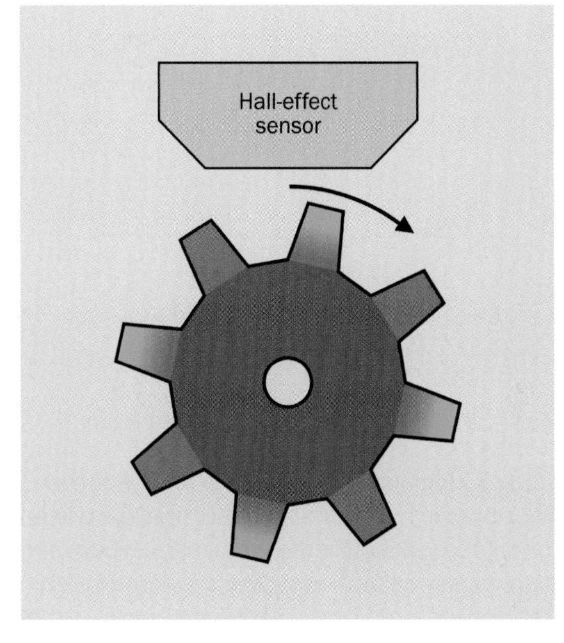

Figure 30-5. *A Hall sensor can measure the speed of rotation of wheel with teeth of alternating magnetic polarity.*

In your test experiment, try turning your magnet 90 degrees and then slide it past the sensor, so that it is affected by first one pole and then the other. Once again you should see that the LED makes a clean on-off transition—and displays some hysteresis, because the sensor tends to stick in its on and off states.

The hysteresis can be graphed as in Figure 30-6. Compare this diagram to Figure 6-8, illustrating the behavior of a comparator. The difference is that the comparator responds to changes in voltage, while the sensor responds to changes in magnetic field.

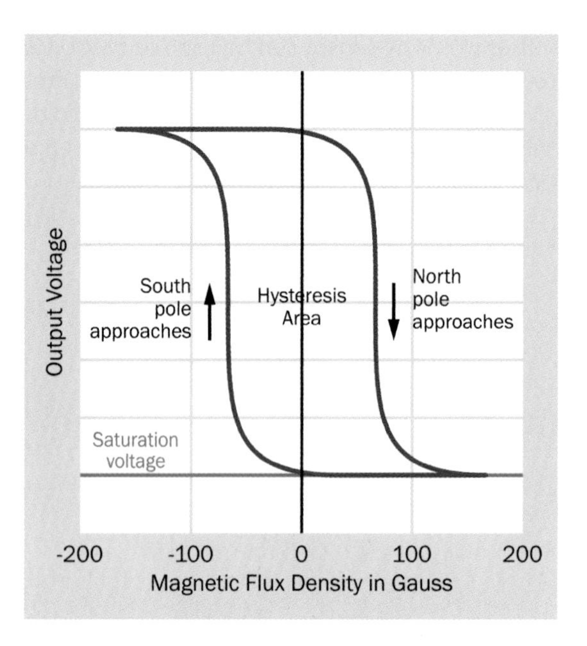

Figure 30-6. *The hysteresis of a bipolar Hall sensor.*

You could make an odometer or a speedometer for a bicycle by mounting a Hall-effect sensor on one of the forks and attaching a magnet between two spokes in one of the wheels. The output from the sensor would go to a microcontroller that would convert each rotation of the wheel into a distance value. The microcontroller could then calculate speed by dividing distance by time.

Quick Facts About Hall Sensors

- The through-hole versions of Hall effect sensors are usually sealed in a little piece of black plastic, about 0.1" x 0.1" x 0.05", with three leads. This is similar to, but smaller than, a generic TO-92 transistor package.
- When a datasheet refers to the "front" or "top" of a sensor, it means the face that has a part number printed on it. This face has bevelled edges, while the "back" of the sensor does not. Usually the sensor is designed to respond to a magnetic pole that is brought toward its front, bevelled face.
- The part number is usually abbreviated to three digits. An additional code below the three digits usually refers to the date of manufacture.
- The power-supply voltage often ranges from around 3VDC to 20VDC, allowing use with a 9V battery. However, some sensors will only tolerate 3VDC to 5.5VDC. Check datasheets carefully!
- Like a phototransistor, a Hall-effect sensor is very often encapsulated along with an NPN transistor that has an open-collector output. The absolute maximum current that sinks into this output can be 20mA to 25mA.
- When a pullup resistor is placed between the open-collector output of a Hall sensor and the positive side of the power supply as in Figure 30-3, the output will appear to be low when the sensor is activated and high when it is deactivated.
- Many Hall sensors also include a Schmitt trigger, which is a circuit that creates a clean on-and-off response with some hysteresis.
- Different variants of sensor may be activated by either a magnetic south pole or north pole. The datasheet should provide this information.
- Hall sensors are free from the contact bounce that is a problem in reed switches. This makes them useful for providing an input to a logic gate.

Hall Types

There are four common types of Hall-effect sensors.

The *bipolar* type has just been described. It requires magnetic fields of opposite polarity to switch it on and off.

A *unipolar* Hall sensor will switch on in response to proximity with one magnetic pole and will switch off when the magnet is removed. It doesn't need an opposite magnetic pole to switch it off.

Unipolar sensors are available in versions that are activated either by a magnetic north pole or south pole. Like the bipolar sensor, they use a Schmitt trigger to achieve a clean on-off response.

Remember that in its "off" state, the open-collector output of a Hall sensor has a high resistance between it and negative ground so that the output voltage provided through a pullup resistor will appear to be high. When the sensor goes into its "on" state, the output voltage appears to be low. This is the same behavior you saw with a phototransistor.

A so-called *linear* Hall sensor does not contain a Schmitt trigger and creates a voltage (amplified by the internal transistor) which varies in proportion with external magnetic field. In the absence of a magnetic field, the output from the sensor is half of its supply voltage. In response to one magnetic pole, the output from the sensor diminishes almost to 0VDC. In response to the other magnetic pole, the output increases almost up to the limit of the supply voltage.

Linear Hall sensors are also known as "analog" sensors. The output pin is usually connected with the emitter of an internal NPN transistor, rather than its collector. A minimum 2.2K resistor must be applied between the output pin and ground to limit the sink current.

The variable output can be interpreted to measure the distance between the sensor and a magnet. The perceived magnetic field decreases with distance, so the sensor will not respond where the distance becomes significant (more than 10mm in most cases).

The *omnipolar* type of Hall sensor closely resembles a reed switch. It is switched on by a magnetic field of either polarity and switches off in the absence of a magnetic field. This sensor actually contains two Hall detectors and a logic component that responds to the voltage difference between them. Because of the additional internal circuitry, and because there is less demand for this type of sensor, it tends to be slightly more expensive (a bit more than $1, at the time of writing, while the other three Hall sensor versions are available for less than $1).

Sensor Ideas

A Hall sensor is easily interfaced with digital logic. A 10K pullup resistor can be used, as in Figure 30-7. The higher value of resistor is acceptable because the inputs of a logic chip have much higher impedance by comparison and require a negligible flow of current. Of course, the voltage should be checked to make sure that it is within the appropriate range.

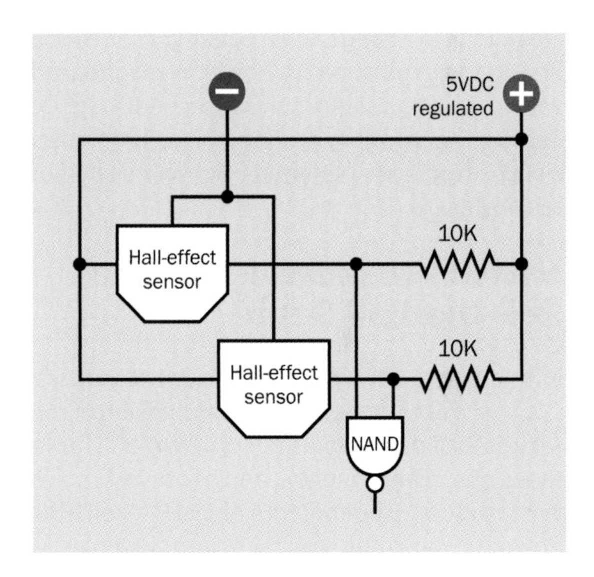

Figure 30-7. *The NAND gate in this circuit will have a normally-low output that only goes high when either or both of the Hall sensors are activated.*

In this circuit, the NAND gate will deliver a high output when either or both of the sensors are activated. Remember, the voltage provided by a pullup resistor with a Hall sensor is low when the sensor is "on," while a NAND gate has a low output when both of its inputs are high and a high output when either or both of its inputs are low. Unipolar sensors may be appropriate in this application.

If a Hall sensor is being used to respond to a handheld magnet whose polarity is unknown, the omnipolar type is appropriate.

A bipolar sensor can behave like a unipolar sensor if a biasing magnet is installed behind the component. In other words, if a sensor is normally activated by a north magnetic pole in front of it and deactivated by a south magnetic pole in front of it, a small magnet can be placed behind it to insure that the sensor returns to its "off" state. When the north pole of a magnet is positioned in front of the sensor, its field can overwhelm the weaker field from the other magnet, causing the sensor to turn "on."

When magnets are used in this way, they must be of high coercivity, meaning that they resist "coercion" to change their polarity. That is, the first magnet must resist being re-magnetized with opposite polarity by the second magnet. Because neomydium magnets have high coercivity, they are preferred for this kind of application.

Make Even More: Miniature Roll-the-Ball Game

In the midway of a traditional carnival or state fair, you're likely to have seen some version of the roll-the-ball game. You sit at one end of a ramp that slopes away from you and has several holes in it at the opposite end. Your object is to roll balls down the ramp and sink them into the holes as quickly as possible, bearing in mind that the holes farther from you will give you a higher score.

If you miss all of the holes, the ball disappears down a slot at the end of the ramp and doesn't score anything at all. The basic layout is shown in Figure 30-8.

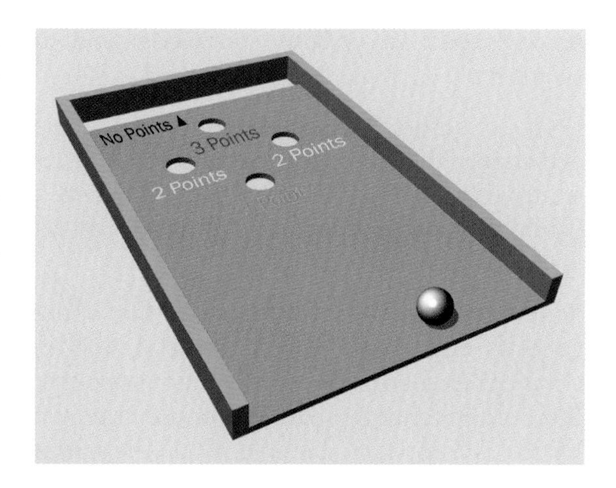

Figure 30-8. *The traditional carnival midway ball-rolling game.*

Usually a lot of people will be competing at adjacent ramps, and the person who reaches a high score first will receive some kind of prize.

You could make a miniature version of this game using ball magnets and omnipolar Hall effect sensors. Assuming you make it as a single-player game, you'll need a time limit in addition to a score counter.

Now, how can you control the addition of one, two, or three points to the score? I think the easiest way is to have multiple sensors along the return paths from the higher-scoring holes. The balls will trigger the sensors by rolling past them.

You can do this by making miniature wooden channels, but I think an easier and better way would be to use plastic tubing.

Plastic Bending

Tubing formed from PETG (polyethylene terephthalate) is readily available online and doesn't cost much. It will keep the returning balls completely under control, even if someone jogs or nudges the game. The only challenge is to bend the tubing smoothly so that it doesn't get any kinks in it; and the way to do this is by inserting a spring and then using a heat gun.

I'm straying from the business of making electronics, here, but now that I'm dealing with sensors, we have to think about ways in which they can interact with the physical world.

Figure 30-9 shows a piece of PETG tube that has been bent by inserting a spring. The next step will be to pull the spring out, which can be difficult. You may have to rotate the spring, causing its diameter to diminish slightly.

Figure 30-9. *A piece of polethylene tube that has been bent into a smooth curve by inserting a spring before applying a heat gun.*

Smaller tubing can be formed using this same system. For the ball-rolling game, you might want to use 3/8" magnet balls inside tubing with a 1/2" internal diameter and 1/16" thick walls. But where do you find the tubing and a long spring to push inside it?

I suggest McMaster-Carr, which stocks probably the biggest range of hardware on the planet. Right now they're selling six feet of PETG tube of the size that I just suggested for about $1.50 a foot. They also sell "cut-to-length" extension springs for around $3.50 apiece. The idea is that you can cut them to whatever length you choose, but for tube-bending purposes, you won't need to cut them at all.

Of course, you may prefer to work entirely with wood, or you may want to build some other project that requires less shop work than this. I realize that not everyone shares my odd interests, such as bending plastic tubing and rolling balls through it. Still, I've come this far, so I'll finish describing the project.

Rolling-Ball Electronics

If you have one sensor on the return tube from the one-point hole, two sensors for each of the two-point holes, and three sensors on the single three-point hole, that's eight sensors altogether. You could connect all eight of them through an eight-input NAND chip to a counter, as shown in Figure 30-10, which shows ball-return tubes as they could be laid out under the ramp.

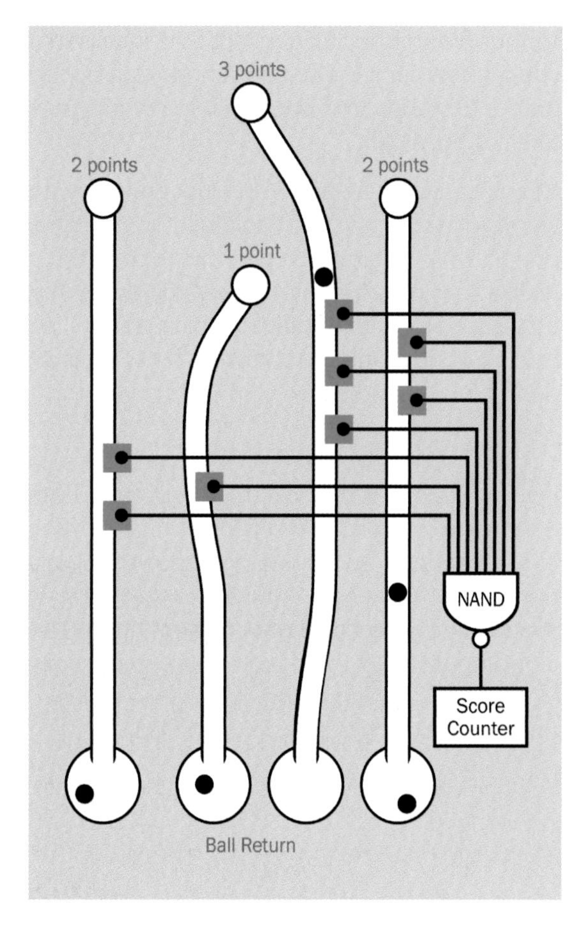

Figure 30-10. *Hall-effect sensors (shown here as green squares) to count points in the rolling-ball game.*

You'll remember from Figure 30-7 that a NAND is appropriate for use with Hall-effect sensors, as its output goes high each time a sensor pulls one of its inputs low.

Omnipolar sensors would be appropriate for this application, as they should respond to the magnetic field created by a rolling ball regardless of its changing polarity. I'm hoping that the sensors will give noise-free signals, which is a primary reason for using them instead of reed switches. There's no reason to debounce a reed switch (using the debouncing techniques that I mentioned in *Make: Electroncs*) when you can use sensors that should be bounce free.

The output from the NAND chip can connect with a 4026B decade counter. This chip is specifically designed to power a seven-segment LED numeral. The carry from this chip can go to a second 4026B and a second numeral, so you can count up to 99 points. The 4026B is another item that I described in *Make: Electronics* (for use with a reaction-timer game).

Add a reset button that will zero the counters and begin a single 30-second pulse from a 555 timer, which will stop the counters at the end of its cycle. And that's it! Your very own arcade game with rolling magnetic balls—like an early, primitive pinball machine.

Wait—could you build an entire pinball machine with neodymium ball magnets?

Maybe so, but quite apart from the complexity, you'd have to make sure that the balls never come into contact with each other. In the ball-rolling game, pulling magnets apart that have accidentally come into contact will just add to the tension as you try to beat a high score.

Experiment 31: Electronic Optics

31

Two basic types of sensors are triggered by variations in light. There is the active type and the passive type.

The Chronophotonic Lamp Controller that I described in Experiment 7 used a phototransistor, which is a passive sensor. It just sits and waits, measuring light from outside sources and changing its effective internal resistance accordingly.

Another common passive light-sensing component is a PIR motion sensor, where the P in the acronym stands for "passive" and the IR stands for "infrared." These are commonly used to switch on lights or trigger alarm systems when they sense body heat from a person moving around.

PIR motion sensors are useful, but the complete units that you find in hardware stores don't leave much room for experimentation. If you want to play with this kind of sensor, it would be better to buy one on a breakout board, as shown in Figure 31-1. Hobby-electronics sources, such as Sparkfun, sell them for less than $10 at the time of writing, and they are designed for direct connection with microcontrollers. For a truly excellent tutorial on using them, visit *http://www.ladyada.net*. I won't be dealing with PIRs here, as I think active sensors have more interesting possibilities.

Figure 31-1. *A passive infrared (PIR) sensor mounted on a breakout board to facilitate experimentation.*

Active Light Sensors

Instead of just sitting and watching the ambient light around it, an active sensor emits some light of its own—almost always infrared, although a few ultraviolet sensors exist. Active sensors are found in copy machines, where they detect paper jams; in industrial automation, where they sense the progress of a product through a manufacturing process; or in robotics, where they verify the positions of moving parts.

Usually the light beam is generated by an infrared LED that operates in a narrow frequency band and may be modulated to distinguish it clearly from other sources. Some circuitry behind a nearby detector (usually, a phototransistor) is tuned to the same frequency.

The general term for a combination of light emitter and light sensor is an "emitter-receiver" combination, and there are two variants:

The *reflective* type of emitter-receiver:

- The LED and the phototransistor are mounted side by side, pointing in roughly the same direction. An example is shown in Figure 31-2.

Figure 31-2. *This reflective emitter-receiver consists of an infrared phototransistor mounted beside a matching infrared emitter. Light from the emitter bounces back from a reflective object positioned a short distance away, and the phototransistor measures any variation caused when something moves to interrupt the beam.*

- You have to position a reflective surface, such as a piece of silvered mylar or a white object, to bounce the light from the LED back to the phototransistor.
- The range is usually very limited. Many reflective sensors expect the reflecting surface to be about half an inch away. There are exceptions to this rule, but they are likely to cost more.

The *transmissive* type of emitter-receiver:

- The LED and the phototransistor face each other across the gap in a single U-shaped mount, as shown in Figure 31-3.

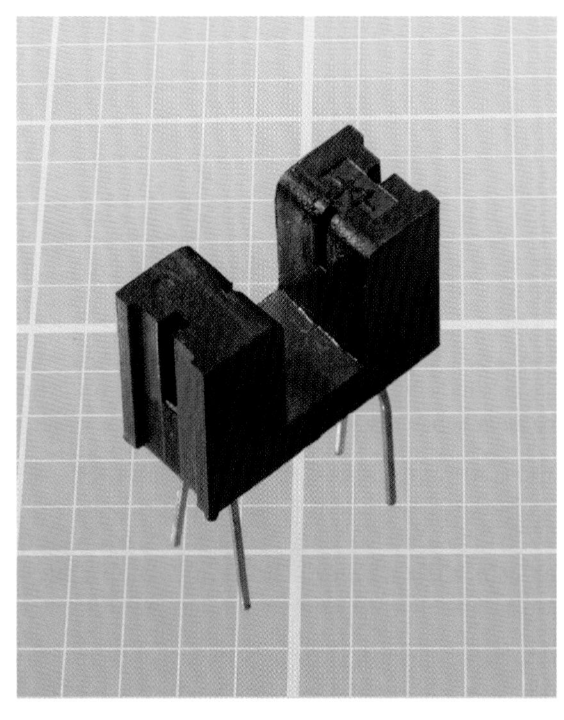

Figure 31-3. *In this Everlight ITR9606-F transmissive sensor, light is transmitted across the gap to a phototransistor facing it in the other half of the component. The schematic symbol for a diode is just visible, molded into the right-hand section of the plastic. An infrared LED is mounted in this part.*

- If an object breaks the light beam, the phototransistor changes its output.
- In many components of this type, the air gap is about a quarter inch wide.
- A transmissive type of emitter-receiver is less versatile than a reflective type but very easy and convenient to install.
- This type of sensor is sometimes referred to as an "optical switch" or an "opto-interrupter."

In the upcoming experiment, I'm going to show you how to use a transmissive sensor, such as the

Everlight ITR9606-F. After you test it, I'll suggest some applications. But first—

Warning: Slow Sensor Death!

Active IR sensors are "always-on" devices. So long as the system is powered up, the infrared LED is emitting light.

This entails some power consumption (usually around 10mA to 20mA), but that isn't the major issue. The sad fact is that infrared LEDs tend to deteriorate with usage. Some datasheets may warn you that the light intensity will diminish by 50 percent over a period of about five years. Other datasheets say nothing about this, but it is still likely to happen. The deterioration may be heat-related, or may be linked with the particular chemical process that is used in manufacturing. My reading on the subject suggests that deterioration is still not thoroughly understood, but everyone agrees that it happens.

If you use an active IR sensor, try to adjust your circuit with a large margin of error so that it will continue to work for as long as possible while the light from the LED diminishes with time. Also try to minimize the current that passes through the LED.

The Numbers

The infrared emitter in a transmissive sensor is an LED that requires a relatively low forward voltage —typically around 1.2V, and not greater than 1.5V. You'll need to add an external series resistor to protect the LED, because no resistor is included inside the component. The resistor value must be chosen according to the voltage of your power supply.

- Check the current consumption of the infrared LED, as well as its forward voltage. Adjust the series resistor so that the current is close to the value described as "typical" in a datasheet. You don't need (and don't want) the current to be up near the "absolute maximum."

The infrared receiver is usually a phototransistor with an open-collector configuration—just like a Hall-effect switch. As before, you are expected to use a pullup resistor with the open collector. The only question is, what value of pullup resistor is appropriate? The datasheet doesn't always say, but it will give a maximum value for current that can be allowed to sink into the open collector. A "maximum" value may be 20mA, but a "typical" value may be much lower. Consequently, you should not expect to drive a (normal, visible-light) LED directly from this type of sensor.

Infrared Sensor Test

Now you have the orientation, let's test this thing! If you are using the ITR9606-F, the pinouts are shown in Figure 31-4. But how do you know which way around the sensor is? The Everlight ITR9606-F makes it easy for you by embossing the symbol of a diode in the plastic. You can see it on the right in Figure 31-3 if you look carefully.

Many other sensors have similar specifications, and their datasheets will show you the functions of the leads. Some of them are the same as on the ITR9606-F, but in others you'll find one pair of leads is reversed.

The experiment here should work with any transmissive optical sensor, provided you proceed methodically, get the pin functions right, and avoid passing excessive current through the component.

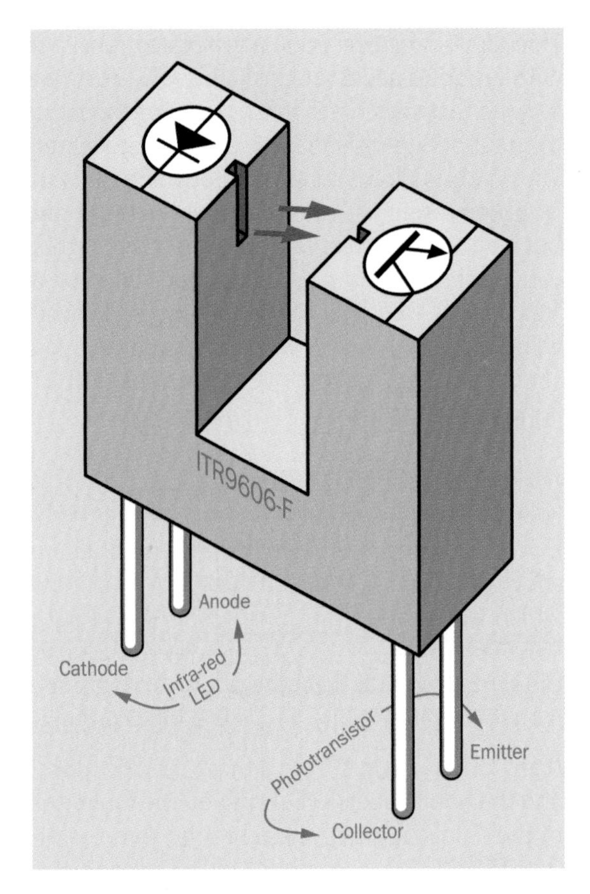

Figure 31-4. *Pin functions of the Everlight ITR9606-F. Other sensors with similar specifications may have the same pin functions, or one pair of pins may be reversed. Check datasheets for details.*

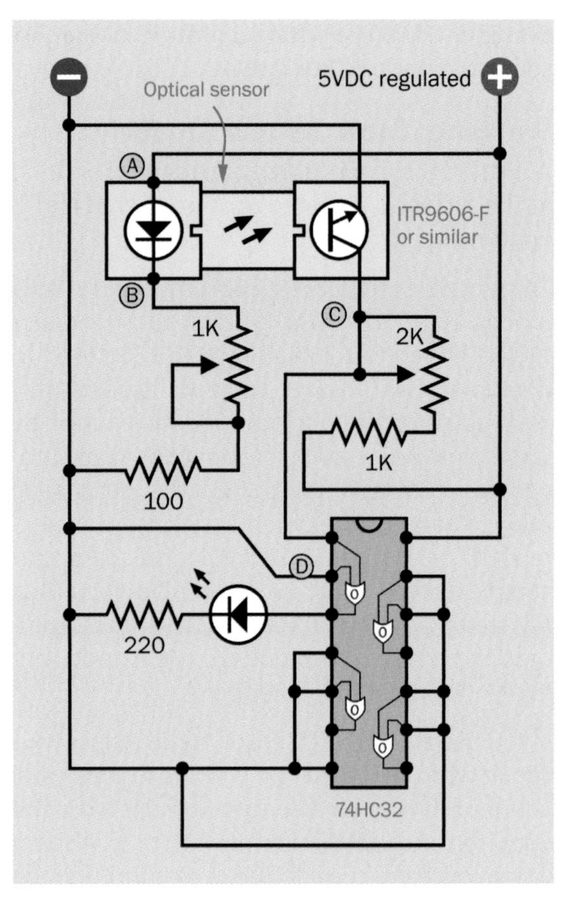

Figure 31-5. *Schematic for testing a transmissive optical sensor in conjunction with an OR logic chip.*

In Figure 31-5 I'm suggesting that you test the sensor with a quad two-input OR logic chip. The reasons for this will become clear as we go along. A picture of the breadboarded circuit is in Figure 31-6.

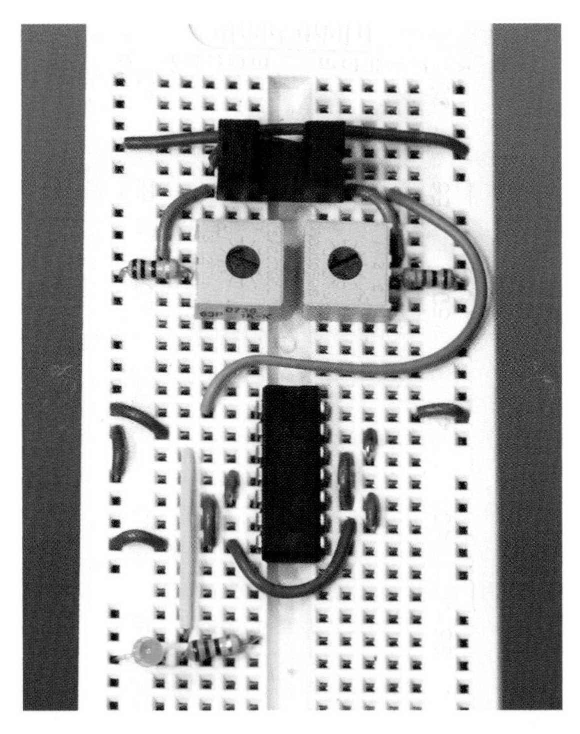

Figure 31-6. *Breadboarded version of the test circuit for a transmissive optical sensor. The U-shaped sensor is at the top, just above the trimmer potentiometers.*

Note that I'm using a 5VDC regulated power supply, because the OR chip requires this.

When you wire the circuit, leave the positive power supply to the chip unconnected initially, and also leave the wire to the collector of the phototransistor unconnected. The first step is just to power the infrared LED.

Infrared LED Test

Remember, you aren't going to see any light emitted from the LED inside the sensor, because it's outside the visible spectrum. You must rely on your meter to tell you that it's working.

In Figure 31-5, adjust the 1K trimmer to its maximum resistance. (Check it with your meter before you plug it into the board, if you are in doubt.) Now measure the voltage between points A and B. It should be between 1VDC and 1.2VDC.

You also need to measure the current that the infrared LED is drawing. Remove the jumper that connects point A in the circuit and the positive bus. Set your meter to measure mA and use the meter to connect point A to the positive bus. As you turn the trimmer, you will see the meter reading change. You want a reading of about 10mA. The LED is rated for more, but because of a future application that I have in mind, I don't want this circuit to draw too much power.

Remove the trimmer from the circuit and measure the resistance between the wiper and the top end. In my test I found that it was about 350Ω. So, the total series resistance for the infrared LED was 100 + 350 = 450Ω, which is relatively high for an LED powered by 5VDC, but my circuit worked with this value.

- You can now remove the 1K trimmer and the 100Ω resistor and substitute a series resistor of 450Ω, or whatever value gave you the 10mA current flow.

Phototransistor Test

Now that the LED is taken care of, connect the collector of the phototransistor and measure the voltage between point C and negative ground. This voltage is supplied through the 1K resistor and the 2K trimmer, which together form the pullup resistor for the open collector.

While you are measuring it, insert a piece of card in the sensor to block the infrared light beam, and then remove the piece of card. By adjusting the trimmer, you should be able to achieve a high voltage that exceeds 4.5VDC when the card is removed, and a low voltage that is less than 0.5VDC when the card is inserted. I'm betting that a total resistance of about 2K will be appropriate.

- You can now remove the 2K trimmer and the 1K resistor and substitute an appropriate pullup resistor that creates the desired voltage range at point C in the circuit.

Logic Test

Because a 74HC00 likes to have a high logic input greater than 3.5VDC and a low logic input less than 1VDC, the output from the sensor should be acceptable.

I've seen some transmissive optical sensors that recommend a pullup resistance on the phototransistor as low as 100Ω. If you happen to be using one of these, verify the current that is sinking into the sensor. The way to do that is to set your meter to measure milliamps, disconnect the emitter of the transistor from the circuit, and insert your meter between the emitter and negative ground. You shouldn't need a current higher than 4mA.

Now connect point C with the top-left OR gate in the logic chip, as shown. The other input of this OR gate, at D, is tied to negative ground for the time being.

When you move a piece of card into the sensor, you should see the LED lighting up without any hesitation or flickering.

Because the input impedance of a 74HC00 series logic chip is so high, you should find that it doesn't reduce the open-collector voltage significantly.

While you are testing the circuit, experiment with different objects blocking the infrared LED in the sensor. You'll probably find that a thick piece of cardboard will provide a higher open-collector voltage than a piece of white paper. Bear this in mind if you use a sensor for a practical application.

Options

You could take this experiment a step further by adding another optical sensor. Disconnect the grounding wire from the second input to the OR gate at point D, and connect this input to the output from the second sensor. Now you should get a high output from the OR gate when you block the infrared light beam in either of the sensors. Naturally you could use an OR gate that has more inputs if you want to check whether any of a set of sensors is triggered.

You could substitute an XOR gate for the OR gate. Now you'll get a result when only one sensor is triggered. Logic gates enable you to customize your circuit depending on your needs.

Alternatively, a transmissive optical sensor is well suited to drive a comparator, in the same way as we used a phototransistor with a comparator in the Chronophotonic Lamp Switcher (see Chapter 7). The comparator will tolerate a wider range of voltages than a logic chip and will allow you to set a threshold voltage that gives you a large margin of error.

Note that when you were experimenting with a Hall-effect sensor, I suggested using a NAND gate to couple the outputs of multiple sensors. That was because the Hall output from the open collector goes low when the sensor is activated. The optical sensor behaves in the opposite way, because blocking the infrared beam causes the effective resistance of the phototransistor to increase, which makes the open-collector output appear to go high when the sensor is activated.

If you have a circuit where you want to invert the behavior of the optical sensor, you can change it to give you an emitter-follower output:

1. Disconnect the pullup resistor to the collector of the phototransistor. Disconnect the ground wire to the emitter of the phototransistor, and substitute the resistor between the emitter and ground.
2. Connect the collector of the phototransistor directly to the positive side of the power supply.
3. Move the connection to the OR chip from point C to the emitter of the phototransistor.

Now the output from the sensor will be normally high and will go low when the infrared light beam is blocked.

Quick Facts About Transmissive Optical Sensors

- Typically the internal infrared LED requires a voltage no greater than 1.5V (preferably around 1.2V). It may consume 10mA to 20mA.
- In an open-collector circuit, the phototransistor needs a pullup resistor between its collector and the positive side of the power supply. The value varies widely for different types of optical sensor. The collector voltage will be low when the sensor is unobstructed, and high when the sensor is obstructed.
- An emitter-follower circuit inverts the behavior of the sensor. A resistor is placed between the emitter of the phototransistor and the negative ground side of the power supply. An emitter connection should be high when the sensor is unobstructed and low when the sensor is obstructed.
- Transmissive optical sensors are not designed to drive significant loads. The output should go to a high-impedance component, such as a comparator or a logic gate.
- Remember that the output of an infrared LED will diminish over time. Try to design your circuit so that it will still work with a reduced light intensity. A comparator will tolerate a wider range of input voltages than a logic gate.
- Because infrared light is not visible, you won't see anything while the component is active. Don't leave the component switched on by mistake!

Better Slots

What are we going to use this thing for? Well, how about a coin detector in the Hot Slot game? You'll need sixteen sensors, but they're quite cheap, and by the time you read this, they may be even cheaper.

To do this, I need to address two issues: how to wire the sensors, and how to mount them in a box suitable for coin insertion. I'll deal with the wiring first.

Proof of Concept

You'll remember that in the original version of the game, a multiplexer provided power through one of its sixteen outputs (see Chapter 21). If you need to refresh your memory, check Figure 21-7 and Figure 21-8.

The optical sensor requires so little current, I think the multiplexer can supply both its infrared LED and its phototransistor. This would be a nice arrangement, because all the other sensors can be left in an "off" state, consuming no power and conserving their infrared LEDs.

Another reason I like this arrangement is that instead of having just one LED that lights up when someone wins, each sensor can drive its own individual "win-notification" LED (if I add some kind of amplifier for it). When someone hits the hot slot, the win-notification LED will come on right beside that slot.

The only thing I'm wondering about is the exact sequence of events. It's a little tricky, so I'm going to show what happens to just one sensor.

In Figure 31-7, a sensor is doing nothing yet, because it is receiving no power through the multiplexer. Its infrared LED is dark. Its phototransistor is not powered, either.

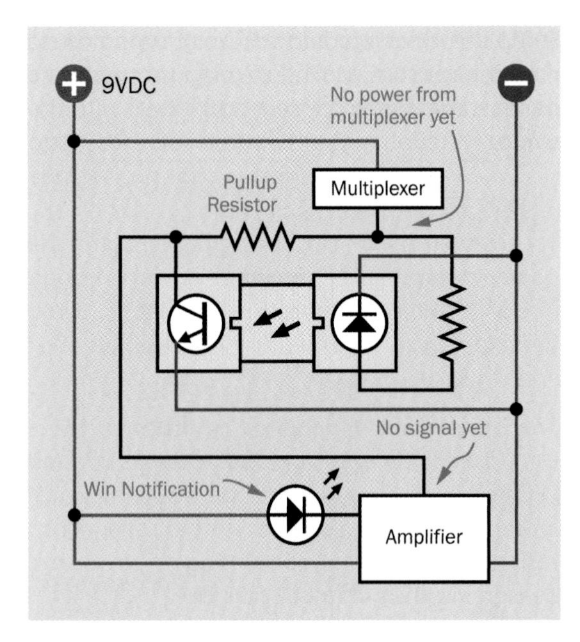

Figure 31-7. *Step 1 in the powering of a coin sensor. See text for details.*

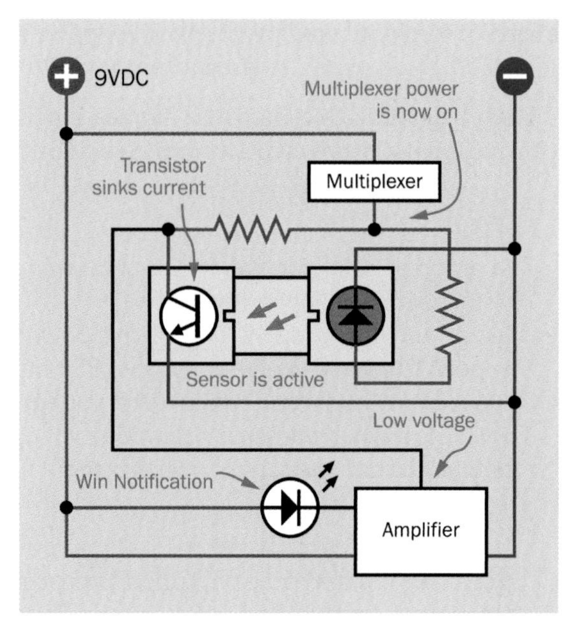

Figure 31-8. *Step 2 in the powering of a coin sensor. See text for details.*

Power is connected to an amplifier for the win-notification LED. The amplifier is our old friend a ULN2003 Darlington, which will work by sinking current through the win-notification LED—but only when a signal from the coin sensor tells the Darlington to do this. There's no signal yet, because the sensor has no power. Consequently, the win-notification LED is dark.

Now take a look at Figure 31-8. The multiplexer has just chosen this sensor to be the Hot Slot. It is powering the sensor's infrared LED and its phototransistor. The infrared LED immediately lowers the effective internal resistance of the phototransistor, which sinks current through a pull-up resistor. Because the phototransistor is sinking current, hardly any of it gets to the amplifier. Consequently, the signal input of the amplifier has a low voltage, and the win-notification LED stays dark.

Finally, in Figure 31-9, someone inserts a coin in the sensor. The phototransistor cannot see the infrared light anymore, so its effective internal resistance rises. Now the voltage on the signal input to the amplifier rises, so the amplifier switches on the win-notification LED.

So—do you think this will work? I wasn't entirely sure when I first drew the circuit. I wondered if there was just a chance that when the multiplexer applies power, the sensor would take a moment to respond, and during that moment, the phototransistor wouldn't sink any current, so the power would sneak around to the amplifier, and the win-notification LED would light up, revealing that this is the Hot Slot.

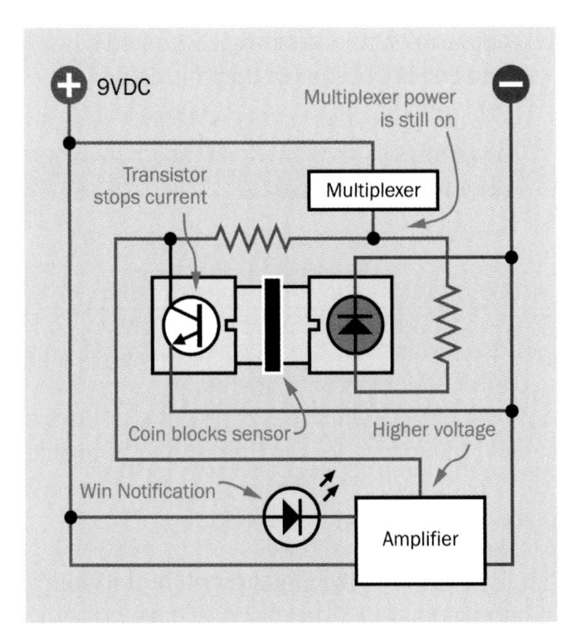

Figure 31-9. *Step 3 in the powering of a coin sensor. See text for details.*

However, when I wired the circuit, I found that the sensor responds so quickly, any current to the amplifier is so brief, the win-notification LED does not make a visible response. So, yes, the circuit does work. But sometimes you have to try these things to be sure. At least, I have to.

That's it for the proof of concept. Now it's just a matter of getting the resistor values right and fitting everything onto a breadboard.

Actually, I can't fit everything onto a breadboard. But I can do eight sensors, which is the most the Darlington array can handle anyway. The breadboard layout can then be duplicated on another breadboard to make a full sixteen slots for the Hot Slot game.

The Schematic

Figure 31-10 shows three sensors with a Darlington array. Because I don't have unlimited space, and because each sensor is wired identically, I'm leaving it to you to add five more sensors to the ones shown in the figure, using exactly the same hookups.

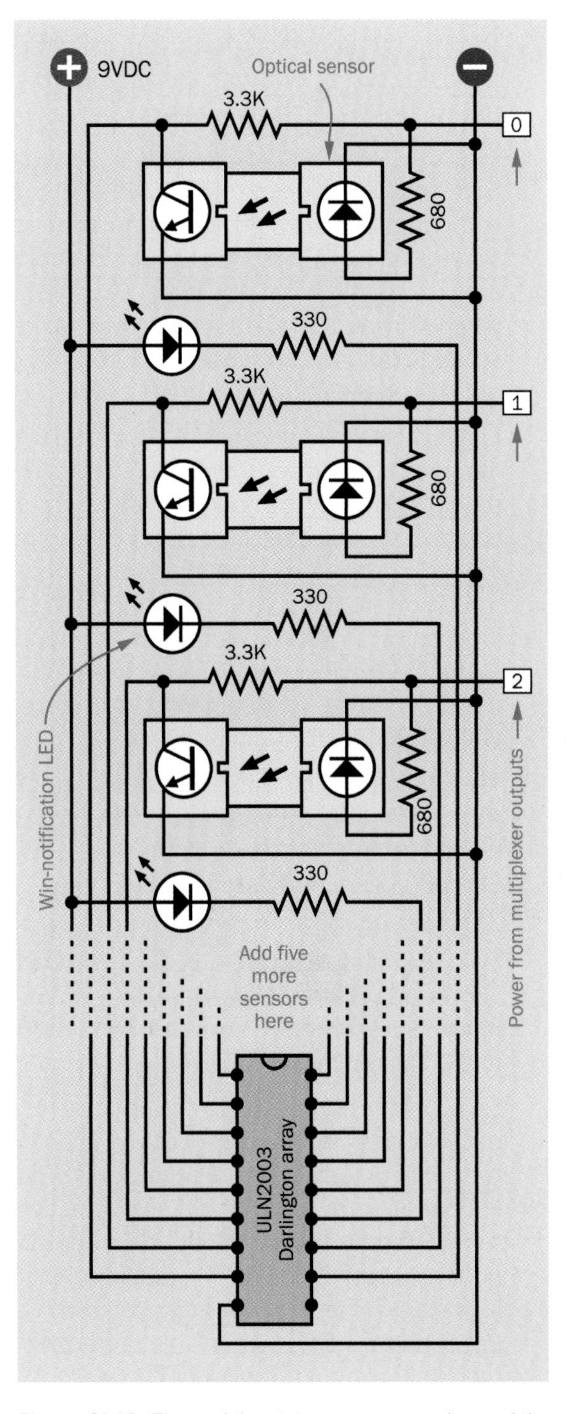

Figure 31-10. *Three of the sixteen sensors and one of the two Darlington arrays that would be needed to implement optical coin sensing for the Hot Slot game.*

- The inputs numbered 0, 1, and 2 on the right-hand side correspond with the outputs from the multiplexer in the original Hot Slot schematic. See Figure 21-7.
- Each sensor has been rotated 180 degrees compared with the way I had them in the test schematic in Figure 31-5. I turned them around to minimize the wire crossovers when driving the Darlington array. Make sure you mount your sensors with the infrared LED on the right, not the left.
- I have increased the series resistor for each infrared LED in the sensors to 680Ω, and each pullup resistor for the open-collector output is now 3.3K. This is because the circuit is going to be driven by a multiplexer using a 9VDC power supply. We're not using 74HC00 logic chips, so 5VDC is not necessary—and might be inadequate, because the multiplexer imposes some internal resistance, which will drop the voltage a little.
- The Darlington chip sinks current (does not source it) so each win-notification LED is connected to the positive side of the power supply and sinks current into the Darlington.
- Although the power for each sensor comes from the multiplexer in Figure 21-7, the Darlington arrays are powered directly from the main power supply. This way, they won't add to the load on the multiplexer.
- Each Darlington pair has a lower input impedance than the OR chip that was used in the previous test. Consequently a Darlington will pull down the output voltage of an active sensor slightly. I tried various values for the pullup resistor on the sensor and found that 3.3K was the best compromise. You can adjust this upward if you don't get correct on-off behavior from the win-notification LED.
- I'm suggesting a series resistor of 330Ω for each win-notification LED. The ones that I used are designed for a forward voltage of 2VDC. If you use different LEDs, you may need to adjust the series resistor. Check the voltage across one of the LEDs when it is on, and check the current that it is drawing.
- A Darlington pair can sink 100mA without any trouble, so you don't need to worry about overloading it.

The Breadboard

Figure 31-11 shows a breadboard layout that you can use for each of the sensors, so that it only occupies four rows, including the win-notification LED, which has to be right beside the sensor. The gray strips are the conductors inside the breadboard. The sensor pins are shown as black circles with white centers.

- The diagonal line linking two of the sensor pins is a jumper that must be inserted in the breadboard before the sensor is inserted above it.

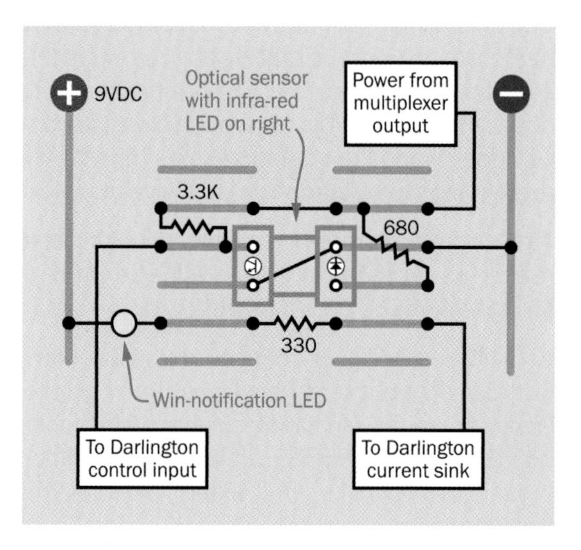

Figure 31-11. *A possible breadboard layout for one of the sensors, occupying a minimal number of rows.*

Because the components are packed tightly together, you have to be very careful about placing them. If you put a resistor or a jumper wire one space away from its proper location, you can reverse the voltage to a sensor and burn it out. In my own circuit, I burned out two sensors while testing it. Impatience, of course, is always the problem. Well—it's my problem.

In Figure 31-12 you will find a photograph of my breadboarded circuit with eight sensors. This circuit would be duplicated to take care of the remaining eight sensors.

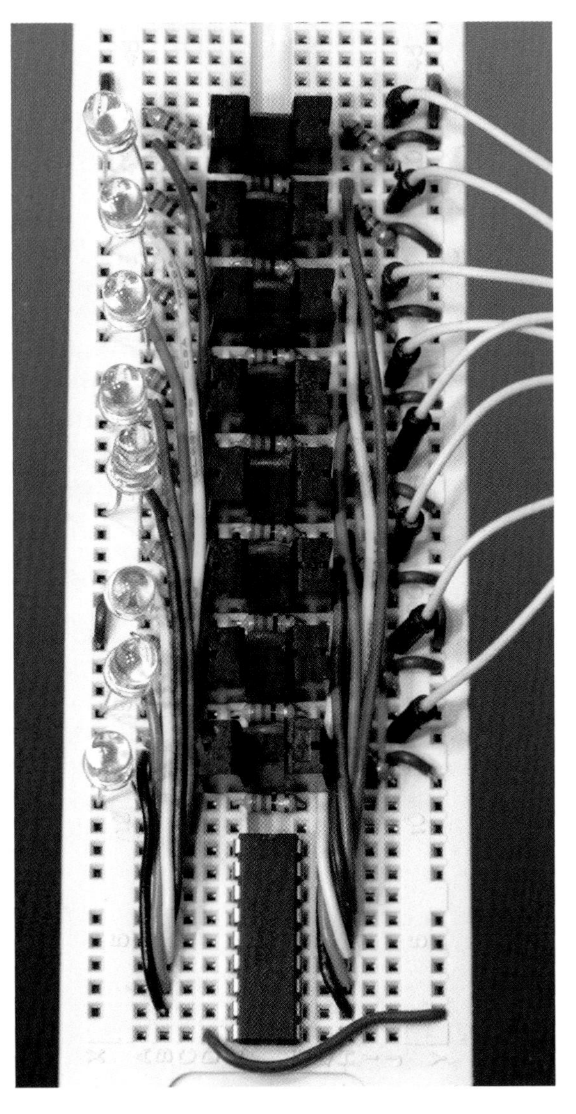

Figure 31-12. *Breadboarded circuit featuring eight of the sixteen coin sensors for the Hot Slot game.*

In the photograph, the white jumper wires coming in from the right are from the multiplexer outputs from the Hot Slot circuit that you can look up in Figure 21-7. Because I wanted to use bright LEDs in this circuit, they're water clear and 5mm diameter. You can see them lined up on the left.

What do you think? Is it worth the extra trouble to create a neater user interface so that coins can be dropped effortlessly into place instead of being forced between some tight metal contacts? Personally I think so, and when you see my idea for a really elegant enclosure (coming right up), maybe you'll be convinced, too.

The Slot Box

The most obvious way to build this circuit permanently would be to solder all the components onto perforated board and position the board underneath slots that you cut in the lid of a box. I'm not too excited by this idea, as it won't control the coins properly.

I think a better plan is to sandwich the sensors between vertical layers of wood or opaque plastic that are glued together.

Figure 31-13 shows the first stage in this process, with four sensors and a quarter-inch-thick piece of wood or plastic that has semicircular coin receivers.

Figure 31-13. *The first step in assembling an enclosure for the sensors.*

Figure 31-14 shows how the sensors sit in notches at the bottoms of the semicircular slots.

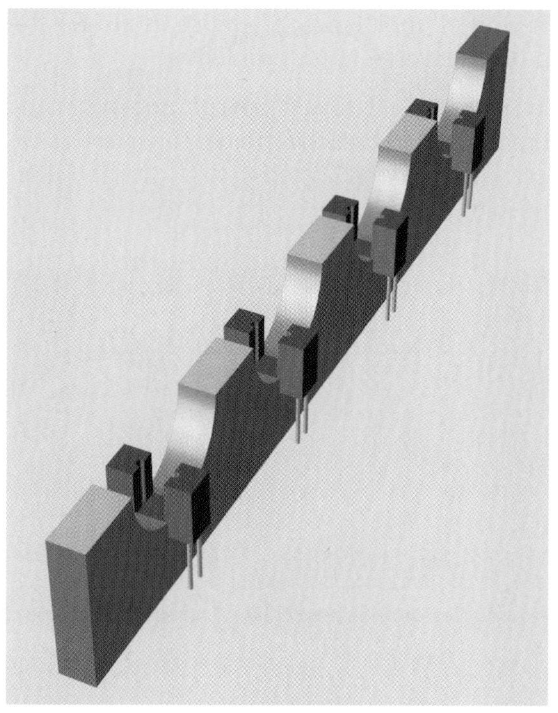

Figure 31-14. *The sensors rest in notches at the bottoms of the coin slots.*

Now in Figure 31-15, a 3/4" spacer is added, with LEDs that have been mounted in holes drilled vertically from top to bottom.

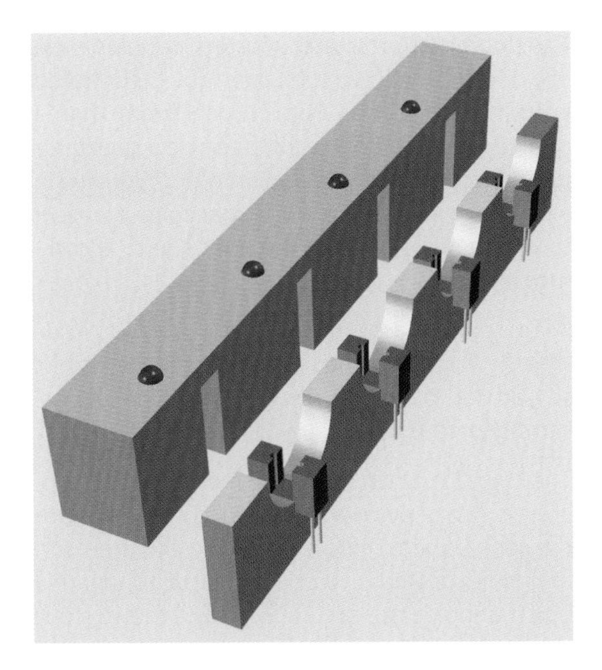

Figure 31-15. *A spacer will hold the sensors in place and prevent any leakage of infrared light from one sensor to another.*

After adding four spacers and four coin-slot segments, all that remains is to add one more spacer on the front and then enclose the whole assembly in a box that will house the electronics in the bottom.

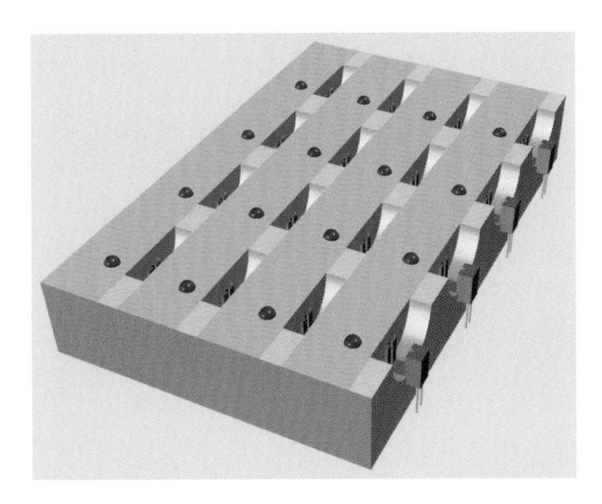

Figure 31-16. *One more spacer will be needed at the front, after which the assembly can be enclosed in a box.*

Cutting the pieces for this enclosure is not a very challenging task. The plans are shown in Figure 31-17. The colors have no special meaning and were added just to clarify the distinction between the sections.

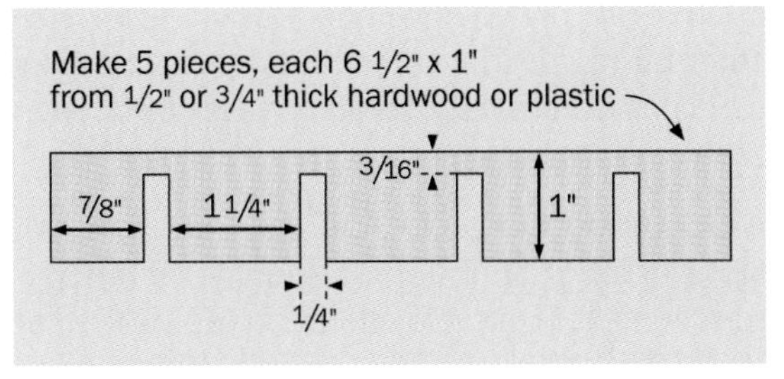

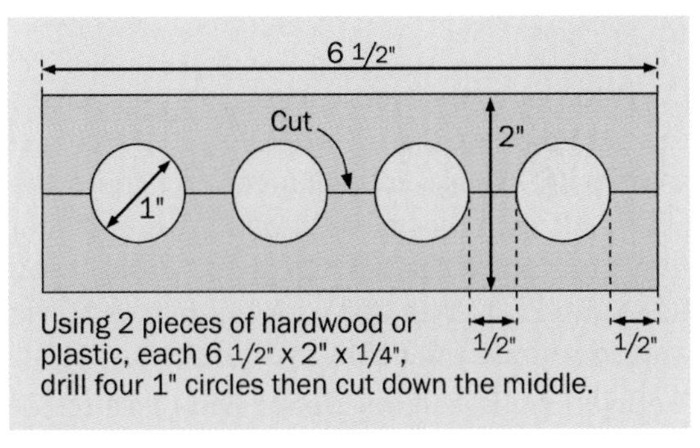

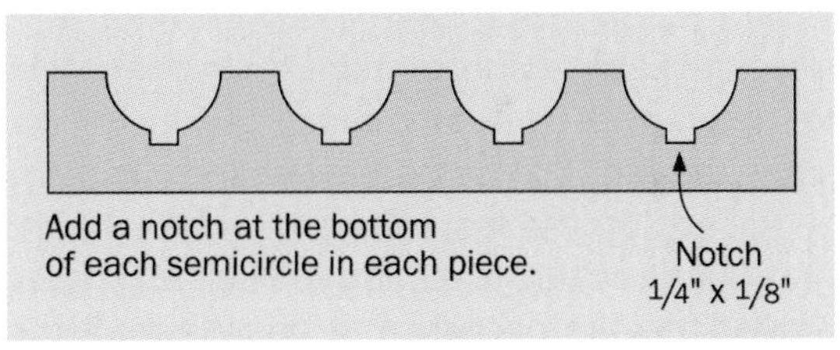

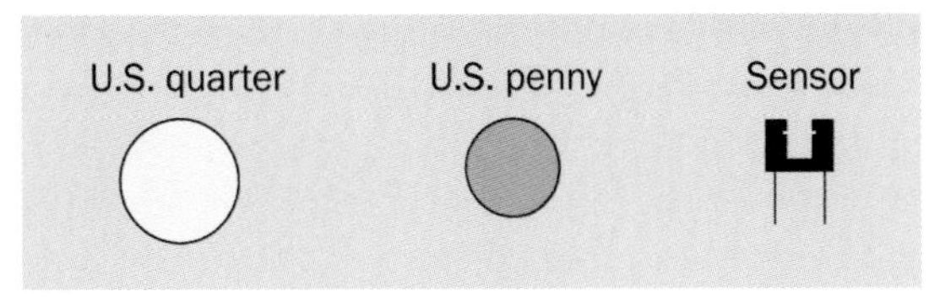

Figure 31-17. *How to cut the sections that can be assembled to enclose the sixteen sensors, coin slots, and win-notification LEDs.*

The spacers are of wood or plastic, 3/4" thick. If you use wood, it should be a hardwood, because there is only 3/16" above each notch, and softwood will provide insufficient strength.

The coin receivers are made by cutting four 1" diameter circles in two pieces of wood or plastic measuring 6-1/2" x 2" x 1/4". The thickness must be 1/4", because it has to match the notch width in the sensors. You may have difficulty finding hardwood in this thickness, in which case a good-quality 1/4" plywood will do. I would use a Forstner bit to cut the circles, but a hole saw may work if you are careful. If you can use plastic such as ABS, that's better.

After drilling the circles, cut each piece in half along its long dimension. The notches at the bottom of each semicircle can be made with a flat metal file with an abrasive edge, or a very small square metal file.

Don't forget to drill vertical holes in the spacers to accommodate the LEDs before you start to assemble the pieces. You will need to extend the LED leads with pieces of 24-gauge wire so that they poke out at the bottom of each hole. If you use 5mm LEDs that don't have a flange at the bottom, they will sit neatly in a hole that is 13/64" diameter. Add a smear of epoxy to hold each one in place.

Use a little epoxy, also, to stabilize each sensor in its notch. Don't get any epoxy on the sensor leads, though. Also, be extremely careful to place all the sensors the same way around. You might draw a little picture to remind yourself which way around they are, because after you assemble the pieces of the enclosure, you won't be able to tell.

The slots are sized so that quarters, pennies, nickels, or dimes will work to interrupt the infrared light in each sensor. Coins much larger than a quarter will not fit in a slot, because a quarter is just a fraction under 1" in diameter. If you live in a country where there are bigger coins, I leave it to you to make holes of the right size to accommodate them.

If you turn the assembly upside down, you'll see that the leads from the sensors are still accessible, so you can build the whole slot-and-spacer assembly before you do any wiring.

The chips in the circuit can be mounted on a separate piece of perforated board, connected with the sensors with ribbon cable. You can then add four sides of thin plywood or plastic to create a box that holds the circuit board and the 16 slots.

This project is high on my "must build!" list, and I regret that I don't have a photograph of a finished version. I'd like to start fabricating it right now, but my primary goal is to finish writing this book so that you can consider building the projects yourself.

In a way, it's good that I don't have my own fabricated enclosure to show you. This way, if you build your own, you won't be influenced (for better or worse) by someone else's work.

Experiment 32: Enhancing Ovid

32

You'll remember from Experiment 23 that I wanted a better way to distinguish between one player's token and the other player's token when two people are playing Ovid's Game. At that time, the best I could do was to suggest that each person should identify himself by pressing a button.

With sensors, this is no longer necessary.

Suppose one player uses magnetic tokens while the other player uses nonmagnetic tokens, all of which will fit into the same holes in the playing board. If each hole is equipped with a Hall-effect sensor (which will respond to the magnetic tokens but not to the others), and if each hole also has a transmissive IR sensor (which will respond to all of the tokens), we should be able to distinguish one type of token from the other type automatically.

This sounds like a logic problem to me, so my first step will be to express it in words.

The Logic Option

Here's how it should work:

If the IR sensor is triggered, AND the Hall sensor is NOT triggered, we must have a nonmagnetic token in the hole.

If the IR sensor is triggered, AND the Hall sensor IS triggered, there's a magnetic token in the hole.

This can be represented by the logic diagram in Figure 32-1. Note that the inverted triangle with the small circle at the bottom is an inverter, which changes a high input to a low input and a low input to a high input. I mentioned inverters in *Make: Electronics* but have not found a reason to use one in this book until now.

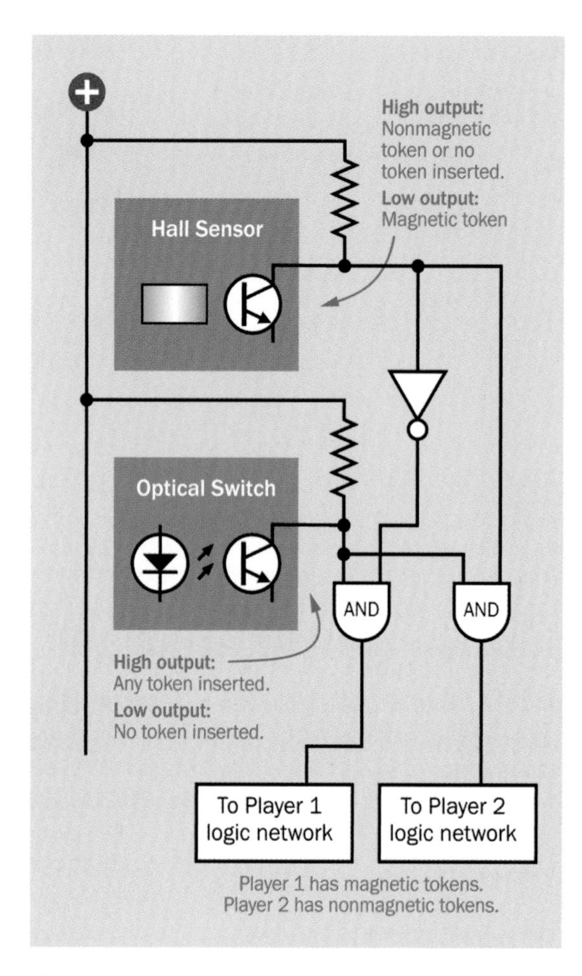

Figure 32-1. *Two AND gates and an inverter can be used to activate the appropriate "Who wins?" logic network in Ovid's Game, depending whether a magnetic token (from Player 1) or a nonmagnetic token (from Player 2) has been inserted in the playing board.*

The output from the righthand AND gate would go to a logic network to detect if the player who is using the nonmagnetic tokens has placed three in a row to win the game. The output from the lefthand gate would go to a logic network to detect a win for the player who is using magnetic tokens. One of those logic networks is shown in Figure 23-4. Each switch in that figure would be replaced by ANDed sensors, as shown here.

One snag is that the sensors won't be activated simultaneously. The Hall sensor is likely to switch on first, because it will be triggered when a magnetic token is still approaching the sensor.

However, I don't think this will matter, because nothing will happen until there is an output from the optical sensor. This is required to activate either of the AND gates.

What concerns me more is that this sensor system is too complicated. The complete game has nine playing positions on the board. Implementing this logic will require five quad two-input AND chips and two hex inverters. That's a lot of wiring. Is there some way to do it more simply?

Switching It Around

As usual, there is no methodical process that can help with optimization. It just requires some creative thought. The first step is to reconsider the roles of the sensors. Really, they perform different functions:

- The optical switch says, "Is any type of token inserted?"
- The Hall-effect sensor says, "Is it a magnetic token or a nonmagnetic token?"

This is really a two-stage operation. Step 1 has a yes-or-no output. That sounds like a logic gate. But step 2 has a "one thing or the other" output. That doesn't sound so much like a logic gate. It sounds more like . . . a double-throw switch! So maybe a Hall-effect sensor isn't really the best tool for the job. How about if I substitute a single-pole, double-throw reed switch?

I mentioned previously that double-throw reed switches do exist. I also said that reed switches create problems for logic gates, because their contacts bounce. Still, the contacts settle very quickly, and if the reed switch is activated first, as the player inserts the token, the contacts should have time to settle while the token slides into place and activates the optical switch. By the time the switch says, "I see a token," the reed switch should be saying what type of token it is, without any hesitations.

Therefore the sensing circuit can be simplified to eliminate logic gates completely. The optical sensor provides an output that is high or low, while the reed switch routes it to the AND gates for Player A or Player B. This is shown in Figure 32-2.

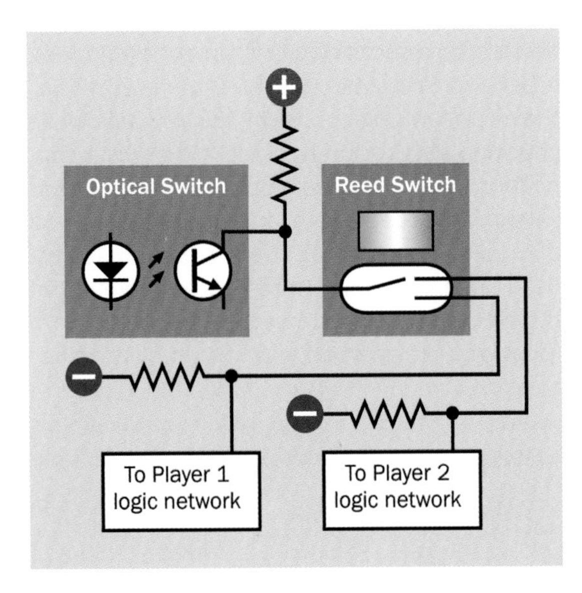

Figure 32-2. *Use of a SPDT reed switch instead of a Hall sensor can eliminate the logic gates that were necessary in the previous version of the token sensing circuit.*

The only snag is that a switch will leave one circuit open, and logic gates must never have an open circuit connected to one of their inputs. Therefore we will have to add pulldown resistors. These resistors must have a relatively high value so that the signal from the optical sensor will be able to overwhelm them and change the input voltage.

In Figure 32-2, the pullup resistor for the optical switch will probably be around 2K. Each pulldown resistor for a logic network would be 10K.

This looks promising. But in Ovid's Game, a player can remove a token to reposition it in another slot. Will we get switch bounce, once again, when the magnetic field is released?

Yes, but by the time the removal of the token allows the reed switch to flip back into its other position, the output from the optical switch will have changed from high to low. And internal hysteresis in the optical switch insures that it has a clean output.

Magnetic Issues

Another issue that has to be resolved, here, entails matching the magnet with the reed switch. The magnet has to be strong enough to operate the switch reliably, but not so strong that it affects adjacent switches. I'm thinking that the holes in the game board will have to be spaced at least an inch apart, and the reed switches must be located a safe distance underneath so that none of them will be activated if a player simply drops a token on the board.

The token could be like a peg with a magnet at the tip. Suppose you use little rectangular magnets, about 1/4" × 1/2" and 1/16" thick. You could cut a slot in the bottom of a wooden or plastic peg and glue the magnet into the slot.

We also have to be sure that the reed switch and the magnet are properly oriented. While a SPST reed switch isn't very fussy about its placement relative to a magnetic field, the reed in a DPDT reed switch must be "pushed" or "pulled" from one contact to the other. I'm thinking that the game tokens must be shaped so that they won't fit into a hole the wrong way around. They could be T-shaped in cross section, to fit T-shaped holes in the playing board.

This leads me to yet another possibility. If the peg was sufficiently wide, it could have a magnet mounted on one side, but not the other. Player 1 could have magnets on the left sides of the pegs, for instance, while Player 2 could have magnets on the right sides of the pegs. You could then get rid of the optical sensors and have two reed switches on opposite sides of each hole. One set of reed switches would be activated by the pegs used by Player 1, while the other set of reed switches would be activated by the pegs used by Player 2. Figure 32-3 shows how this could work.

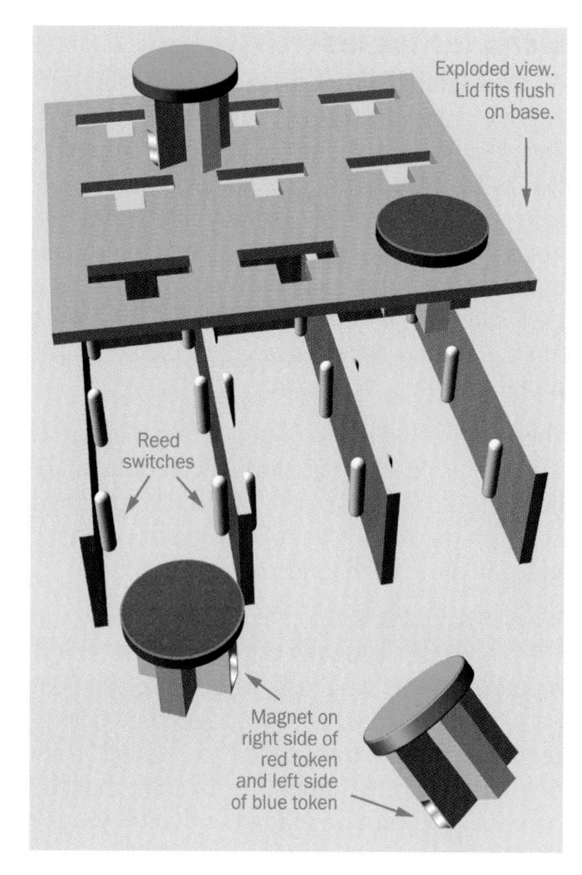

Figure 32-3. *Simplified rendering of a version of the game with one set of reed switches activated by magnets on Player 1's tokens, and a second set for Player 2. To avoid simultaneously triggering two reed switches on opposite sides of a wooden divider, the divider should probably be thicker in a functional version of the game.*

The advantage of this configuration is that you could just replace the manual pushbuttons in the original schematic for Ovid's Game with SPST reed switches. Job done!

Careful testing would be required to make sure that the system would be reliable. This is the down side of using sensors: you have to deal with mechanical attributes of the real world.

Make Even More: Microcontrolling It

Ovid's Game is difficult to adapt to a microcontroller, because there are nine positions on the playing board, and each of them can have three states: empty, occupied by Player 1, or occupied by Player 2. That represents a lot of inputs.

However, the number of inputs can be greatly reduced by matrix encoding—the same system that is often used to detect if someone has pressed a key on a key pad. In a three-by-three matrix, the microcontroller polls one row at a time while checking each column to see if there is a connection at the intersection. The basic idea is shown in Figure 32-4. Using this system, instead of requiring nine inputs, the microcontroller just uses three inputs and three outputs. Note that the diodes are necessary to prevent current from following incorrect pathways if multiple switches are closed.

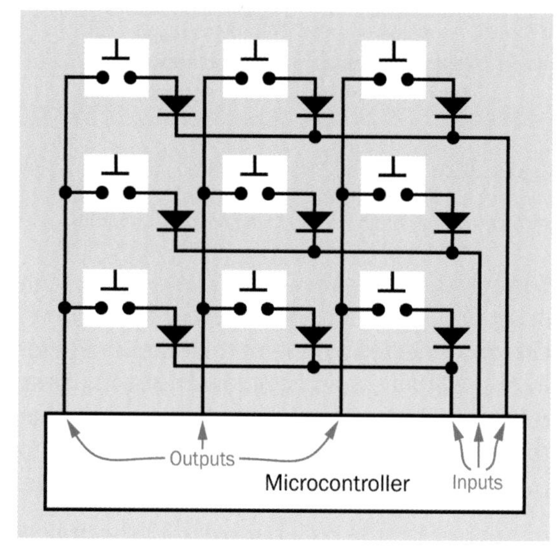

Figure 32-4. *The basic concept of matrix encoding.*

In Ovid's Game, if two sets of reed switches are used for the two players, matrix encoding can be modified as shown in Figure 32-5. The two sets of switches have been colored pale red and pale blue in this figure to indicate that one set will be operated by one player's tokens, while the other set will be operated by the other player's tokens.

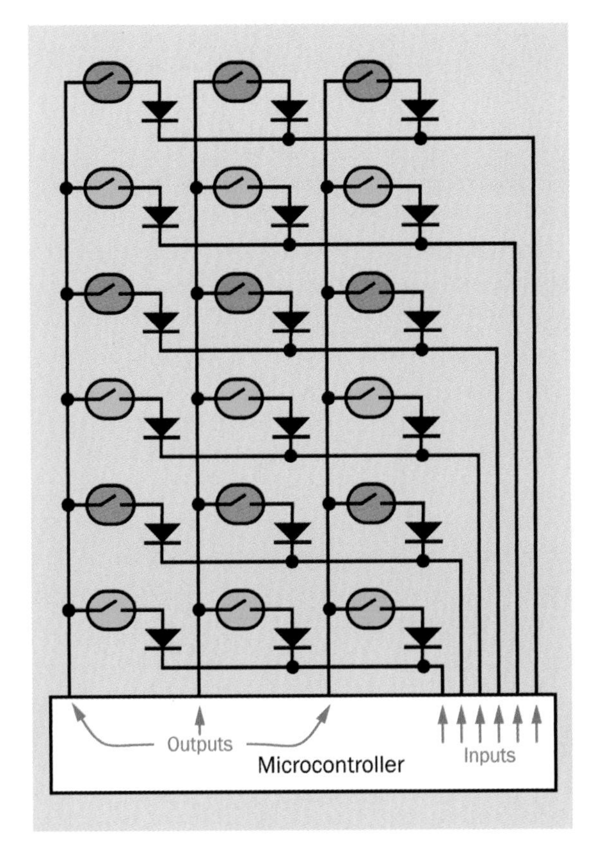

Figure 32-5. *Matrix encoding of Ovid's Game where the two players activate separate sets of reed switches, colored pale red and pale blue.*

Alternatively, could a phototransistor that uses visible light distinguish between two sets of player tokens, if one was matte black and the other was glossy white?

Or would a phototransistor be able to distinguish between two colors of tokens—say, red and green—if they were illuminated by different colors of LEDs? The color green appears bright in green light and dark in red light, while red appears bright in red light and dark in green light. You could use a comparator to determine if a green token or a red token was in the slot.

Probably you can think of other methods for token identification. The best will be the one that is reliable, cheap, easy to use, aesthetically pleasing, and reasonably easy to build. I'd love to see any ideas that you come up with.

But now that I have kept my promise to suggest some ways to improve the user input for Ovid's Game, I want to move on to other types of sensors.

Experiment 33: Reading Rotation

33

Everyone is familiar with that old-school favorite, the potentiometer. In this book alone, I have used maybe twenty trimmer potentiometers in the various circuits.

In consumer electronics, it's a different story. Your car stereo probably still has a volume-control knob, but it will turn a full 360 degrees—and keep on turning. This suggests that something other than a potentiometer may be hidden behind it.

What component has taken its place? The answer is a rotational encoder, also known as a rotary encoder, or an incremental encoder, or an electromechanical encoder (because it has mechanical contacts inside it). A selection of them is shown in Figure 33-1. Their appearance is reminiscent of potentiometers—indeed, most of them even have three terminals. But their behavior is very different.

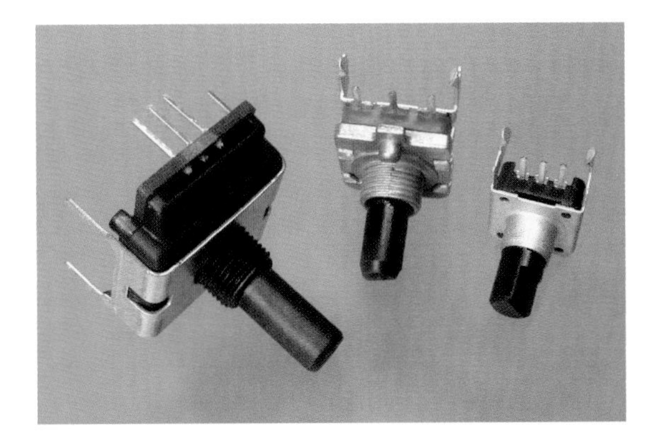

Figure 33-1. *A selection of commonly used, inexpensive rotational encoders.*

Defining a Rotational Encoder

First we have to distinguish clearly between different types of encoder. Obviously I'm not talking about solid-state encoder chips. We already explored them in detail, and you must have noticed that they didn't have any knobs on them.

A rotational encoder has a shaft and at least two terminals, and generates a stream of pulses from internal contacts while its shaft is rotating. Another component (usually, a microcontroller) must interpret that stream and decide what to do in response. It can adjust the sound in an audio system, or cycle through some prompts on a screen, or perform any other task specified by its program.

Originally a rotational encoder was a high-end kind of component, often using optical methods to measure rotation very precisely (more than 100 intervals within 360 degrees). That definition has changed. Any component containing contacts that generate a pulse train when you turn a knob is now likely to be called a rotational encoder.

Specification

For this experiment, you need a rotational encoder with the following attributes:

- "Quadrature" output (I'll explain what this means below).
- A "resolution" of at least 24 changes-of-state in each complete turn of 360 degrees. This may also be expressed as "pulses per revolution," or PPR.
- The same number of "detents" as the resolution. Detents are the little clicks that create momentary resistance when you turn the shaft.

The Bourns ECW1J-B24-BC0024L is an example, but there are more than a hundred alternatives with 24 PPR and 24 detents. Because the field is evolving rapidly, any encoder I specify today is liable to be replaced by a slightly different one tomorrow. Don't be afraid to make a substitution, so long as you read the specifications carefully.

- Some decoders are not of the "quadrature" type and may have a number of terminals greater or less than three. I won't be dealing with those types here.

The Pulse Train

Demonstrating the behavior of an encoder is very easy. Although the contacts inside it will not tolerate much current, they can deal with the few milliamps required to light an LED.

Set up your encoder as shown in Figure 33-2. Many encoders have pins or leads that are spaced 0.1" apart and can plug directly into a breadboard, as shown in Figure 33-3.

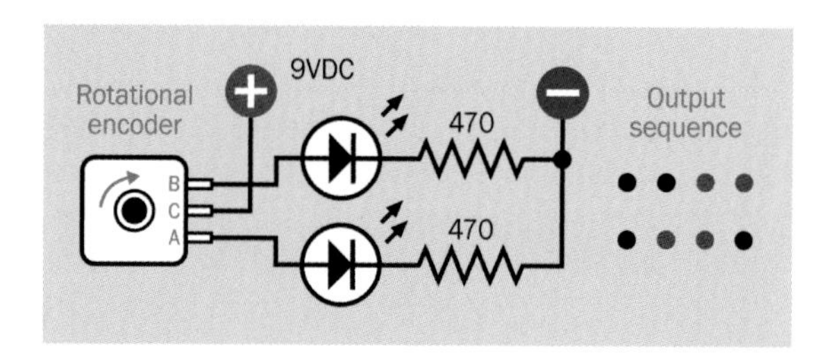

Figure 33-2. *When a rotational encoder is wired as shown, its two outer terminals should run through the output sequence shown here.*

Figure 33-3. *Many encoders, such as this one, can be plugged into a breadboard for testing.*

Turn the shaft very slowly, and you should see the LEDs lighting up in the output sequence that is shown in Figure 33-2. Now turn the shaft in the opposite direction, and you should see that the pattern reverses itself. This is the pulse train I referred to earlier.

Note that the black dots in the diagram do not represent a logic-low state. An encoder output is either "high" or "off" because it has little switches inside. When they are "off," they are open circuit and not connected to anything.

An encoder may be wired with negative ground attached to the center terminal, instead of the positive side of the power supply that I have chosen to use. In that case, the output can be thought of as a series of negative and off pulses, instead of positive and off pulses.

Warning: Mediocre Encoders

If the output sequence created by your encoder doesn't look like the sequence of red and black dots shown in Figure 33-2, or if you see hesitations or flickering, this may be because you are using a cheap encoder. Paying a little more can get you a cleaner, more regular output. This is desirable for the experiment that follows.

Inside the Encoder

The pulses that you have seen are created by two pairs of contacts inside the encoder, mounted slightly out of step with each other (that is, they are out of phase). Figure 33-4 shows this concept.

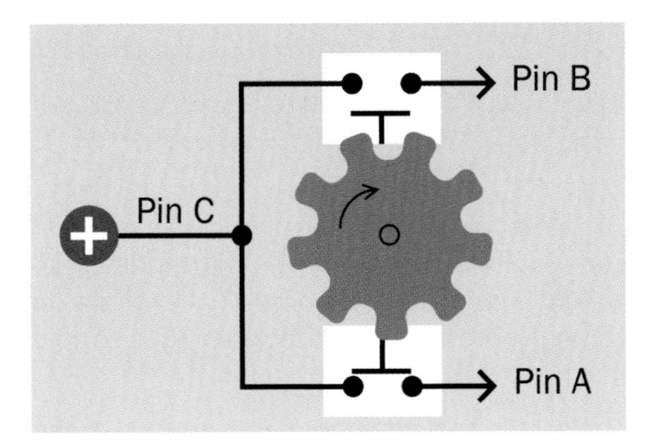

Figure 33-4. *In this diagram, the pushbuttons represent pairs of contacts inside a quadrature encoder.*

A graphical representation of the output is shown in Figure 33-5. Each of the dashed white lines represents a detent, and the A and B output combination changes from each detent to the next.

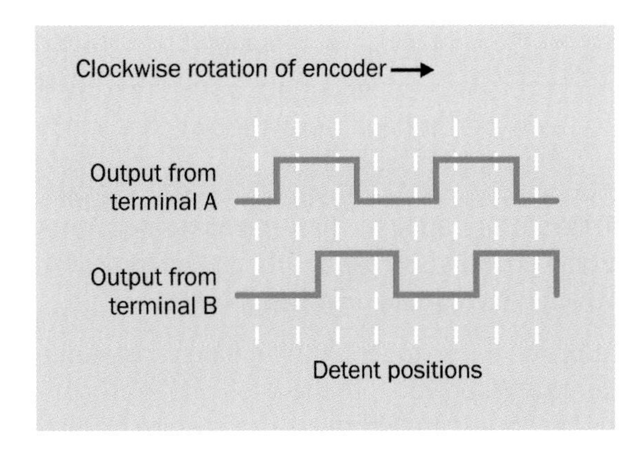

Figure 33-5. *Output of an encoder, where the number of detents is the same as the resolution.*

The way in which encoders are described can be confusing. Here are the rules:

- An encoder of the type that I have been describing, with two output terminals that have four possible on-off combinations, is called a quadrature encoder.
- The resolution is the number of transitions of each terminal, both up and down, multiplied by the number of terminals (two, in the case of a quadrature encoder), within a complete rotation of 360 degrees.
- Therefore, the resolution is the same as the number of state changes during one revolution.
- The term "PPR" counts "off" pulses as well as "on" pulses, from both terminals. Therefore PPR means the same thing as resolution.

Encoder Usage

The encoder is a very simple-minded, lazy device. All it does is click its little internal switches, leaving some other component to do the intelligent work of interpreting their sequence.

That other component could be a logic chip—but if you plan to connect a rotational encoder with any type of digital logic, there is a very important point to remember:

- You must always use pulldown or pullup resistors on encoder outputs so that they have a defined voltage when the switches inside the encoder are open.

This rule also applies when you are using a rotational encoder as an input to a microcontroller, which is its most common application.

Suppose the encoder is functioning as a volume control. When you turn the knob, the microcontroller figures out which way you're turning it by comparing the pulse trains from the two terminals. For example, if Switch A closes a moment before Switch B, this indicates that the direction of rotation is clockwise. If Switch B closes before Switch A, the direction is counterclockwise.

Having determined that, the microcontroller then counts the number of pulses to find out how much you want to turn the volume up or down.

Programming a microcontroller to respond in this way is more complicated than it seems, because in addition to the other chores, the program should ignore the contact bounce that a rotational encoder creates. Fortunately I don't have to worry about all this, because I'm not going to be using an encoder with a microcontroller. I have my own peculiar plans for it.

Incidentally, you may be wondering what a rotational encoder is doing in a section of this book devoted to sensors. Surely it's an input device, not a sensor?

This is true. It is not a sensor. It is an input device —but it's a useful thing to know about, and I'm going to use it as a sensor.

It Can Be Random

If an encoder is symmetrically designed, and if someone turns the knob to a random position, you will have an equal chance of ending up with outputs A and B both low, or A high and B low, or A low and B high, or A and B both high.

Hmm—those combinations sound like every possible combination of two binary digits—and indeed, they can be used this way.

I explored this idea briefly in a column that I wrote for *Make* magazine about a "Magic 8 Box" that would function similarly to the old Magic 8 Ball toy. My idea was to attach a lead counterweight to an arm mounted on an encoder shaft, as shown in Figure 33-6.

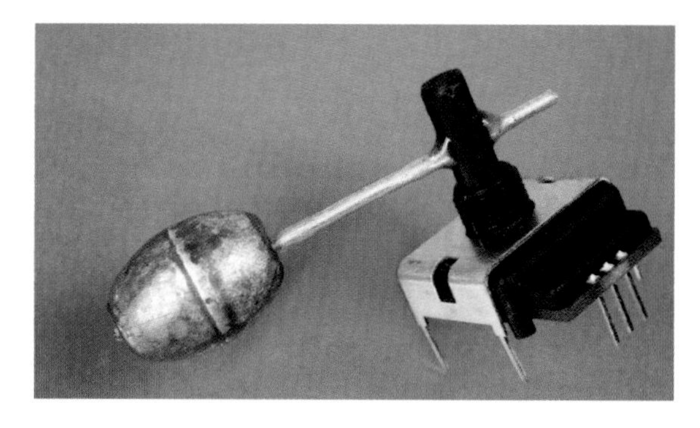

Figure 33-6. *A lead sinker weight, normally used by people who go fishing, has been added to this encoder on the end of a stiff piece of galvanized wire.*

I drilled a hole through the plastic shaft of an encoder, drilled another hole through the little lead weight, inserted a piece of #14 galvanized wire through both holes, and used epoxy glue to hold it in place.

The wire is a standard item in hardware stores, and I bought a little package of one-ounce lead weights in Walmart. They're used as "sinkers" by people who catch fish. I don't go fishing, myself, so I'm a bit sketchy on the exact purpose of sinkers, but you don't need to know how they are normally used in order to use them for something else. You'll find them in the fishing section of the sports department.

Now, suppose the rotational encoder is secured firmly to the inside of a box, which also contains a circuit that you have built. Suppose someone turns the box in various random directions. Why would a person do this? Well, maybe you could mount an on-off switch for your circuit on the underside of the box so that the person who wants to use it must turn it over to find the switch. The inertia of the lead weight on the shaft of the encoder will cause the shaft to rotate—and where it stops, nobody knows.

To make things more interesting, suppose we have a second rotational encoder mounted on an inner wall of the box, at 90 degrees to the first. This way, the motion of the box will be sensed along two different axes. Since each encoder delivers a pair of outputs that can be thought of as generating a two-bit binary number in the range 00, 01, 10, and 11, we can combine the outputs from the two encoders to get a four-bit binary number ranging from 0000 to 1111, and it will be totally unpredictable. This can be supplied to the input of a decoder chip or the control pins of a multiplexer, to choose a random number from 0 through 15.

Other ideas are also possible—and indeed, I have one right here.

Rotational Decider

You can reduce the four encoder outputs to a single output that can be either high or low. This could function as a "decision maker" toy, like a very dumbed-down version of the I Ching. Think of a question, pick up the box, shake it, put it down, press a button, and it lights either a "yes" LED or a "no" LED.

Part of the appeal of this toy is that you can feel the encoder detents clicking and the weights shifting when you move the box. It feels as if something really complicated and mysterious is going on in there. (If someone asks you about this, you can tell them that it's just too complicated and mysterious to explain.)

Perhaps you're wondering why we need two encoders, each creating four possible states, if all we need is a yes-or-no answer. The reason is that the device should be as unpredictable as possible, and the more inputs you can throw in, the less predictable the final result will be.

But how do you reduce four outputs to one? By XORing them. This is shown in Figure 33-7. Note the four pulldown resistors that insure there is always a defined voltage on the encoder outputs.

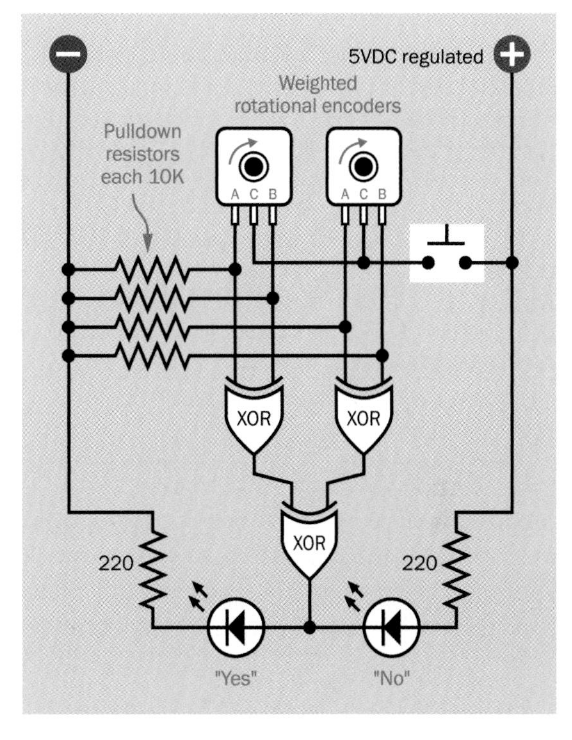

Figure 33-7. *This circuit produces a random "yes or no" answer from a couple of weighted rotational encoders.*

Trace the logic through this circuit and you'll find that the output from the XOR gate at the bottom has an equal chance of being high or low—so long as the rotational encoders are behaving in an even-handed fashion.

One nice aspect of this simple circuit is that it needs no on-off switch. The encoders consume no power, so you can shake them into a position and put down the box before you press the button. Depending whether the output of the lowest XOR gate is high or low, the button will power the "Yes" LED or sink current from the "No" LED. Remember, a 74HC00 series logic gate can sink the same amount of current as it will source.

- You must use an HC series chip in this circuit. It won't work with older series of chips that have insufficient output power or cannot source as much current as they sink.

A schematic for this circuit is shown in Figure 33-8.

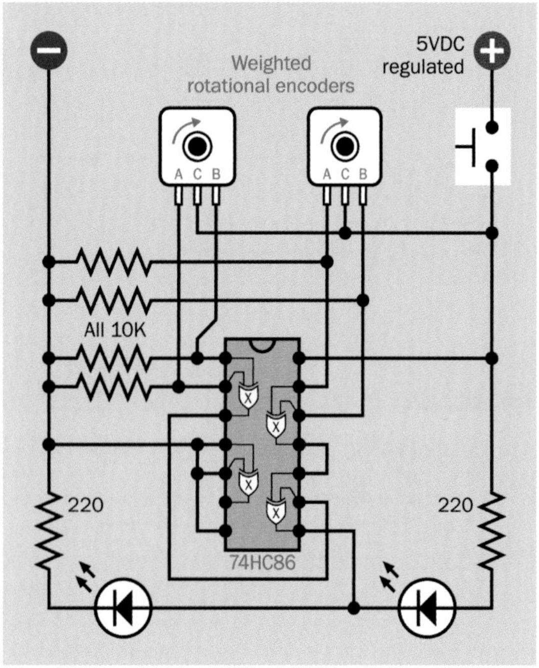

Figure 33-8. *Schematic for the encoder-based yes-or-no decider toy.*

Rotational Equivocator

Equivocation is what you do when you don't feel willing to make a decision. Elected legislators do this all the time. For example, if I were a legislator and someone asked me, "Can we adapt the Rotational Decider to become a Rotational Equivocator?" I could say, "Well, there are two sides to this issue; I think something is to be said for both of them, and the topic may need further study."

Since I'm not a legislator, I won't say that. I'll say, "Yes, of course we can build a Rotational Equivocator."

The logic diagram for it is shown in Figure 33-9. It will light the "Maybe" LED if the left XOR has a high output and the right XOR has a low state. It lights the "Maybe not" if the states are reversed.

But what if both XORs have the same output—either both low or both high? In that case, the third XOR has a low output and sinks current through the third LED, which says "I'm not quite sure."

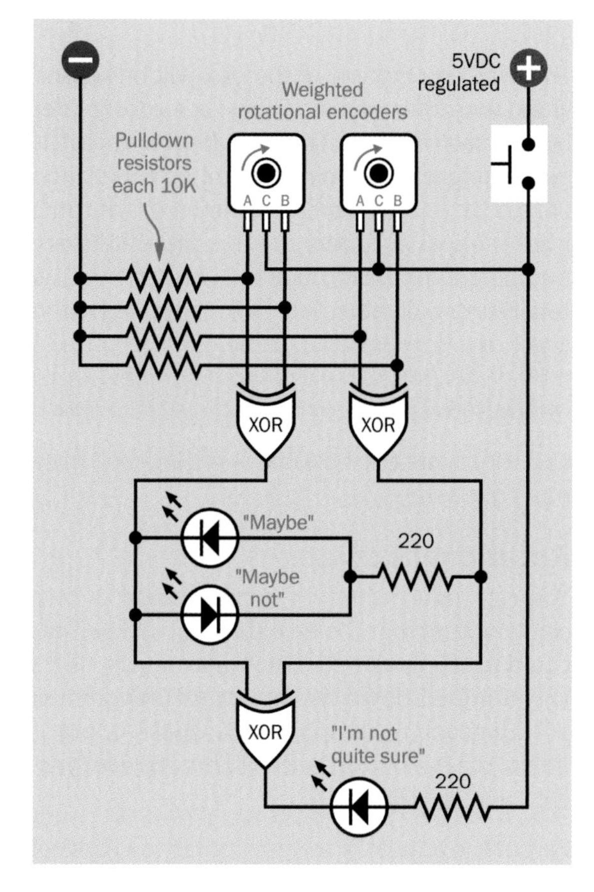

Figure 33-9. *The previous yes-no circuit has been modified to give equivocal answers.*

The schematic is shown in Figure 33-10, while a photograph of the circuit appears in Figure 33-11. In the photograph, LEDs have been used that contain their own resistors. The rotational encoders have been placed far apart to allow their lead weights to rotate without hitting each other.

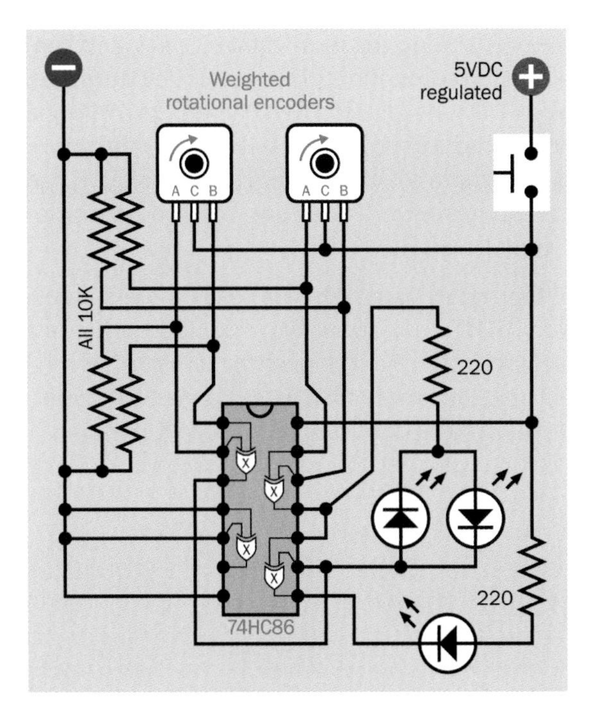

Figure 33-10. *The Rotational Equivocator schematic requires only a very small modification of the previous yes-or-no schematic.*

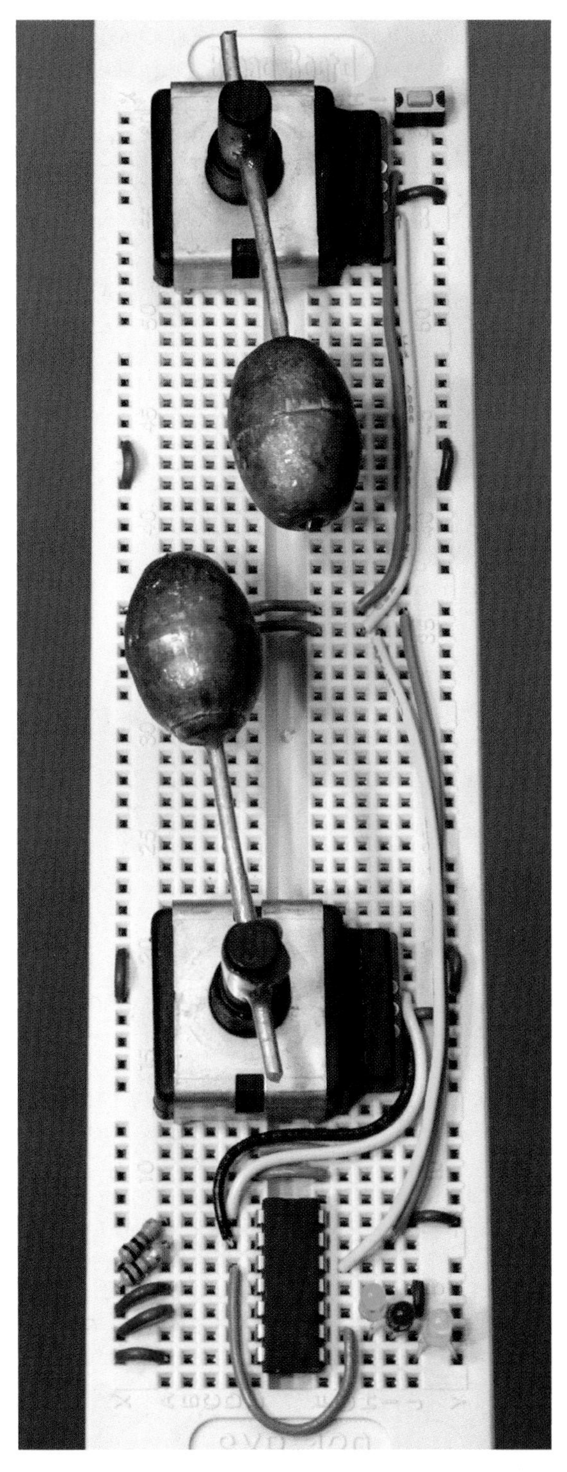

Figure 33-11. *Breadboarded version of the Rotational Equivocator. No series resistors are included with the LEDs, as they have their own internal resistors.*

If you follow the logic in Figure 33-9, you'll find that the "Not sure" output is twice as likely as either of the other outputs. This insures that the toy will be as unhelpful as possible.

If you have a friend in politics, maybe you could give this person one of these as a campaign contribution.

Seriously Random

I included this little digression into rotational encoders because I wanted to include an easy project among the more difficult ones, and I enjoy using components for unintended purposes. Also, I like the idea of a toy that requires only one logic chip, a couple of LEDs, and a pushbutton. If you include a 5VDC voltage regulator, you can power it from a 9V battery that should last for several years.

Other types of sensors are really more suitable for creating randomicity. I'm going to suggest several options in the next section, after which I will explain a way to achieve "perfect" randomicity—and I'll discuss what this really means.

Experiment 34: Ambient Sensing

34

In this section, I'm returning to the basic concept that I have used before, of choosing a random number by stopping a fast timer at an arbitrary moment.

This was the technique in the Ching Thing in Experiment 28, and also in the Hot Slot game in Experiment 21. However, both of those games required the player to do something that would stop the timer. What I'm going to describe here is a way for a sensor to inject a random factor without any user input.

First I'll recap the basic setup of a slow monostable timer, which starts and stops a fast asynchronous timer, which drives a counter. I'll be using this in all the sensor-based randomizing applications below.

Then I'll pause for a moment to show you how a timer can be wired to restrict the random number sequence. Instead of counting from 0 through 15 or 0 through 9, it can count to any lower number. You can even make it count from 0 through 1.

Finally, I'll throw in the sensors.

One Timer Controlling Another

In the circuit in Figure 34-1, the timer at the top is wired in the same way as the timer in Figure 27-1, to produce a single clean pulse when the DPDT pushbutton is pressed.

The output from the first timer lights an LED (the little yellow circle) just to confirm that it's working. It also connects with pin 4, the reset pin, of the second timer. When the reset pin is high, the timer is allowed to run. So, the high output from the first timer unlocks the second timer (for a limited time only).

The second timer sends a stream of pulses to the ring counter, and you'll see the LEDs labelled 0 through 9 counting in sequence.

When the first timer reaches the end of its cycle, it pulls down the reset pin of the second timer, stopping it and leaving one LED illuminated.

Press the Counter Reset button, and then press the Rerun button—which is not debounced, because for testing purposes, I'm expecting you to let go of it before the pulse from the slow counter is over.

You should find that at the end of each cycle, the same LED is left glowing every time. No randomness, here—the components are doing what we expect them to do. They are behaving consistently.

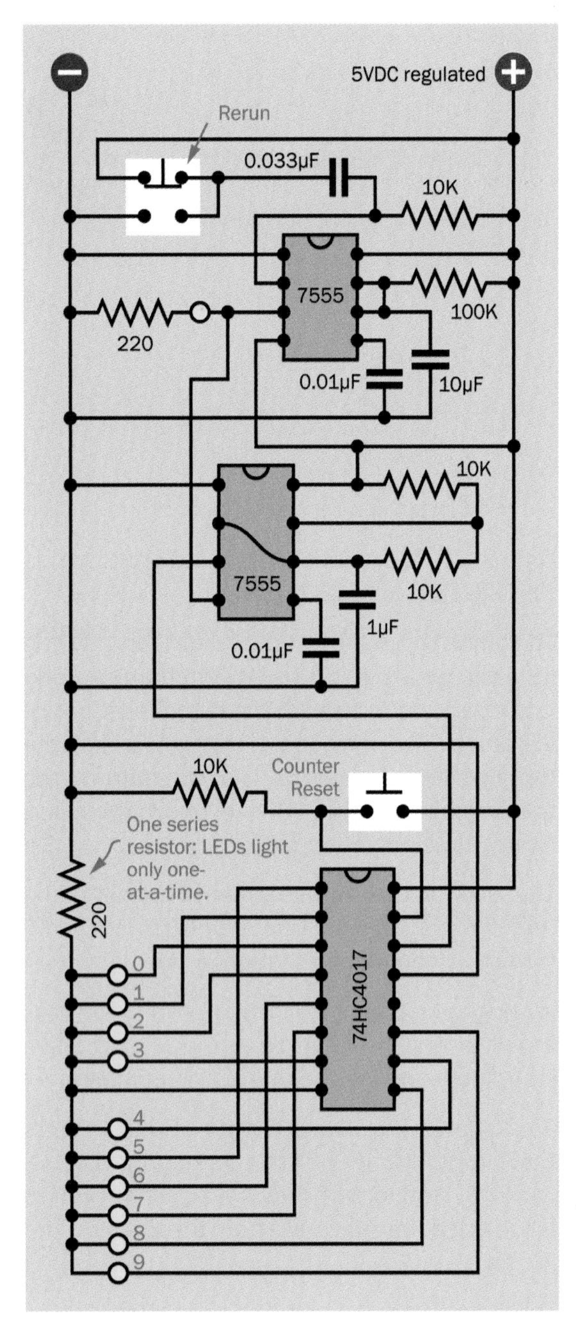

Figure 34-1. *This basic circuit illustrates the concept of a slow timer enabling a fast timer for a limited period. The ring counter controlled by the fast timer should stop in the same state every time, so long as you press the Counter Reset button before each run.*

Temperature Control

Now we're going to make things a little more interesting by using a thermistor, which is like a resistor, except that it changes its resistance with temperature. A photograph of a thermistor is shown in Figure 34-2. It's very tiny because the smaller it is, the more rapidly it will respond to changes in temperature. It has long leads because the longer they are, the less heat they will transport to or from the thermistor itself. For the same reason, the leads are very thin.

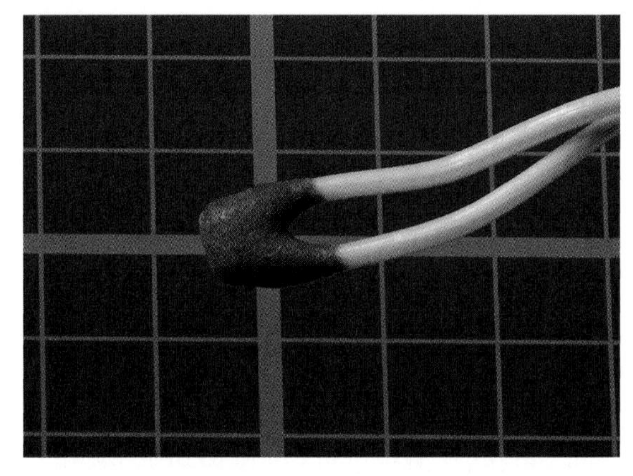

Figure 34-2. *A high-quality thermistor. Its small size allows it to respond rapidly to changes in temperature.*

The thermistor that I'd like you to use is rated at 100K, which is the baseline resistance it should have at 25 degrees Celsius. A thermistor has no polarity, so you don't have to worry about which way around you connect it.

To insert it in the circuit, simply remove the 100K resistor from the first timer and substitute the thermistor.

Repeat the cycle. So long as the temperature of the thermistor remains constant, you should end up with the same LED illuminated every time, because the first timer will give the second timer the same window of opportunity in which to do its counting operation.

Run the test a few times. Don't forget to press the Counter Reset between each test and the next. Now grip the thermistor between your finger and thumb to increase its temperature. Run the test again. Does the result vary?

Random Factors

In addition to the changing resistance of the thermistor, various factors may have an effect:

- The timers themselves change their performance slightly as they warm up with use.
- The Rerun pushbutton may not perform exactly the same way every time.
- You may have variations in your power supply.
- Connections on your breadboard have some resistance, which may vary if you touch any of the wires or components.
- There could be some other ambient factor that I haven't thought of.

Automating the Randomizing Circuit

We can speed up the testing process by getting rid of the user input.

The first step is to change the first timer from one-shot to astable mode. It can run for one second, then pause for one second, then run for another second, and so on. This way you can just sit back and watch the results instead of constantly having to press the Rerun button. Figure 34-3 shows the rewired circuit.

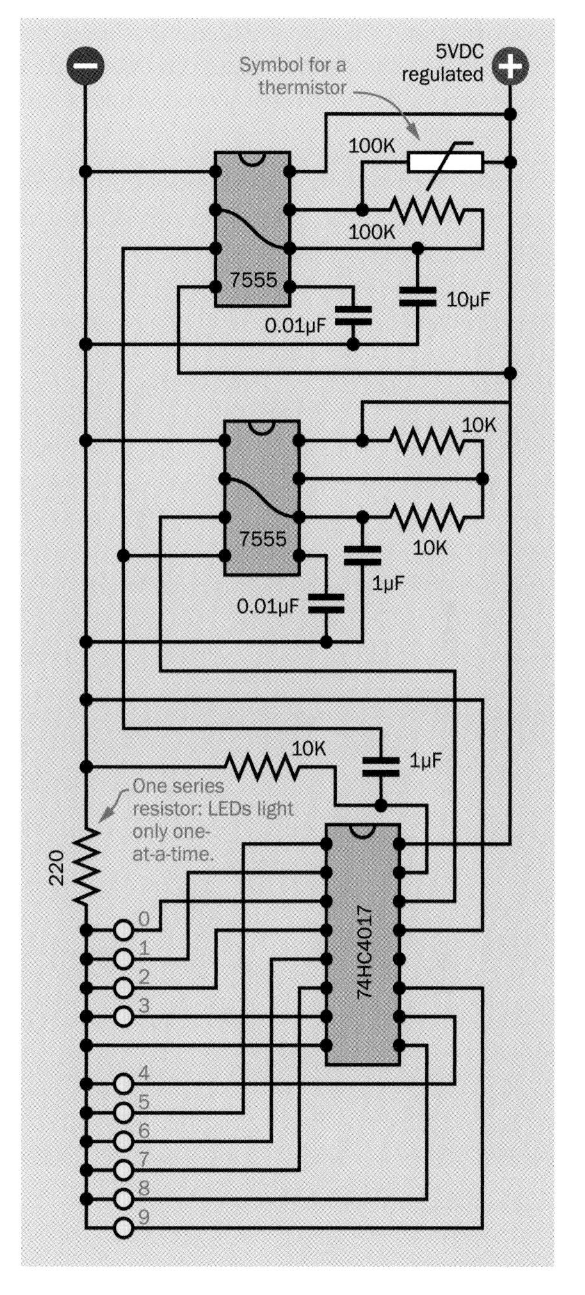

Figure 34-3. *The previous version of the randomicity test circuit, rewired so that it runs automatically, without user input.*

The second step is to eliminate the Counter Reset button and fix the counter so that it resets itself. The 74HC4017 counter's reset pin responds to the rising edge of a transition from low to high. Well, the output from the first timer rises from low to high at the beginning of each cycle, so let's connect it to the counter reset pin—through a capacitor, so that the reset pin only goes high momentarily.

I have incorporated this modification in Figure 34-3. You can also see a breadboarded version of the circuit in Figure 34-4.

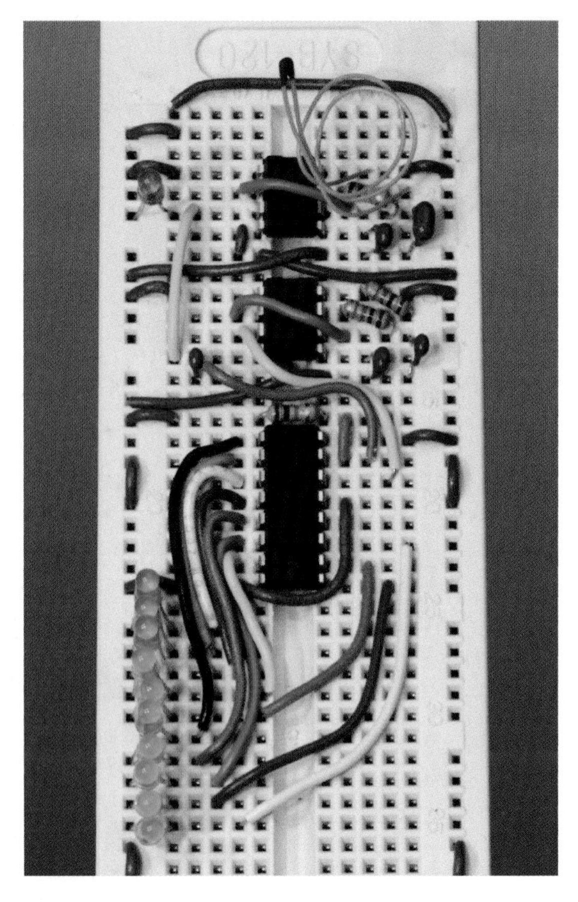

Figure 34-4. *The original thermistor testing circuit described in the text has been modified to run itself without user input.*

Now the system should run itself.

Background: Lower Counting

I'm going to mention this important detail before I leave the subject of the ring counter.

- You can easily modify most counters so that they count to a lower number. The ring counter is no exception.

Pin 1 of the 74HC4017 timer normally goes high to light LED number 5. What do you think will happen if you put a jumper between Pin 1 and Pin 15, the reset pin?

The counter will illuminate LEDs 0, 1, 2, 3, and 4 as usual. Then it gets to Pin 1, which normally lights LED 5, and the high output feeds back to the reset pin. Within a fraction of a second, the counter resets itself to 0 and stops its output to pin 1. Consequently, LED 5 never has a chance to emit any visible light.

The counter then resumes counting from 0 upward, so long as clock pulses are still arriving. It will repeat the 0 through 4 sequence indefinitely and has been transformed from a decade counter into a half-decade counter.

- Connecting an output pin back to the reset pin is standard procedure to change the cycle length of a counter.

This is particularly easy in the case of a counter with decoded outputs, as you can choose any number as your cutoff in the count cycle. When using a counter with binary weighted outputs, you have fewer options. For example, suppose you connect the third binary output pin back to the reset pin. The third pin has a value of 4, so the counter will restart itself after counting from 0 through 3. What if you want the counter to restart at, say, a value of 6? That will be a little more difficult.

You can address the problem by using logic gate(s) to select a combination of pins corresponding with the cycle length that you want. An AND gate between the second and third pins will generate a positive output when the counter reaches binary number 110, which is 6 decimal. Connect the AND output to the reset pin, and your binary counter now runs from 000 to 101 (0 through 5 decimal) before repeating:

- Limiting a counter is a valuable technique when you want a counter to choose a random number outside of the usual range, for a game.
- To convert a 74HC4017 counter so that it only counts from 0 to 1 before repeating, just run a jumper from pin 4 (the pin that has value 2) back to the reset pin.

Now, I'll get back to sensors and randomicity.

Speed Adjustment

If you're not achieving much variation from your automated randomizing counter output, you should increase the running speed of the second timer. 50Hz is very slow, and I chose it only so that you could see the LEDs flashing in sequence. Remove the 1µF timing capacitor from the second timer and substitute a 0.1µF capacitor, to get 500Hz. Or use a 0.01µF capacitor to get 5,000Hz.

The faster the second timer is running, the more its interrupted state is likely to vary because of small changes in the run time. This is explained in Figure 34-5.

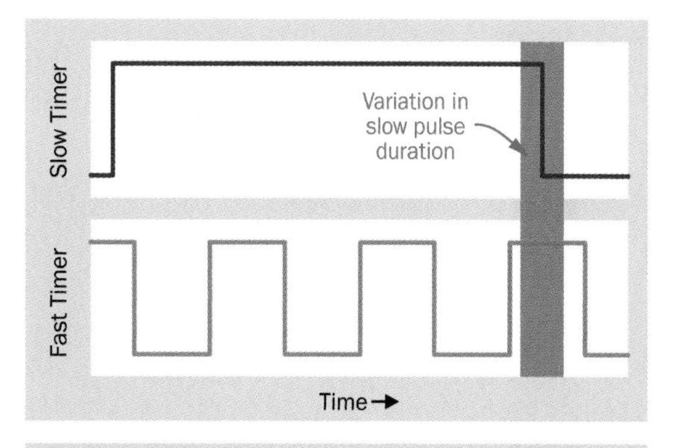

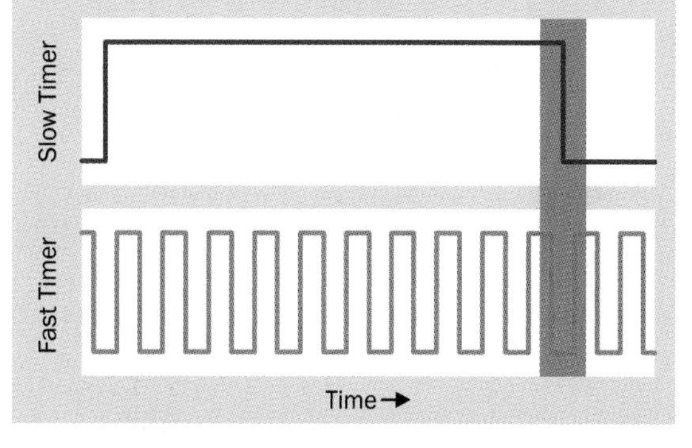

Figure 34-5. *When a slow timer controls a second timer, variations in the slow timer speed (shown by the darker blue rectangle) will stop the second timer at a wider range of pulses when the second timer is running faster.*

In the upper half of the figure, the slow timer pulse length (purple) can vary between the limits shown by the darker blue rectangle, as a result of sensor variations in its timing circuit. Because the second timer (orange) is not running very fast, it will be stopped at the same number of pulses, both at the low end and the high end of the slow timer's range.

In the lower half of the figure, the second timer is running faster. Consequently, it will stop at two different possible pulses at the low end and the high end of the slow timer's range.

If you use a 1nF capacitor to create 50,000Hz, I'm willing to bet that the second counter will stop in a different place almost every time—at least, so far as you can tell.

So, have we solved the whole problem of automatically randomizing the behavior of games to make them seem unpredictable?

Well, maybe.

First I want to go a bit farther into the topic of thermistors.

Quick Facts About Thermistors

Thermistors change their resistance in response to temperature. The NTC type decreases its resistance as the temperature rises and has a linear response over a reasonably wide range (usually from around -40 to +125 degrees Celsius). The PTC type increases its resistance quite suddenly when its temperature goes up.

NTC stands for Negative Temperature Coefficient, while PTC stands for Positive Temperature Coefficient. The PTC type is often used as a substitute for a fuse, to block current from a power supply when it exceeds a maximum value.

The NTC type is the one we're using here.

Thermistors are cheap. Many of them cost less than 50 cents, and they are available in a wide variety of basic resistance, which is usually measured at 25 degrees Celsius,

To test a thermistor, set your meter to measure kilohms and touch the probes firmly against the leads of the component. Don't let it touch your hand, as the contact will affect not only the temperature of the thermistor but also the resistance that you measure. Press the probes against the thermistor leads on a firm, insulating surface, and wait for the meter reading to stabilize.

Now move your other hand close to the thermistor, without actually touching it, as that could affect the contact being made with the meter probes. You should see the thermistor resistance gradually changing as it feels heat radiating from your skin. Physically small thermistors will respond more rapidly than larger thermistors, as they have less mass to be heated or cooled.

Making a Thermistor More Random

Many factors can affect the temperature around a circuit, but heat is also generated by the circuit itself. You could put a thermistor inside a box with your circuit. To make the thermistor more responsive, you could tape it to a 220-ohm resistor, wired straight across your power supply. The resistor would consume about 100mW at 5VDC, well within the usual quarter-watt operating range, but sufficient to generate a little heat. Naturally, this wouldn't be a very smart idea for battery-powered devices. You should also be careful not to push the thermistor to one extreme end of its range, where it will no longer respond to changes in temperature—although this is unlikely.

Alternatively, you can mount a thermistor so that it is exposed to ambient air at the back of the box, where it will respond to local temperature fluctuations. Better still, use two thermistors, in series or in parallel, with one inside the box and one outside the box.

If this isn't random enough for you, maybe you'd like something a little more exotic.

Humidity Sensor

Humidity in a room usually changes very slowly, unless you have a bathroom or a kitchen in the immediate vicinity. But you could factor those slow changes into your timer control circuit.

The Humirel HS1011 humidity sensor is available from Parallax (and other suppliers) for less than $10. It only has two leads, and the capacitance between them will change as the humidity changes.

Yes—the capacitance. That's a novel concept. How can we use it in the timer control circuit?

Easily! Just substitute the humidity sensor for the timing capacitor with the first timer in Figure 34-4. The Humirel datasheet tells me that the capacitance of the sensor varies from 177pF to 183pF. Those are small numbers, so you'll have to use a lower-value timing resistor.

Humidity Control

The humidity sensor could potentially be used to switch a humidifier on or off, if you like the idea of a humidity-controlled environment. Books, papers, old audio tapes, and human sinuses tend to do better when the amount of moisture in the air is controlled.

One way to achieve this would be by using the variable capacitance to adjust the running speed of a fast astable timer, while a second timer activates it for a brief period. The astable output is connected with a four-bit binary counter. The counter is connected with a four-bit multiplexer, with its sixteen input/output pins wired in series as a long voltage divider. The voltage selected by the multiplexer becomes the variable voltage input for a comparator. The reference input to the comparator is set by a potentiometer, which is your humidity control. The output from the comparator goes to a solid-state relay that turns the humidifier on and off.

Is that complicated enough for you? It sounds like at least an all-day project, to get everything working right. Maybe you would do better to buy an off-the-shelf humidistat, which works just like a thermostat. But that would be so boring! Wouldn't you rather have the aggravation of building your own version, followed by the elation when it finally works, and the mystified looks that you will get from your friends, who may not understand that it's a whole lot of fun to build gadgets that are unnecessary?

I would definitely prefer the aggravation-elation-mystification cycle myself, if only I had time for it. Right now, I have to move along to another sensor option.

Accelerometer

I wrote about accelerometers for a column in *Make* magazine. They're getting quite cheap these days. Because they respond to a force in any direction, they can sense which way up they are by measuring the force of gravity. If you have a handheld device, an accelerometer will vary its output resistance as the person shifts his grip on the device slightly. So, here's another possible source of random resistance values.

The actual accelerometer is likely to be a tiny surface-mount device, but it's also likely to be available mounted on a breakout board, like the one in Figure 34-6. This makes it really easy to use.

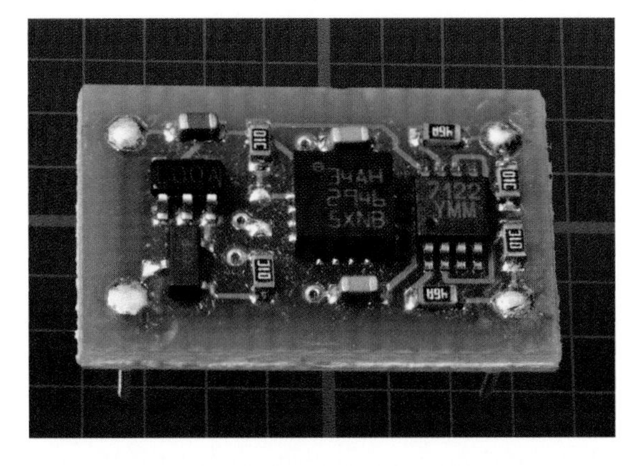

Figure 34-6. *A surface-mount accelerometer mounted on a breakout board. The accelerometer measures any force, including the force of gravity. Consequently, its output changes depending which way up you hold it. The output is converted to a simple change of resistance between pins on the board. Graph squares are 0.1" x 0.1".*

Touch Sensor

This, I think, is the most promising device of all. It's a thin sandwich of flexible plastic layers, with a pressure-sensitive resistor between them. You'll find a picture of it in Figure 34-7. I'm not sure exactly how it works, but it seems to be reliable, and it has a very wide range of resistance values. When you don't touch it at all, the resistance is almost infinite. When you push hard, its resistance drops to around 1K.

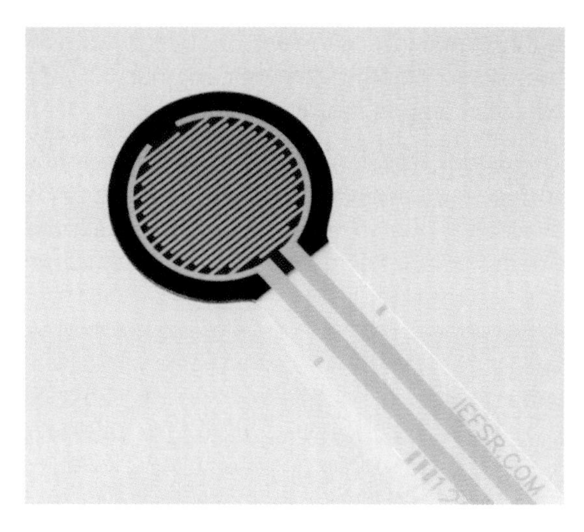

Figure 34-7. *A touch-sensitive pad that changes its resistance when you press it with your finger tip.*

You could use it as a substitute for a Start button that will also run a timer. In other words, the touch sensor can do dual duty as a high-tech on-off switch and a randomicity generator. Its very wide range of resistance values should provide you with an equally wide range of randomicity values.

But, I have to admit, I'm speculating about this. I don't have time to put the sensor through, say, a thousand cycles, to check if it gives a good variety of values. Even if I did have time, there might be something about the way I touch it which tends to create some resistances more than others. Even if the sensor worked well for me, there might be someone else who could touch it in such a consistent way, it wouldn't be so random.

It seems I'm forever doubting the performance of these components and saying "What if" this, and "What if" that.

I admit that I tend to be a skeptical kind of person, but the real problem is that I have been running these experiments on an *empirical basis.*

Empirical Issues

An "empirical" study means that the results are based on observation or experience.

Well—what's wrong with that? It's certainly better than just sitting around and wondering what might happen.

That's true. Hands-on work is more valuable than a vague prediction of what may happen (indeed, that's what this book is all about). Most research uses observations to confirm the conclusions. But there is an additional aspect of research: it often has a theoretical basis.

For instance, when astronomers very carefully measured the apparent position of the planet Mercury as it disappeared behind one side of the sun and reappeared from the other side, they weren't just doing it because they had some spare time and wanted to test their equipment. Einstein's theory of relativity predicted that the sun's gravity would bend the light reflecting off Mercury. Scientists made their observations because they wanted to find out if he was right. (Indeed, he was.)

But if you don't have a theoretical basis, and you're hoping that your observations today will still give you the same results tomorrow, there are no guarantees—especially if you have a lot of uncontrolled factors, such as different people pressing a sensor in different ways.

You may feel that a thermistor, or a humidity sensor, or an accelerometer, or a pressure sensor should behave similarly with different users, and they should create a good spread of random values, but you cannot be certain of that. If someone asked you, your correct response would be:

- The output seems to be random today, and I don't see why it shouldn't be random tomorrow. But I can't prove it.

Well, what if I told you that you can actually connect some components together that will generate an unpredictable stream of numbers, all on their own, without any external factors interfering? And what if you could guarantee mathematically that the sequence of numbers will be the same every time? And what if the sequence could continue for so long before it repeats, no human brain would ever be able to predict which number comes next?

That sounds like the perfect pseudorandom number generator, so long as you don't always begin the sequence in the same place. But it also sounds complicated, and you may be wondering if it's necessary.

It's not very complicated. Whether it's necessary depends on what you want to use it for.

How Random Is Random?

In a typical electronic game, we just need an input that seems pretty much random. If we are generating arbitrary numbers from 0 through 15, and number 13 comes up just a bit more often than the others during a series of hundreds of games—probably it doesn't matter.

In the Hot Slot game, the player who goes second is hoping to make a 12.5% profit. Suppose the random number generator in the game chooses one slot maybe 0.5% more than another—probably it doesn't matter.

But in some experiments, we require numbers to be precisely evenly weighted. If we are generating a random sequence of ones and zeroes over a long period of time, we may need to be absolutely positive that the series will contain 50% ones and 50% zeroes, not a ratio of 50.1% to 49.9%.

Maybe you feel you're unlikely to need such accuracy. But think back to the Telepathy Test in Experiment 15. Suppose we change it from a two-person experiment to a single-person experiment.

The challenge will be for someone to use psychic powers (if such things exist) to guess whether an LED is on or off. A circuit will turn the LED on and off in a seemingly random sequence, but during, say, 254 tests, we must be absolutely sure that the LED will be on 127 times and off 127 times. Otherwise, if the person makes correct guesses slightly more than half the time, we won't know how to evaluate his performance.

In any paranormal research, even a small variation from a median value can be significant, and therefore it really matters if one number is generated more often than another.

Could we rebuild the Telepathy Test in a single-player version that could produce really solid results? Actually, I don't think it's too difficult, and I want to give it a try.

I'm going to approach the challenge in two stages. First, I'll show you how to create a perfectly pseudorandom number generator. Then I'll incorporate it into the test.

Experiment 35: The LFSR

35

Let's suppose we have a black box containing some kind of circuit that generates a stream of numbers, without any influence from the environment. How would we decide if the stream is random? I think it should satisfy two requirements:

- The stream should be relatively free from repetition. I say "relatively" because any self-contained number generator will repeat itself if we allow it to run long enough. The objective is to create a sequence that is so lengthy or complicated, it exceeds the capabilities of human memory or attention span. (I am assuming that the number generator is on a large enough physical scale for quantum effects not to be relevant.)
- If we can specify a range of numeric values that are supposed to be equally likely to occur, they should be evenly weighted so that they all have an equal chance of appearing in the sequence and no values are omitted.

There is a circuit that can almost satisfy both of these requirements. It's called a linear feedback shift register, or LFSR. Its output sequence can be of any length (almost), and its output is absolutely evenly weighted (almost). After you build an LFSR, I'll show you how to make those "almosts" small enough to be ignored.

Getting to Know Your LFSR

Begin with the simple test circuit shown in Figure 35-1. This is very similar to the shift register test circuit in Figure 27-2. However, the Data Input switch has been omitted, because this circuit will now generate its own data by recycling it.

The outputs from memory locations E, F, G, and H have been left unconnected, because the function of the circuit will be easier to understand with just A, B, C, and D active. It will also be clearer if you put the LEDs (little yellow circles at bottom-left) in a line, as I have indicated. Of course, their input pins can't share the same row on the breadboard, because they have to be activated separately. But you can bend the wires so that the LEDs line up pretty well.

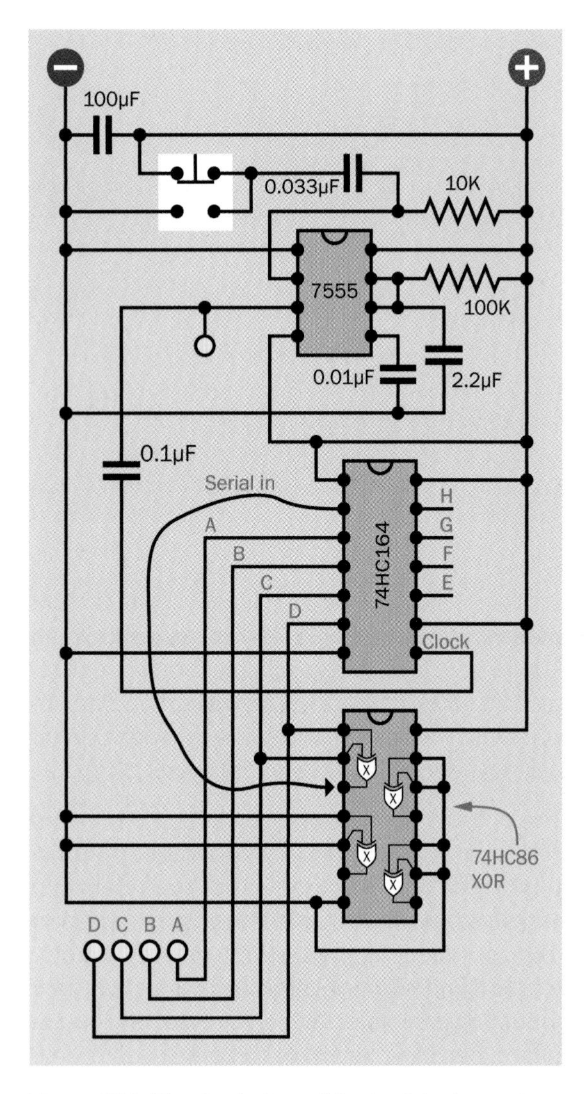

Figure 35-1. *The simplest possible circuit to demonstrate a linear-feedback shift register. Negative ground connections for LEDs are not shown, and a series resistor for each LED must be added if the LEDs do not contain their own resistors.*

A photograph of the breadboarded circuit appears in Figure 35-2.

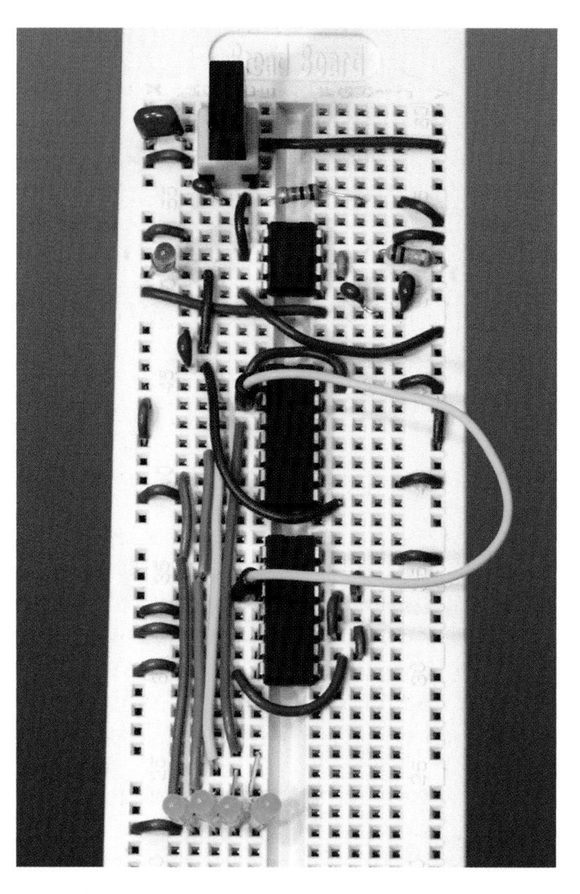

Figure 35-2. *Testing the linear-feedback shift register.*

Instead of a manual data input, we now have an XOR gate that taps into the outputs from the C and D memory locations and feeds them back to the serial data input. I've shown this connection as a wavy line, indicating a flexible jumper wire—the type with a plug on each end. Initially, both ends of this jumper are connected.

The states of the C and D memory locations recirculate. The only problem is, if you apply power cleanly to the shift register, its memory locations will all have low states. Also, when the XOR gate has two low inputs, its output will be low, too. Consequently, the feedback won't do anything. The low states will circle around and around, and the LEDs will remain dark.

Often when you apply power to a circuit, there is an initial surge or some noise in the supply that can result in some random values being loaded into the shift register. The 100µF capacitor in the schematic is intended to stop this from happening, although it may not be entirely effective.

If you do get some initial random values, some of the LEDs will be illuminated.

Regardless of whether this happens or doesn't happen, I want to load the shift register entirely with high states. To do this, remove the bottom end of the jumper wire from the XOR chip and apply it to the positive bus so that the serial input to the shift register will be high. Now hit the pushbutton four times to load the memory locations A, B, C, and D.

When all four LEDs are lit, reconnect the jumper to pin 3 of the XOR chip. Now press the button sixteen times, and you'll see a sequence that should be the same as shown in Figure 35-3. It will repeat after fifteen steps, and if you changed the timer from one-shot to astable mode, the sequence would be self-sustaining.

Quick Facts About a LFSR

Summarizing what you just learned:

- If the LEDs display low states corresponding with 0000 binary, they will remain stuck in that state, as the LFSR merely recirculates zeroes.
- If the LEDs display any value other than 0000, they should cycle through fifteen distinct patterns before the sequence repeats. All values from 0001 through 1111 will appear, but not in numerical order. No value will be omitted (except 0000), and no value will be repeated until the whole sequence repeats.

The snag is that this sequence is so brief, your human eye and brain will soon recognize that it's repeating.

Perhaps if we use all eight memory locations in the shift register, and increase the number of LEDs from four to eight, we will have a greater variety of patterns, and repetition will take longer to occur. This seems plausible, but before I try that, I want to show exactly what is happening here.

Bit-Shifting in Closeup

In Figure 35-3, the red boxes at the top represent memory locations in the shift register. The shifting in this figure is going to occur from right to left, as you can tell by the identifying letters A, B, C, and D. If you explore this topic elsewhere, you may find bits being shown moving from left to right, but I don't want to do this, because I'm going to be using the shift-register outputs to represent a binary number. Moving bits from right to left makes better intuitive sense for this purpose.

Remember how a shift register works. The data input is stored in an input buffer until a clock pulse moves everything along to make room. So, when we XOR the current states of locations C and D and feed them back to the input, they stay in the input buffer until the next clock signal clocks them in. At that point, the XOR of the two new states in C and D are copied into the input buffer, until the next clock cycle—and so on.

If the memory locations are assigned the usual binary place values of 8, 4, 2, and 1 (reading from left to right), their initial decimal value will be 8 + 4 + 2 + 1 = 15. But feedback from the XOR gate will change the rightmost bit to 0, because the XOR is receiving two high inputs, which always makes it produce a low output. So on the second line, the memory locations add up to 8 + 4 + 2 = 14. On each line in the figure, the total decimal value of the binary digits is shown as black numerals in a white box.

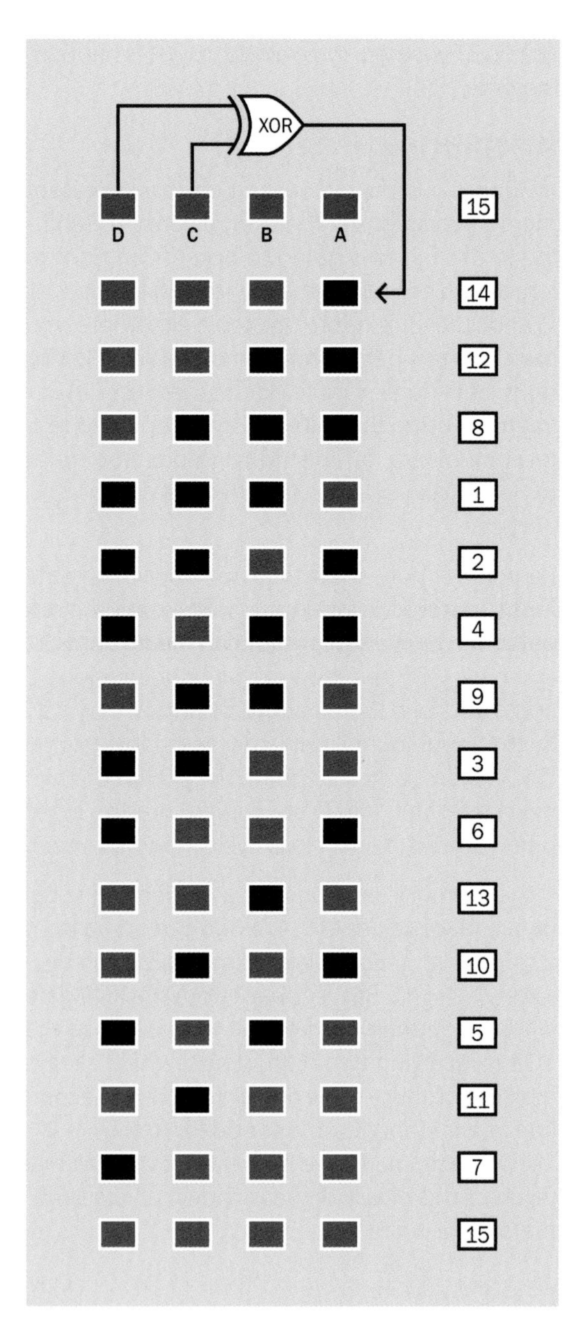

Figure 35-3. *The behavior of a four-bit linear shift register.*

The Problem with Zeroes

To deal with the problem of the LFSR seeming unresponsive when it is powered up with all low values in its memory locations, a text book will typically suggest preloading it with other values. This can be done with components that will run the clock of the shift register briefly while holding its data input high. But I'd like to avoid this requirement.

There's an easier answer: rewire the circuit with an XNOR gate instead of an XOR gate. XNORs are not often used, but you can still find them in either the 74HC00 series or the 4000B series. (Another option would be to invert the output from the XOR gate, but this of course would require an inverter as an additional logic gate.)

Back in Figure 15-5, in Experiment 15, I showed that the output from an XNOR gate is the inverse of the output from an XOR gate. It responds to two low inputs by delivering a high input—so it will work in the linear feedback shift register, even if the shift register starts from 0000.

The sequence using an XNOR gate instead of an XOR gate is shown in Figure 35-4. This includes every value from 0000 through 1110 binary (0 through 14 decimal). With a little thought, you can see that 1111 will lock up the circuit with an XNOR in it, just as 0000 locks up the circuit with an XOR in it. The XNOR gate simply leaves out 1111 instead of 0000. Still, it will perform the task of starting automatically when power is first applied.

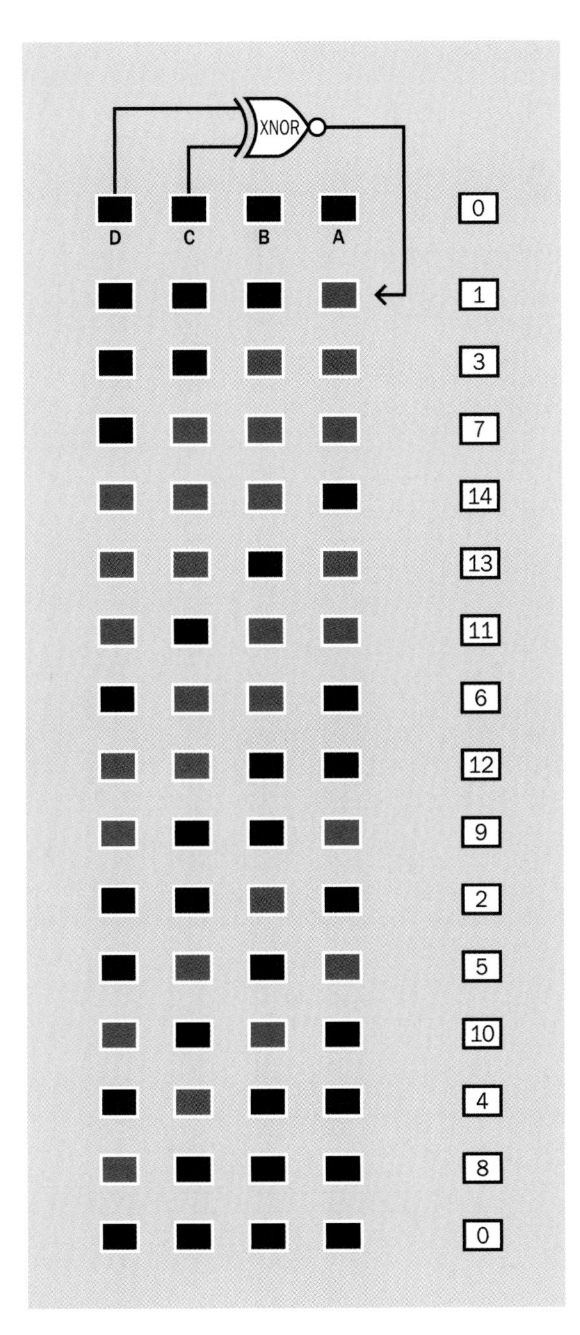

Figure 35-4. *The linear-feedback shift register sequence when an XNOR gate is used to process a four-bit value beginning with 0000.*

The Need to Be Nonrepeating

Before you rebuild the test circuit with a quad two-input XNOR chip, I need to get back to the issue of repetition. We really need more than fifteen steps before the sequence repeats.

Well, if we use all eight memory locations, the output values can range from 00000000 through 11111110 binary (0 to 254 decimal), and they will take 255 steps before they repeat.

This sounds more promising, but where should we put the XNOR to provide the feedback to make this happen? Will it work if we just tap into locations G and H and feed back to the input buffer?

No, if you try this, you'll find that it does not work. Now that we are shifting eight bits, we need three XNORs, as shown in Figure 35-5. (The locations of gates would be the same if they were XORs instead of XNORs. The only difference would be that the shift register could not start containing all zeroes.)

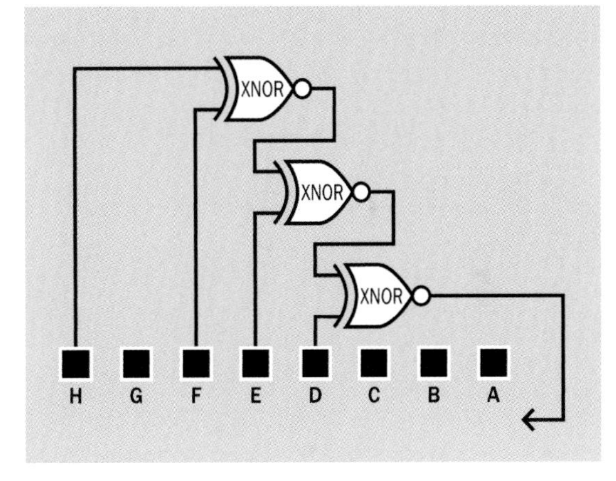

Figure 35-5. *Three XNOR gates are required to make an eight-bit linear-feedback shift register.*

Why do we now need three XNOR gates, and why are they in those particular locations?

Well, if we had fewer of them, or more of them, or if they were in different locations, the circuit probably would not generate a sequence including every value from 0 through 254, with no omissions and no repeats.

But how do I know this?

It can be proved mathematically. This proof is not very simple, though. It gets us into areas such as "primitive polynomials" and "finite field arithmetic" that are difficult to understand and would require many pages to explain—even if I were qualified to explain them.

If I can't prove it to you, how can you be sure it's right?

The answer is, we can test the theory by making an observation. Figure 35-6 shows the linear feedback circuit that replicates the logic in Figure 35-5. You can trace this out very easily.

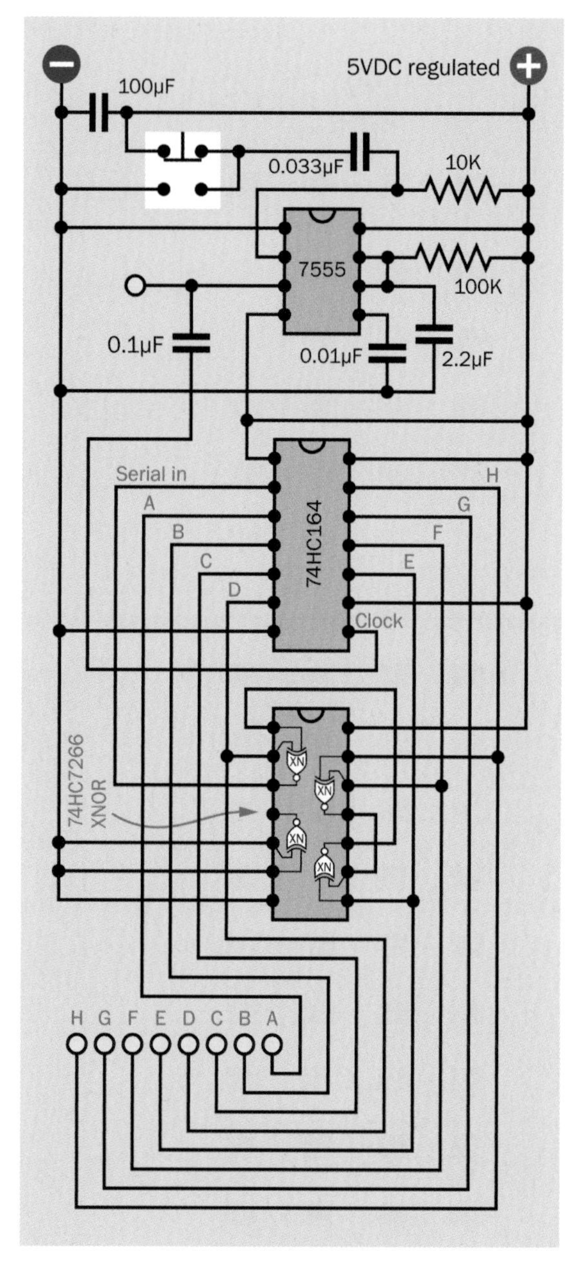

Figure 35-6. *Schematic for an 8-bit linear feedback shift register test. Negative ground connections for LEDs are not shown, and a series resistor for each LED must be added if the LEDs do not contain their own resistors.*

Shift register outputs H and F are tapped to provide inputs to the top-right XNOR gate in the XNOR chip. The output from that gate goes to an input on the gate below it, which also receives and input from output E. The XNORed output from that gate goes around to the top left gate, which receives an input from output D. The XNORed output from that gate feeds back to the serial input of the shift register.

At the top of the circuit there is a debounced pushbutton so that you can step through the sequence at your leisure.

The breadboarded version of the circuit is shown in Figure 35-7.

Figure 35-7. The 8-bit linear-feedback shift register, breadboarded.

Warning: XNOR Idiosyncracies

Make sure you wire the XNOR chip correctly. Its internal connections are totally different from those of all other logic chips. Just for the record, the pinouts are shown in Figure 35-8. If you make the mistake of wiring it like an OR gate or an XOR gate, you may damage it permanently.

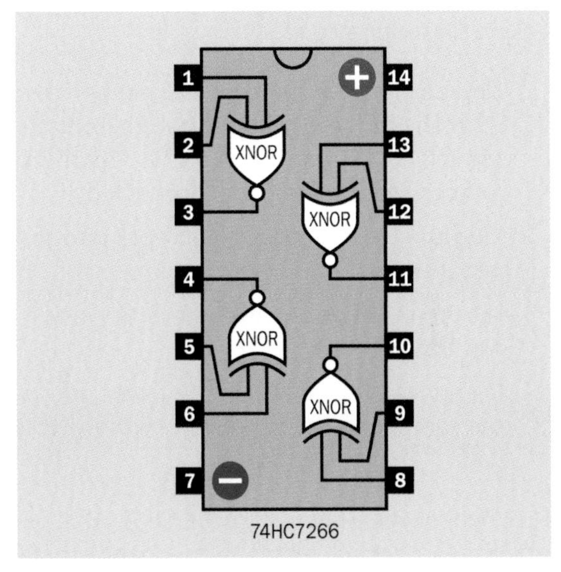

Figure 35-8. Pinouts of a quad two-input XNOR chip. The internal connections of this chip are different from those of all other chips containing logic gates.

Another thing to watch out for is that there is a variant of the XNOR chip under the part number 74HC266, which is just one digit different from the 74HC7266. The 74HC266 version of the XNOR chip has open-drain outputs, which are intended to control loads of up to 100mA and must not be connected with other chips. The Texas Instruments SN74HC266N is an example. Don't buy it by mistake!

The 74HC7266 is what you want, but it costs a bit more than most logic chips. You can alternatively use the 4077 (the old CMOS version) which is one-quarter the price. The pinouts are exactly the same.

Running the Test

Because I want your results from this test to be exactly the same as my results, you need to start with your shift register in the same state that I used with mine. I made sure that it had a low state in every memory location. If you don't see this when you power up, you'll have to take care of it manually, as you did in the four-bit LFSR test.

Follow these steps carefully:

1. Unplug the lower end of the jumper that normally connects with the output from the top-left XNOR gate. This is the "Serial In" jumper. Leave the top end connected.
2. Plug the lower end of the jumper into the negative bus.
3. Hit the pushbutton eight times to clock in eight zeroes.
4. Veeery carefully reattach the lower end of the jumper to the output from the XNOR gate.

Now when you start using the LFSR, the sequence you get will match the sequence that I am about to show you.

This sequence is listed in Figure 35-9, where 0 represents an LED that is off and 1 represents an LED that is on. Each time you press the button, the next line in the figure should match the LEDs on your breadboard.

00000000	11001000	01111001	10000100
00000001	10010001	11110010	00001000
00000011	00100011	11100100	00010000
00000111	01000110	11001001	00100000
00001111	10001101	10010011	01000000
00011110	00011011	00100111	10000001
00111101	00110111	01001110	00000010
01111010	01101111	10011100	00000101
11110100	11011111	00111000	00001011
11101000	10111110	01110000	00010110
11010000	01111101	11100001	00101100
10100001	11111010	11000011	01011001
01000011	11110101	10000110	10110011
10000111	11101010	00001100	01100110
00001110	11010100	00011000	11001100
00011100	10101001	00110001	10011001
00111001	01010010	01100011	00110010
01110010	10100100	11000110	01100101
11100101	01001001	10001100	11001010
11001011	10010010	00011001	10010101
10010111	00100101	00110011	00101011
00101111	01001010	01100111	01010111
01011111	10010100	11001110	10101110
10111111	00101001	10011101	01011100
01111111	01010011	00111010	10111001
11111110	10100110	01110100	01110011
11111101	01001101	11101001	11100111
11111011	10011010	11010010	11001111
11110111	00110100	10100101	10011111
11101110	01101001	01001011	00111110
11011100	11010011	10010110	01111100
10111000	10100111	00101101	11111000
01110001	01001111	01011011	11110001
11100011	10011110	10110111	11100010
11000111	00111100	01101110	11000101
10001110	01111000	11011101	10001010
00011101	11110000	10111010	00010101
00111011	11100000	01110101	00101010
01110110	11000001	11101011	01010101
11101101	10000010	11010110	10101010
11011010	00000100	10101101	01010100
10110100	00001001	01011010	10101000
01101000	00010010	10110101	01010000
11010001	00100100	01101010	10100000
10100011	01001000	11010101	01000001
01000111	10010000	10101011	10000011
10001111	00100001	01010110	00000110
00011111	01000010	10101100	00001101
00111111	10000101	01011000	00011010
01111110	00001010	10110001	00110101
11111100	00010100	01100010	01101011
11111001	00101000	11000100	11010111
11110011	01010001	10001000	10101111
11100110	10100010	00010001	01011110
11001101	01000101	00100010	10111101
10011011	10001011	01000100	01111011
00110110	00010111	10001001	11110110
01101101	00101110	00010011	11101100
11011011	01011101	00100110	11011000
10110110	10111011	01001100	10110000
01101100	01110111	10011000	01100000
11011001	11101111	00110000	11000000
10110010	11011110	01100001	10000000
01100100	10111100	11000010	00000000

Figure 35-9. *The sequence of 255 outputs from an 8-bit linear feedback shift register, plus a repeat of the initial state.*

You may be wondering how I generated the listing. Did I laboriously type it with one hand, while pressing the debounced pushbutton with the other? Well, actually, no. I wrote a little computer program that emulated a linear-feedback shift register, and I saved the output. However, I did check the output against the performance of the actual circuit—and now, so can you.

Binary numbers are difficult for the human brain to interpret, so I also generated the same sequence in decimal values. Here it is:

0, 1, 3, 7, 15, 30, 61, 122, 244, 232, 208, 161, 67, 135, 14, 28, 57, 114, 229, 203, 151, 47, 95, 191, 127, 254, 253, 251, 247, 238, 220, 184, 113, 227, 199, 142, 29, 59, 118, 237, 218, 180, 104, 209, 163, 71, 143, 31, 63, 126, 252, 249, 243, 230, 205, 155, 54, 109, 219, 182, 108, 217, 178, 100, 200, 145, 35, 70, 141, 27, 55, 111, 223, 190, 125, 250, 245, 234, 212, 169, 82, 164, 73, 146, 37, 74, 148, 41, 83, 166, 77, 154, 52, 105, 211, 167, 79, 158, 60, 120, 240, 224, 193, 130, 4, 9, 18, 36, 72, 144, 33, 66, 133, 10, 20, 40, 81, 162, 69, 139, 23, 46, 93, 187, 119, 239, 222, 188, 121, 242, 228, 201, 147, 39, 78, 156, 56, 112, 225, 195, 134, 12, 24, 49, 99, 198, 140, 25, 51, 103, 206, 157, 58, 116, 233, 210, 165, 75, 150, 45, 91, 183, 110, 221, 186, 117, 235, 214, 173, 90, 181, 106, 213, 171, 86, 172, 88, 177, 98, 196, 136, 17, 34, 68, 137, 19, 38, 76, 152, 48, 97, 194, 132, 8, 16, 32, 64, 129, 2, 5, 11, 22, 44, 89, 179, 102, 204, 153, 50, 101, 202, 149, 43, 87, 174, 92, 185, 115, 231, 207, 159, 62, 124, 248, 241, 226, 197, 138, 21, 42, 85, 170, 84, 168, 80, 160, 65, 131, 6, 13, 26, 53, 107, 215, 175, 94, 189, 123, 246, 236, 216, 176, 96, 192, 128, 0

To me, that looks satisfactorily pseudorandom. My program also checked to make sure that each value is included once, and only once, without any omissions or repetitions.

Ones and Zeroes

Having verified that the logic diagram in Figure 35-5 really will work as advertised, the next step is to decide how we reduce this sequence so that it generates just a single 1 or 0 in each cycle. Remember, we want to use this for the Telepathy Test, where the circuit will switch on an LED, or switch it off.

One way would be XORing the outputs from the shift register, just as I did with the outputs of the rotational encoders in Experiment 33. That would work, and you could use a "tree" of three levels of XOR gates, each averaging the outputs of the previous layer, to get down to a 0 or 1 output.

However, this isn't necessary. All we need to do is tap into the "A" memory location of the shift register, and use that.

At first this makes no sense. We included more outputs from the shift register to get a sequence that runs through more states before it repeats. And now we're just going to throw away those extra digits?

Not quite. All eight memory locations are still being used in the feedback process. The eighth, sixth, fifth, and fourth locations will still be XNORed as before, so we will still have a sequence that runs through 255 states before it repeats. I'm just suggesting we can subsample it.

The entire pattern of ones and zeroes will still take 255 steps before it repeats.

Perhaps you find this hard to believe. In fact, I wasn't totally convinced of it myself. Theory told me that it should be true, but I decided to confirm it with another observation. I modified my computer program to sample only the rightmost digit from each of the 255 steps in the linear feedback process, and got this sequence of 1s and 0s:

```
01111010000111001011111110111000
11101101000111111001101101100100
01101111101010010010100110100111
10000010010000010100010111011100
10011100000110001100111010010110
11010110101011000100010011000010
00000101100110010101110011111000
1010101000000110101111011000000
```

Then I ran the actual circuit, looking only at the rightmost LED. I got exactly the same series.

When you predict something like this, and it turns out to be precisely correct, that provides excellent evidence that it should happen every time.

You may object that this sequence doesn't seem entirely random, because it has patterns such as 0000000 and 1111111 in it. Yes, indeed it does, but a random sequence actually should contain some repetitions of this type. Remember, when you toss a coin, it can come up heads or tails several times in succession. In fact, the probability of multi-digit repetitions increases as the number of trials increases.

So, the presence of some repeated numbers in the sequence above is not a matter for concern. Of course, if the sequence consisted mostly of 11111111 and 00000000 combinations, that would be a different matter; but the distribution of patterns is actually very good. I counted the frequency of repeated sequences of 1s and 0s, and they look like this:

```
0 alone – 33 times

00 – 16 times

000 – 8 times

0000 – 4 times

00000 – 2 times

000000 – 1 time

0000000 – 1 time
```

Total: 128 instances of value 0.

```
1 alone – 32 times

11 – 16 times

111 – 8 times

1111 – 4 times

11111 – 2 times

111111 – 1 time

1111111 – 1 time
```

Total: 127 instances of value 1.

The Problem of Weighting

There's only one thing that doesn't look quite right in the list that I just presented. There are 33 single zeroes, but 32 single ones. What? This sequence was supposed to be absolutely evenly weighted!

No, I said it would be absolutely evenly weighted, "almost."

Here's the problem (which will then lead us to a solution). The entire sequence of all possible values would be 00000000 through 11111111 before it repeats. But you'll recall that using XNOR gates for feedback, the XNOR-based shift register is unable to display 11111111, so it skips it. Because that value has a 1 at the end, there is a 1 missing from our tabulation of digits at the end of each number.

Suppose I try to get around the problem by using the penultimate digit instead of the last digit. No, that won't solve the problem, because 11111111 is still missing, so there will be one more 0 than 1 in the final tabulation.

I will address this issue.

Skipping 254

One answer would be to add more shift registers. Suppose I have four of them chained together, providing enough memory locations to store a binary sequence of thirty-two digits. A complete sequence from this LFSR will run through more than four trillion values before it repeats. In that sequence, every pattern of thirty-two ones and zeroes will be represented except for the last, consisting of thirty-two ones. That missing single value will now be buried among more than four trillion others—which I think should be acceptable for just about any purpose.

Unfortunately, building that circuit would entail more work. Do you want to wire four shift registers and an unknown number of XNORs to create the linear feedback, plus additional components, to make the output from the Telepathy Test more perfectly balanced? If you do, that's great! I'd love to see someone do this! Personally, however, I don't have time, because—well, I already told you, I have to finish writing this book, to give you an opportunity to read it.

So how about an easier alternative? We can simply tell the shift register (somehow) to skip an eight-bit pattern that would have created a 0 as the last bit, to make up for the missing 11111111 that has a 1 as the last bit. Then the outputs will be even.

Since 11111111 is already missing, we could get rid of 11111110 as well (254, in decimal notation). That will be easy: just use a seven-input AND gate to look for 1111111 in places B through H, and when its output goes positive, use it to advance the clock to the next value. The complete logic diagram is shown in Figure 35-10.

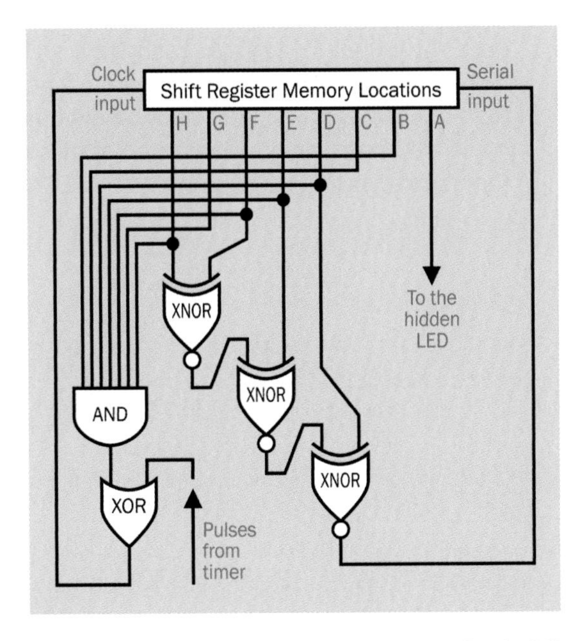

Figure 35-10. *Logic diagram for a linear feedback shift register with an evenly weighted binary output, achieved by forcing the register to skip the state 11111110.*

Is there such a thing as a seven-input AND gate? No, but you can buy one with eight inputs. If the eighth input is tied to the positive side of the power supply, the remaining seven can be ANDed. This chip is in the old 4000B series, not the 74HC00 series, but it will work.

Sharing the Clock Input

You'll notice that the seven-input AND does not feed back directly to the clock input of the shift register. That's because the clock input must also be used by a timer (or other input) that advances the shift register on a regular basis. I'll call this the "regular clock" to distinguish it from the "AND input."

To deal with the situation, I have added an XOR gate. Here is the imaginary sequence of events:

1. The XNOR feedback has reached the value 11111110, which we want to skip. It is applied to the input buffer.
2. The regular clock moves the 11111110 forbidden state into the shift register.
3. The eight-input AND gate immediately detects this and emits a high output. This feeds into the XOR gate.
4. The regular clock pulse hasn't ended yet, so the XOR gate has two high inputs. It switches to a low output.
5. The regular clock pulse ends. But the 11111110 forbidden state is still generating a high output from the AND gate.
6. The XOR now has one high input from the AND gate, and a low input from the regular clock. So the XOR output goes high.
7. The high XOR output triggers the shift register to move on to the next combination.
8. The AND gate doesn't sense the 11111110 forbidden state anymore, so its output goes low.
9. The shift register is now stable until the next regular clock pulse comes along.

You see from the above, there will be a moment while the forbidden 11111110 state is active, but only for the remainder of a clock pulse. If the pulse is very brief, the 11111110 state will be skipped almost immediately. During the Telepathy Test, it the state won't last long enough to light the LED.

At least, I think this is how it will work. In the next experiment, we'll confirm that it works.

Any Other Options?

Blocking one input pattern is not a very elegant solution to the problem. Maybe there's a better way, but if there is, I'm not sure what it is. I even consulted some friends in the encryption business, where big LFSRs generate pseudorandom numbers, but they just told me to use more shift registers—which I had already decided would be too much trouble.

So, my solution to the problem has to be skipping an input, for better or worse.

Seeding

One thing that still has to be taken care of is that so long as there are no voltage spikes, we will always start with a value of 00000000 in the shift register.

Obviously if we want to make a game seem unpredictable, it won't achieve its goal if it starts in exactly the same way every time. We must start at an unknown point in the sequence. How can we deal with this?

The answer is very simple. It is standard practice to "seed" a random-number generator with a value that will not be the same every time. In a computer program, often the system clock in the computer is used for this purpose, because the time represented by the clock is constantly changing. In the XNOR Randomizer, ideally you should run it for an arbitrary number of cycles before you use its output in a game.

Well—no problem! We can use the system I mentioned in Experiment 34, where a resistor-capacitor combination triggers a single-pulse, slow-running timer when power is switched on. The duration of this timer pulse will be adjusted by a sensor of some type. During the pulse, the slow timer allows a fast asynchronous timer to run. The asynchronous timer cycles the clock input of the LFSR. At the end of the pulse from the slow timer, the LFSR stops in an unknown state, and we're ready for some perfectly balanced pseudorandomicity.

Make Even More: Other Games and Other Numbers

Before I get to the Telepathy Test, I want to show you how many ways there are to use the output from an LFSR. I'm going to call it an All-Purpose 8-Bit Randomizer, with applications such as the ones below.

Twinkling LEDs

Suppose you have eight LEDs: two red, two green, two blue, and two yellow. Each pair of LEDs is powered by two outputs from a linear-feedback shift register. One output goes through a 330Ω resistor, while the other goes through a 1K resistor. Each color will now have four states: off, low, medium, and high intensity. Figure 35-11 shows how this could be done, using a Darlington array so that you can have bright, high-current LEDs.

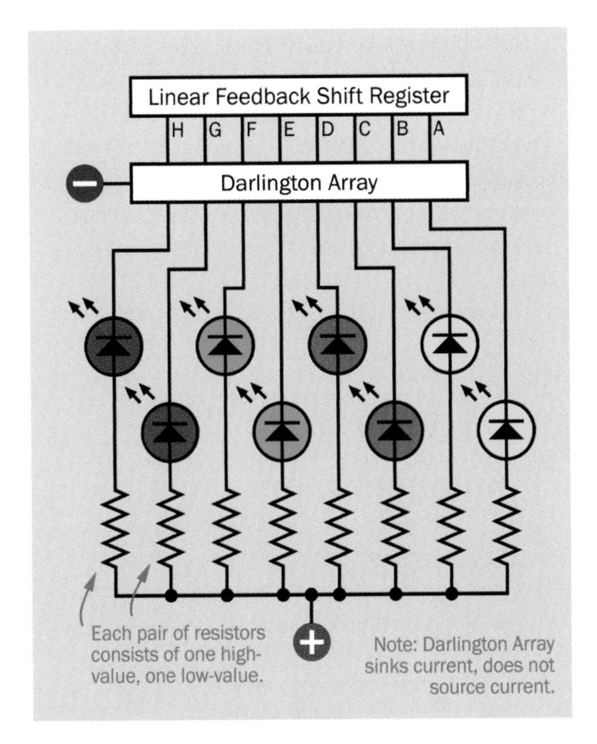

Figure 35-11. *Using the linear-feedback shift register to create pseudorandom twinkling LEDs.*

If you mount the LEDs behind a diffusing panel (like the plastic diffusers that are used with recessed fluorescent tubes), you'll get an attractive twinkling effect when the shift register is running at a fairly high clock speed.

Slot selector

The linear-feedback shift register can run the multiplexer in the Hot Slot game. Just connect four of the register outputs to the multiplexer control lines.

Ring counter variant

In Experiment 26, a ring counter flashed sixteen LEDs sequentially. We could use an LFSR to flash them randomly, powering them through a multiplexer. If there are also sixteen buttons, one beside each LED, the challenge will be to press a button while the LED beside it is still illuminated.

The same voltage that powers the LED can be applied to the contacts in the pushbutton, and the other sides of the pushbuttons will connect with a beeper, so that if you press the button while the LED is on, you'll hear a beep and your score will increment. See Figure 35-12.

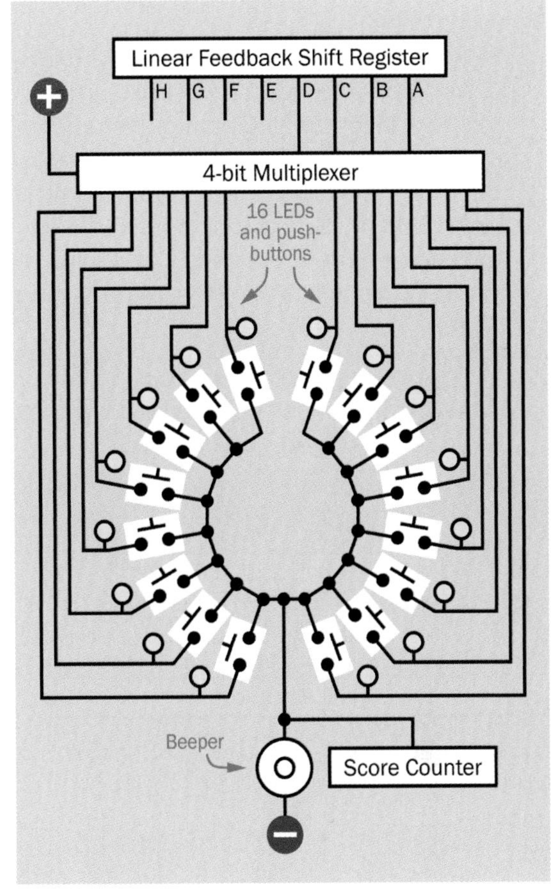

Figure 35-12. *The game in Experiment 26, using a ring counter, could be remade with a linear feedback shift register to flash LEDs randomly. The player tries to press the button beside each LED while it is illuminated.*

If you use reed switches instead of pushbuttons, you could put everything, including the LEDs, behind a thin membrane of translucent plastic. A reed switch would be hidden beside each LED. The player would point to the LED with a magnet-tipped stylus, thus activating the reed switch beside it. This would be an elegant user interface, and it would eliminate the problem of a player cheating by touching multiple switches. Of course, you would have to choose the strength of the magnet carefully so that it would activate one reed switch reliably without affecting the other switches either side.

Random tones

Still using four outputs to control a multiplexer, it could connect one of sixteen resistors to a timer running at audible frequencies. If you use trimmer potentiometers instead of fixed-value resistors, you can tune each of them to produce a note in the diatonic scale. (You would need a keyboard instrument for reference to get the tuning right.)

The result would not be music, as such, but if the pitch changes occurred slowly, they could be interesting. See Figure 35-13.

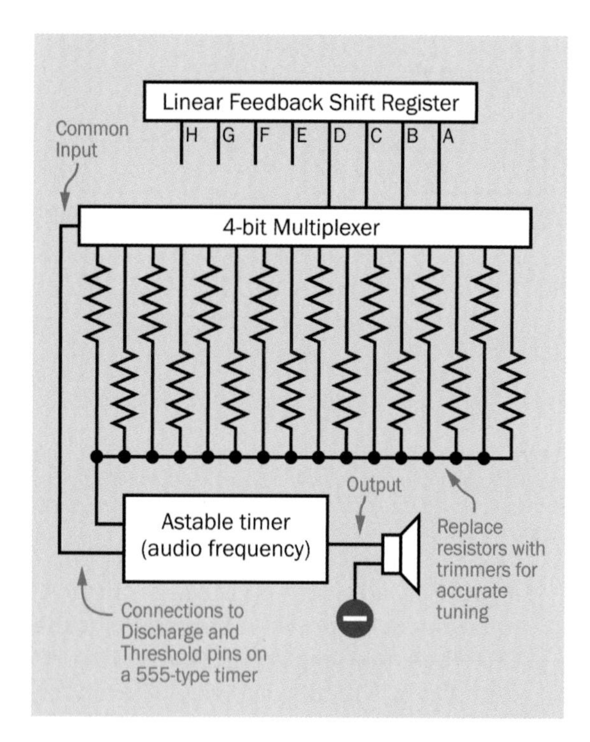

Figure 35-13. *Using an LFSR to create sixteen random musical tones.*

An alternative would be to use two multiplexers, each using four of the eight outputs from the shift register. One multiplexer would control pitch, while the other could set sound quality by inserting resistor-capacitor combinations between the output and negative ground. See Figure 35-14.

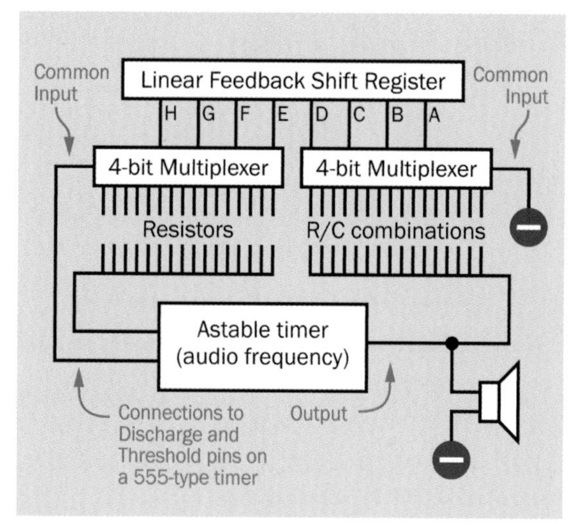

Figure 35-14. *Resistor-capacitor combinations could be selected randomly by a second multiplexer to add tonal variety to a random-tone generator.*

Ching input

Instead of the fingertip-resistance randomizer that I suggested for the Ching Thing (see Chapter 27) you could use four outputs from the linear-feedback shift register to control the decoder chip.

Make Even More: Microcontroller Randomicity

The concept of a linear feedback shift register is used in computer languages to create pseudorandom number sequences. This includes the high-level languages that are installed in some microcontrollers. Depending which microcontroller you are using, some sort of statement should allow your program to generate a seemingly random number on demand.

But will the distribution of values be evenly weighted? I have to say that when I investigated the random number function in the version of BASIC implemented on a PIC microcontroller (the so-called PICAXE), I was not impressed. Depending on the range that I specified, some values came up significantly more often than others.

If you're wondering if C language on the Arduino is more likely to give you a better simulation of randomicity, I encourage you to test it to find out. Note that it is common practice to read the value of an unconnected pin on the chip, via the internal analog-to-decimal converter, to start the internal random-number generator with an unpredictable value.

Experiment 36: The One-Person Paranormal Paradigm

36

Here's the plan. There will be a single LED behind a screen. An electronic circuit will switch the LED on or off, and will then prompt the player to guess which state it is in—by using his psychic powers, if he is fortunate enough to possess them.

The player will press a righthand button if he thinks the LED is on, or a lefthand button if he thinks the LED is off. The circuit will tell him whether the guess was correct or incorrect, and the cycle will repeat.

I think we now have all the necessary knowledge to build this circuit. What we don't have is a way to assess the results. How many correct guesses do you think a player needs to make, compared with his incorrect guesses, to achieve a result which seems too unlikely to be merely a matter of chance? I will explain how this can be assessed, after the circuit has been constructed.

The Last Logic Diagrams

In the previous experiment, I included a logic diagram for the XNOR Randomizer, wired to deliver a 0 or 1 output. I have adjusted this slightly, as shown in Figure 36-1, to make it compatible with the Telepathy Test circuit that I have in mind.

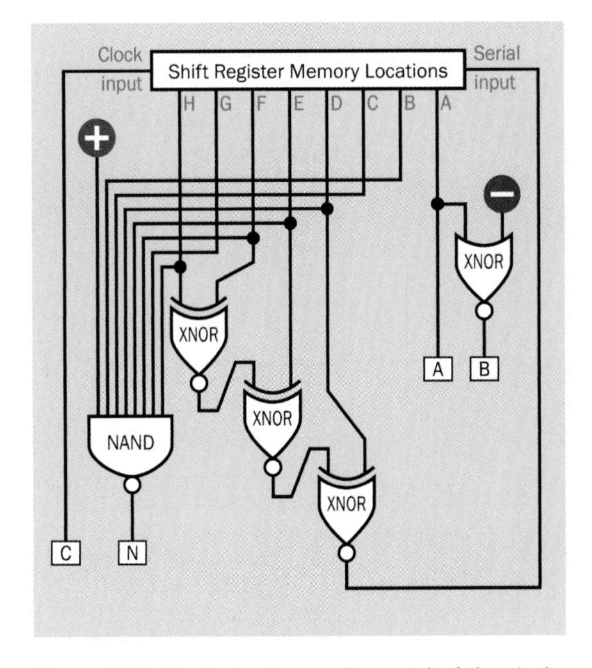

Figure 36-1. *The logic diagram for part 1 of the single-person Telepathy Test. The input and outputs labelled A, B, C, and N will connect with part 2.*

- I've substituted a seven-input NAND for the seven-input AND gate. Instead of a normally low output that goes high when it detects the disallowed state of 1111111, it has a normally high output that will go low. This will be easier to use in the next part of the logic.

- I added an XNOR gate beside the binary output. When the output is high, the XNOR output will be low, and vice versa. In other words, it functions like an inverter. So why didn't I use an inverter? Because I had one XNOR gate to spare in the quad two-input chip that will process feedback for the shift register. Might as well use the spare gate instead of adding an extra inverter chip.

Figure 36-2 shows the schematic for part 1 of the circuit. It's very similar to the test schematic in the previous experiment. The most visible change is the addition of the NAND gate. The input and outputs identified by letters A, B, C, and N will connect with part 2 of the circuit.

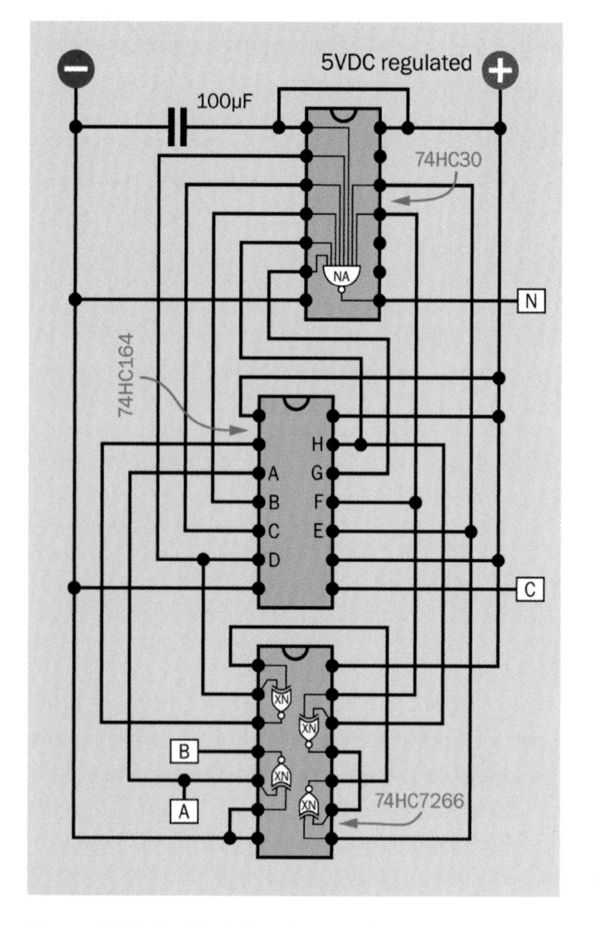

Figure 36-2. *Part 1 of the schematic for the single-person Telepathy Test.*

Looking at Part 2

Part 2 of the logic diagram is shown in Figure 36-3. In this diagram, timers are colored pink for easy identification, and there are four input/output lines labelled to connect with the ones in Figure 36-1.

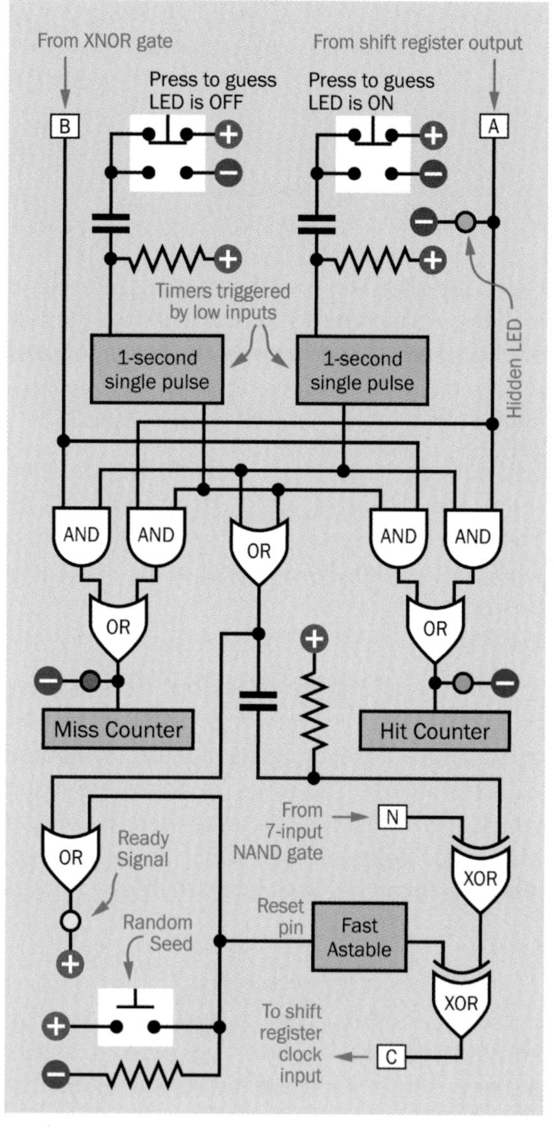

Figure 36-3. *Part 2 of the logic diagram for the single-person Telepathy Test will process user input, provide feedback, and seed the linear-feedback shift register.*

The function of part 2 can be summarized like this:

1. The player sees a "ready" signal.
2. The player presses button A if she thinks the hidden LED is on, or button B if she thinks it's off.
3. The "ready" signal goes off for about a second. During that interval, the circuit gives a "hit" or "miss" indication, and increments a "hit" counter or a "miss" counter.
4. The "ready" signal comes on again, and the player repeats from step 2.

Behind the scenes, it's going to be a little more complicated.

Input Logic

Since accuracy is essential, there must be no risk of contact bounce in the input buttons. Therefore, I am routing each input through a one-second timer in the same configuration that I have used before, to create a clean pulse that is not retriggered if someone's finger holds down a button for too long.

The outputs from the one-second timers must now be compared with the outputs from the previous logic diagram. The shift-register output is on the righthand side and will be high when the hidden LED is on. The XNOR output is on the lefthand side will be high when the hidden LED is off.

There are two ways to make a correct guess:

- The hidden LED is on, AND the player presses the "LED is on" Guess button
- OR the hidden LED is off, AND the player presses the "LED is off" Guess button

Likewise there are two ways to make an incorrect guess:

- The hidden LED is on, AND the player presses the "LED is off" Guess button
- OR the hidden LED is off, AND the player presses the "LED is on" Guess button

The two pairs of AND gates, each feeding into an OR gate, take care of this logic, in much the same way that I used two pairs of ANDs, each feeding into an OR gate, in the original two-player version of the game.

Two LEDs (red and green circles) will tell the player immediately whether her guess is right or wrong, and two (optional) counters can log the number of hits and misses. I'll deal with the counters later.

The rest of the circuit waits until either the input timer on the left OR the input timer on the right is triggered. Therefore, the timer outputs are linked in an OR gate at the center. The output from this OR has to do a couple of things.

The Ready Signal

The OR output is tapped by a wire that runs to the left, where it goes into another OR gate, which controls the ready signal. What's this all about? Well, I want the ready signal to be normally on, but suppressed under two circumstances:

- For one second after the player has pressed a Guess button
- OR during the initial "random seeding" of the shift register, when the circuit is not yet ready

By connecting the yellow ready-signal LED to the positive side of the power supply, as shown, it will satisfy these conditions.

The rest of the time, the lefthand OR gate will have two low inputs, so the ready signal will light up as it sinks current into the gate. Remember, a 74HC00 series logic chip can sink as much current (20mA maximum) as it can source. You can use it either way.

I could have used a NOR gate in this location, with the ready-signal LED conventionally grounded, but that would mean using one more chip just to get the NOR. Since I had one OR gate to spare, I took advantage of it.

Random Seeding

In Experiment 35 I mentioned the need to random seed the linear shift register (see Chapter 35). I'm choosing to have the player do this manually so that only one fast timer is needed instead of a slow one driving a fast one. When the Random-seed button at bottom-left is not being pressed, it applies a low state to the reset pin of the fast astable timer. Remember that a low state on the reset pin stops the timer. When the Random-seed button is pressed, its high state frees the timer to run.

The output from the fast timer passes through an XOR gate at bottom-right and sends pulses back to the clock input of the shift register.

Before beginning a new trial, a player must remember to push the Random-seed button for an arbitrary interval. Since this is a serious tool for paranormal research (!) I'm going to trust the player to remember the random-seed ritual.

Two More XORs

Now for the tricky part. Going back to the OR gate in the center, its output is connected through a capacitor to an XOR gate below it. The capacitor has a much smaller value than any of those which we have used in previous experiments, because a logic chip has such a sensitive input. In my experience, a 68pF capacitor will deliver a pulse that is short enough to trigger the XOR gate without creating multiple "false positives." If you test the circuit and find that it is behaving unpredictably, you can try a higher or lower value for the capacitor. I included 47pF and 100pF alternatives in the shopping lists, to help you in this respect.

- Remember that 1,000pF = 1nF.

The righthand input of the XOR gate is normally high, because of the pullup resistor, and the lefthand input of the AND gate is also normally high, because it is connected with the seven-input NAND gate in the previous section of the circuit, and the NAND has a normally high input, until it detects the disallowed state of 1111111.

When an XOR gate has two high inputs, it creates a low output. Therefore, the output of the first XOR gate is normally low.

The output feeds into the second XOR gate, which has a low left input while the astable timer is not running. So, the second XOR gate has two low inputs, which cause it to have a low output, which connects with the clock of the shift register.

What happens when the player presses a button? Nothing, until the end of the one-second timer pulse. When the pulse ends, the output from a one-second timer goes from high to low. The 68pF capacitor passes that transition as a brief low pulse.

During that brief pulse, the first XOR gate has a low input on the right, while it still has a high input on the left. Consequently, its output goes high. This causes the second XOR gate to have one high input and one low input, so its output goes high and advances the clock of the shift register.

The low pulse from the capacitor is very brief. When it ends, the pullup resistor takes over again, and the two XOR gates return to their initial state. The shift register has been clocked to create a new value, and the ready signal lights up, inviting the player to make another guess.

This is relatively simple, but why do I need two XOR gates to deal with it? Because there are actually three situations where the shift register has to be clocked:

1. When the player has pressed a button. I just described how that will advance the shift register to its next value.
2. When the fast-running timer is cycling the shift register initially.

3. When the shift register reaches the disallowed state of 11111110.

The two XOR gates deal with all the situations where these events can happen simultaneously, or almost simultaneously. This is the difficult part of the circuit to understand.

Timing is Everything

Figure 36-4 shows how the gates will work under normal circumstances when the player presses a button that causes one of the one-second timers to send a pulse. At the end of the pulse, as the timer output goes low, the coupling capacitor passes the transition along, and the XOR logic sends a high pulse to the shift register to advance it to the next state.

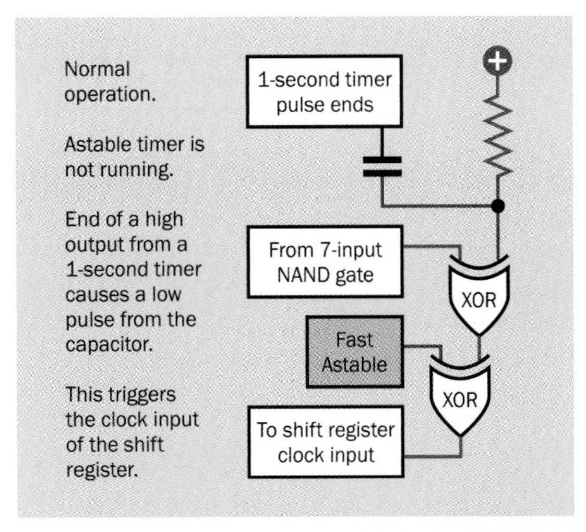

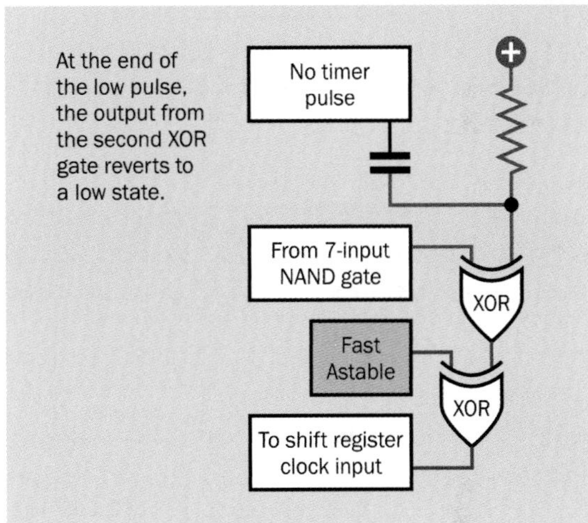

Figure 36-4. *In response to a low transition on the coupling capacitor, the dual XOR gates send a high clock pulse to the shift register.*

But what if the shift register advances to 11111110, the disallowed state? Figure 36-5 shows the sequence of events. The NAND gate in part 1 of the circuit responds to the seven high inputs by taking its output low. But the low pulse from the coupling capacitor hasn't ended yet, so the clock output goes low.

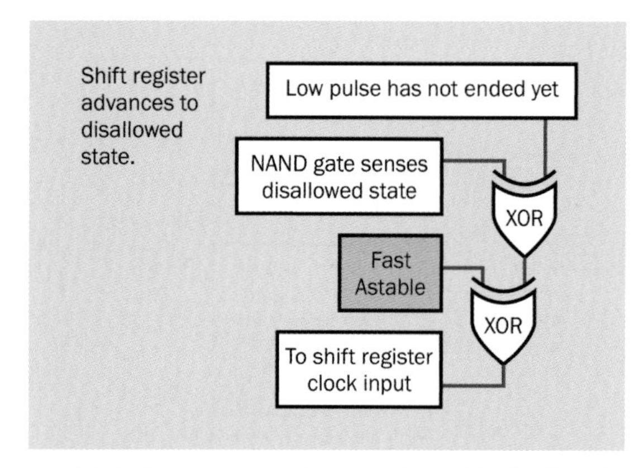

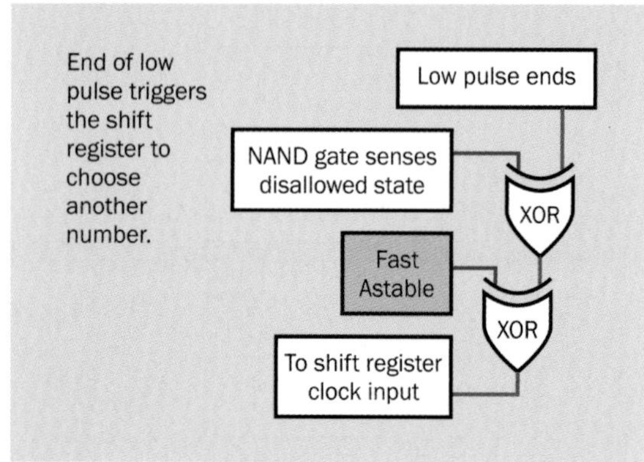

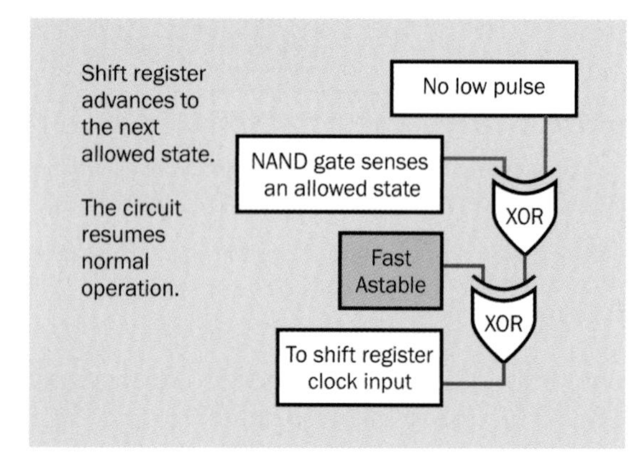

Figure 36-5. *The dual XOR gates process a low signal from the NAND gate in part 1 of the circuit, indicating that the shift register has reached the 11111110 disallowed state.*

As soon as the low pulse ends, the pullup resistor on the first XOR gate takes its input high. This causes a high clock output, advancing the shift register to the next state.

The NAND output returns to its high state, because the new state is not disallowed. The clock output goes low, and the circuit is back in equilibrium.

A similar series of events occurs when the shift register is being seeded by the fast astable timer. While the timer is running, if it advances the shift register to the disallowed state, the pair of XOR gates cause the shift register to skip past it. This will happen even if the timer stops when the disallowed state is reached. The shift register will still move on past it.

The system only fails if two events events happen almost simultaneously. Even in this case, the worst that can occur is that the circuit will lodge in the disallowed state until the player presses a button. This will trigger the shift register, because any change in the XOR gates toggles the output to the clock.

- When feedback is introduced to a logical system, the consequences can be complicated. In a computer, a system clock makes sure that all the chips stay in sync with each other (more or less), and this helps to prevent the kinds of problems that I have been describing. The computer system is "synchronous."
- The single-person Telepathy Test is asynchronous, with all the interesting consequences that this entails.

Making Every Guess Count

How are you going to evaluate your success rate in this experiment? If you don't feel like keeping score manually, and you do have a microcontroller, you can apply the output from each successful guess to one input pin and from each unsuccessful guess to another input pin and run a little program that counts the inputs and displays a running total on an LCD screen.

What if you don't have a microcontroller, but you want to total your guesses automatically? You could buy yourself a couple of off-the-shelf "event counters." Just go to eBay and search for "digital counter" or "digital totalizer." These useful little devices cost maybe $8 each from suppliers in China. You should find that they will make provision for a wide variety of power supplies and inputs. All you need to do is link the counter with your circuit via a common ground and attach the output from your circuit to the counter input through a capacitor (to prevent any flow of DC current).

On the other hand—making marks on paper is not such a bad option. While you're sitting at a desk, pressing one button or the other, it's easy to use a pen with your free hand.

Schematic Part 2

Figure 36-6 shows how to put together components that will perform the logical operations in Figure 36-3. You'll notice that I had to squeeze the components together to fit them into the available space. Still, there are only six chips, which should easily fit on a breadboard, and the connections are relatively simple.

Remember that each of the red, yellow, green, and blue circles represents an LED, and be sure that the yellow one, which is the Ready prompt, goes between an OR gate and the positive side of the power supply, not the negative side. The other three LEDs are conventionally grounded. Naturally you should add series resistors if your LEDs require them.

Points A, B, C, and N are to be connected with points A, B, C, and N in part 1 of the circuit, on an adjacent breadboard.

Figure 36-7 and Figure 36-8 show the breadboarded versions of the circuit. Additional LEDs were included during the testing process.

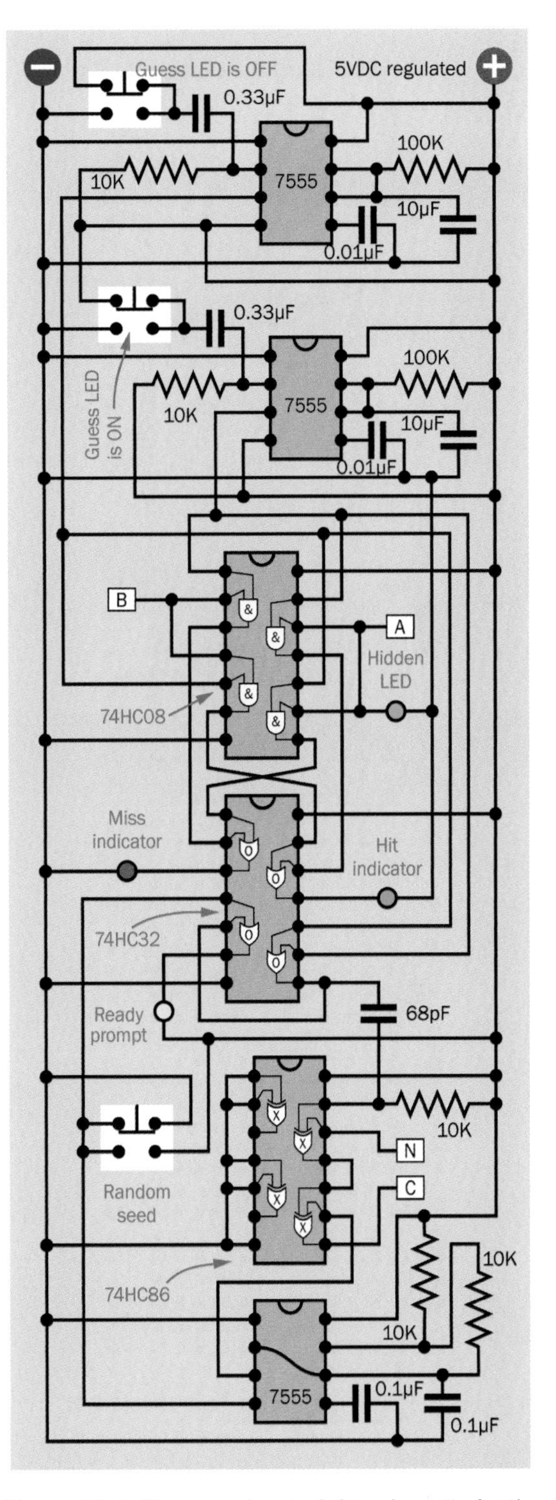

Figure 36-6. *The second part of the schematic for the single-person Telepathy Test.*

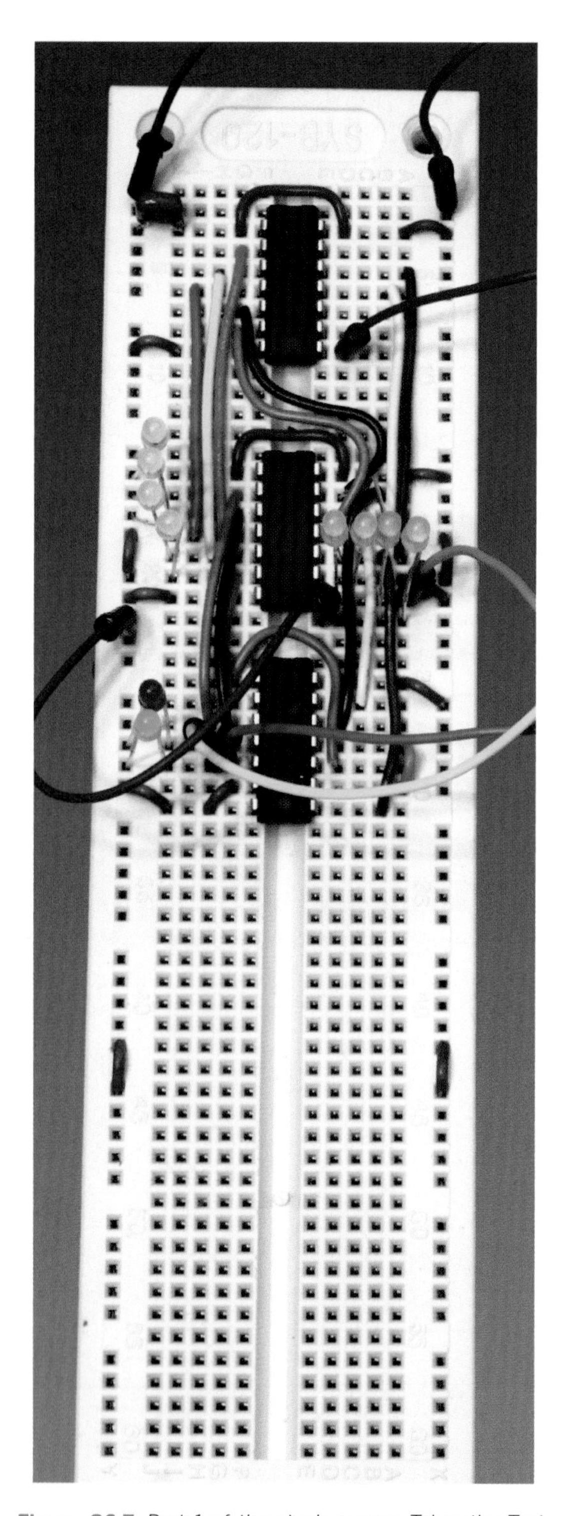

Figure 36-7. *Part 1 of the single-person Telepathy Test, breadboarded.*

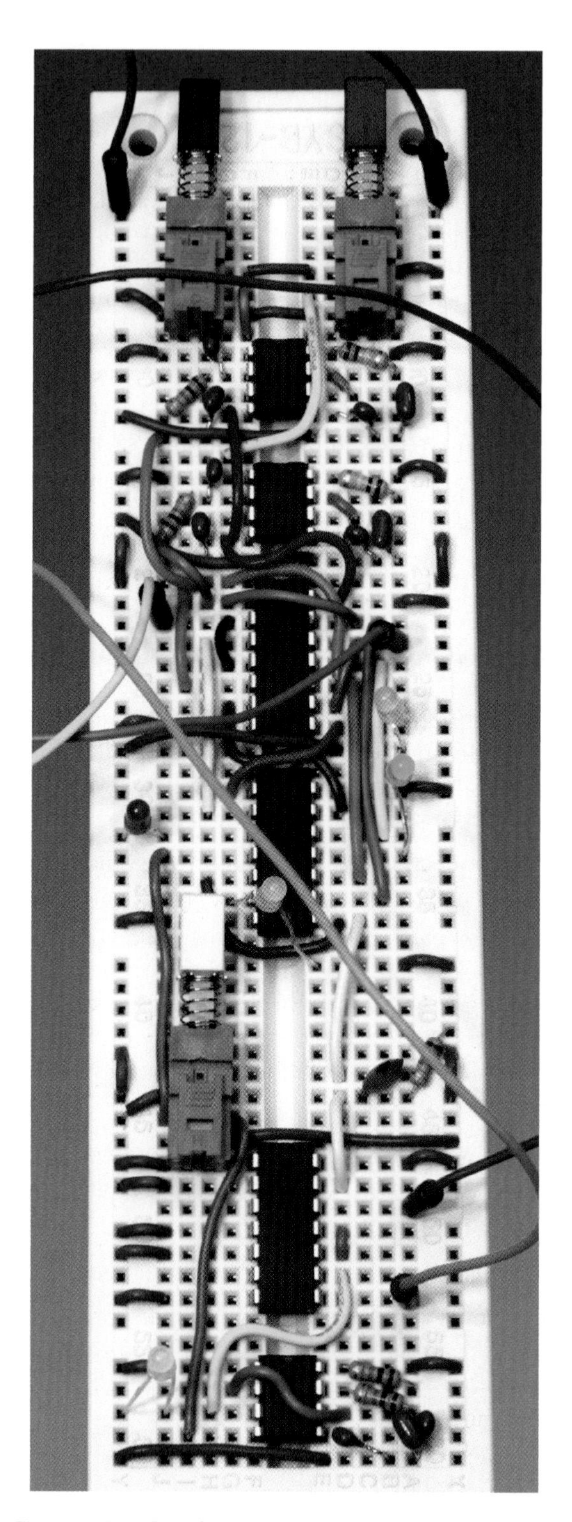

Figure 36-8. *Part 2 of the single-person Telepathy Test, breadboarded.*

Testing the Tester

Initially, slow down the fast-running asynchronous timer at the bottom of the second breadboard by using a 10μF capacitor on its pin 6. Also, add low-current LEDs to the outputs of the shift register on the first breadboard. If you use LEDs containing their own series resistors designed for a 12VDC supply voltage, they will draw so little current that the shift register should still communicate successfully with the NAND and XNOR chips.

Hold down the Random-seed button, and you will see the LEDs attached to the shift register cycling through the linear-feedback sequence that is now familiar to you. If this doesn't happen, you made a wiring error somewhere.

Test the "Guess LED is OFF" pushbutton and "Guess LED is ON" pushbutton, and if each of them generates a pulse lasting about one second, you're ready for the really interesting part of the test, which is to see if the circuit skips the disallowed state.

When you first apply power, the 100μF capacitor on the first breadboard should prevent fluctuations from loading random values into the shift register. The only problem is that if you disconnect power and then reconnect it again, the capacitor may have kept the chips energized sufficiently for them to retain their previous states. Eventually the stored charge will leak away, but this may take a while. To speed up the process, try shorting out the 100μF capacitor (but make absolutely certain the power is disconnected before you do this).

Assuming that you start with the shift register containing eight low states, you can press the Random-seed button quickly, twenty-four times (still with the 10μF capacitor making it run slowly) to get it to the state immediately preceding the one that is disallowed. (See Figure 35-9 to refresh your memory.)

Now if you press the Random-seed button, or either of the Guess buttons, you should find that the shift register skips the 26th state (the disallowed one) and goes straight to the 27th state. This works reliably on the version of the circuit that I built. If you have any problems, your first step would be to try different values for the 68pF capacitor.

Once you have the circuit working, you come to the interesting part: interpreting the results.

How Unlikely Is ESP?

Suppose you run a trial consisting of a thousand guesses. On average, around 500 of them should be correct. But let's suppose you get 510 right, or 520, or 530. How far from the median do you have to go, for the result to look as if it is not just a matter of chance?

This is a complicated question, so I'll start by simplifying it. If you make just four guesses in succession, and we use letter Y to represent a correct guess and N to represent a wrong guess, there can be sixteen possible sequences, all of which are equally likely: NNNN, NNNY, NNYN, NNYY, NYNN, NYNY, NYYN, NYYY, YNNN, YNNY, YNYN, YNYY, YYNN, YYNY, YYYN, YYYY.

We don't care what sequence your guesses are in. We only care about the total number that are correct. This means we need to group the correct and incorrect guesses, regardless of their sequence. Like this:

- NNNN : 1 way to get 0 guesses correct.
- NNNY, NNYN, NYNN, YNNN : 4 ways to get 1 guess correct.
- NNYY, NYYN, YYNN, NYNY, YNYN, YNNY : 6 ways to get 2 guesses correct.
- YYYN, YYNY, YNYY, NYYY : 4 ways to get 3 guesses correct.
- YYYY : 1 way to get 4 guesses correct.

Because there are four ways of getting three guesses correct, but the total number of ways to make guesses is sixteen, the chance of making three correct guesses out of four (in any sequence) is 4/16, or 25%.

But wait—what if you get all four out of four guesses correct? That's even better! So really we should rephrase the question. What are your chances of getting three guesses correct, or even more?

Because we have added another alternative, the chance actually improves to 5/16, or a little over 31%.

If you try to extend this system to calculate the odds for five, six, or more guesses, you find that the number of Y/N combinations gets very big, very quickly. There's a way to look up the number of combinations, though. You can see it in Figure 36-9. Here's how to use this array of numbers:

								1									1
							1		1								2
						1		2		1							4
					1		3		3		1						8
				1		4		6		4		1					16
			1		5		10		10		5		1				32
		1		6		15		20		15		6		1			64
	1		7		21		35		35		21		7		1		128
1		8		28		56		70		56		28		8		1	256

Figure 36-9. *Pascal's Triangle can be used to figure the odds of making any number of correct guesses in a series where a correct and incorrect guess are equally likely.*

The second number in each row is the number of guesses in a trial. For example, the bottom row describes a trial where you make eight guesses.

The numbers in each row tell you how many ways there are (that is, the number of permutations) to make correct guesses, beginning with zero correct guesses (only one way to do this, represented by number 1, at the left end of a row) and ending with all correct guesses (only one way to do this, represented by number 1, at the right end of a row). Between these extremes, the numbers in a row in the triangle tell you how many ways there are to make 1, 2, 3 ... n correct guesses, where n is the total number of guesses (right or wrong). So, for example, the bottom row shows that if you allow yourself eight guesses, and you want to know how many ways there are to get four of them right, the answer is 70. Just count along the bottom row like this: 0, 1, 2, 3, 4 ... and you get to number 70:

- Remember, the first number in any row (always 1) is the number of ways to get no guesses correct. The second number is the number of ways of getting one correct guess—and so on, ending with the number of ways to get all the guesses correct (always 1).
- The white number in the righthand column is the total of all the black numbers in that row of the triangle. In other words, it is the total number of different sequences of right and wrong guesses. Notice that this number doubles with each new row of the triangle.

Now you can calculate the odds. Going back to the previous example, if you want to know the odds of making four correct guesses out of eight, you take 70 as the number of ways to make those four correct guesses (in any sequence), and 256 as the total number of permutations, so your odds of getting exactly four correct guesses are 70/256.

But what if you want to know the odds for a range of guesses? In the bottom row, in a series of eight guesses, what are the odds of getting six right, or more? you would add 28 + 8 + 1 to get 37/256. That's about 14%. You would only expect this to happen, by pure chance, about one time in seven.

Powers of the Triangle

If you have studied much mathematics, you will recognize Figure 36-9 as Pascal's Triangle. I don't have room to go into it in detail here, but the interesting aspect of it is that every number in the triangle (other than 1) can be found by adding the two numbers immediately above it.

In theory, if you want a trial in which you make 1,000 guesses, you just need to extend Pascal's

Triangle downward until you get to a line where the second number is 1,000. Now you will see the odds for making any number of correct guesses, from none out of 1,000 to all out of 1,000.

The only problem is, the numbers in Pascal's Triangle get very large, very quickly. Even a typical computer language is inadequate to handle them. Suppose you have a language that can handle quad integers, meaning whole numbers expressed with 32 binary digits. In decimal notation, a quad integer can have a value of (plus or minus) more than two trillion. But this is only enough to calculate the first 32 lines of Pascal's Triangle. The size of the numbers in a triangle containing 1,000 lines will present a challenge.

John Walker's Probabilities

Fortunately, I don't need to imagine it, because a smart guy named John Walker has already done it for me. Moreover, he has put the results online.

Walker was the founder of Autodesk, which sold the first serious computer-aided design software using MS-DOS. He also happens to have an interest in paranormal phenomena, and pursues that interest in his spare time.

On the page that he created at *http://www.fourmilab.ch/rpkp/experiments/bincentre.html*, you will find a probability table showing how likely it is to make various numbers of correct guesses out of a total of 1,024. For instance, the odds of getting exactly half of the guesses right (512 out of 1,024) is around 2.5%.

Why isn't that number higher? Because you are almost as likely to make 511 or 513 correct guesses. What really matters, once again, is the range.

For instance, what are the odds of getting 562 correct guesses—or more? That is, 50 more than the median—or better? Walker anticipated the need to answer this kind of question, so his table shows the cumulative odds, meaning the chance of a certain number of guesses—or better. For 562 or more correct guesses, he lists the chance as 0.000981032. To change this to a percentage, just multiply by 100, and you see that the odds are about 0.098%. This means you could only expect to achieve it on 1 out of 1,019 trials (according to the table).

Let's suppose that actually happened. Should you think that you must have some paranormal powers? Hmmm, I don't know about that. Achieving a one-in-a-thousand success is unusual, but by definition, just by chance alone, it will happen about one time in a thousand! By the same logic, I would have to think that if someone happens to win a good amount of money in a casino, he must be psychic, too.

However, the odds in Walker's table diminish very quickly as the number of correct guesses increases. For instance, it shows that if you achieved 600 or more correct guesses out of 1,024, the odds of that happening would be 1 in 47,491,007. In other words, you'd have to make almost 50 million trials to expect one in which you made 600 or more correct guesses, just by chance alone. If you can achieve this, I'd be impressed—although, you know, 1 in 50 million is probably about the same as the odds of winning a lottery jackpot. Does that mean I should conclude that jackpot winners are all psychic?

You see the problem. After all the trouble we took to establish an evenly weighted random-number generator, it's still difficult to know how seriously to take any deviation from the norm. Even if you made all of the 1,024 guesses correctly, that could still be a matter of chance—although it is vanishingly unlikely.

Still, testing yourself can be fun. Perhaps 1,024 sounds like a large number, but if you make a guess every three or four seconds, the whole process should only take about an hour. You may never be able to prove beyond doubt that you have psychic powers—but on the other hand, what if your score is about average? That would still be a useful result, because it is a strong indication that you *don't* have paranormal abilities!

This raises an interesting question. Which would you prefer to be: psychic, or not psychic?

After thinking about this, I've decided that I would rather live in a world where paranormal powers don't exist. This is because I prefer to believe that there is a rational explanation for everything.

I'm a big fan of rationality. After all, rational thought processes were behind every valid theory in the history of science. Rational investigations confirmed all those theories, and the ultimate rational discipline of mathematics enabled engineers to apply them. Every bridge, building, car, aircraft, spacecraft, and computer throughout modern history has relied on mathematics for its construction.

When I look at the amazing and wonderful achievements that have been realized through the rational capabilities of the brain, I tend to think that even if some psychic powers exist, they would not be quite so impressive after all.

37 Is That All?

I have finally run out of time and space. No doubt you can think of other topics that might have been included, but I chose the ones that would make an integrated package. All of the primary components and concepts, including phototransistors, voltage dividers, comparators, counters, multiplexers, hysteresis, logic gates, randomicity, timers, and sensors, have recurred from one project to another. With just a little imagination, you may now be able to apply these tools and techniques in applications of your own.

Some people with rather impressive academic credentials have fact-checked everything that I wrote, because personally, I don't have any qualifications. I feel it's only right to tell you that I acquired the knowledge to write this book without formal instruction.

I realize that this method of learning is not suitable for everyone. I also realize that self-education tends to result in an incomplete understanding of a field. Still, if I can acquire sufficient knowledge to write books about a subject, maybe the process that I call "Learning by Discovery" can work for you, too.

We've all had the experience of reading a text book—especially when an exam is imminent—and forgetting a lot of it just a couple of weeks later. When you dig up the information to build something on your own, and you experiment with it, and learn about it by watching what happens, I think the experience is very different. The knowledge becomes embedded in your memory.

Moreover, when you are forced to use your own initiative to solve problems without outside assistance, you develop the ability to innovate.

Tinkering with hardware is a great tradition in the long history of technology. If this book helps you to feel that you can do this—that you can open up a product and figure out how it works, and fix it, or use it for a different purpose, or modify it, or improve it—you will have acquired valuable skills and an empowering mindset, and I will have achieved my purpose.

For me, this is an ending point. For you, I hope it may be a beginning.

—Charles Platt

Bibliography

You can learn anything about anything by surfing the Web, but books on paper are still a very efficient way to acquire information. The following titles and sources were all used in the compilation of *Make: More Electronics*.

An asterisk indicates titles that I think are especially useful.

*123 Robotics Experiments for the Evil Genius** by Myke Predko. McGraw-Hill, 2004.

50 Electronics Projects by A. K. Maini. Pustak Mahal, 2013.

The Art of Electronics by Paul Horowitz and Winfield Hill. Cambridge University Press, 1989.

Basic Electronics Theory by Delton T. Horn. TAB Books, 1994.

Beginning Analog Electronics through Projects by Andrew Singmin. Newnes, 2001.

The Circuit Designer's Companion by Tim Williams. Newnes, 2005.

CMOS Sourcebook by Newton C. Braga. Prompt Publications, 2001.

Complete Electronics Self-Teaching Guide by Earl Boysen and Harry Kybett. John Wiley and Sons, Inc., 2012.

Electronic Components by Delton T. Horn. TAB Books, 1992.

*Electronic Devices and Circuit Theory** by Robert L. Boylestad and Louis Nashelsky. Pearson Education, Inc., 2006.

Electronics Explained by Louis E. Frenzel, Jr. Newnes, 2010.

Fundamentals of Digital Circuits by A. Anand Kumar. PHI Learning, 2009.

*Getting Started in Electronics** by Forrest M. Mims III. Master Publishing, Inc., 2000.

*Practical Electronics for Inventors** (third edition) by Paul Scherz and Simon Monk. McGraw-Hill, 2013.

*TTL Cookbook** by Don Lancaster. Howard W. Sams & Co, Inc., 1974.

Shopping for Parts

B

Depending on your budget and your preferences, you have four different ways to acquire components for the experiments in this book:

1. *Minimum Shopping.* If you disassemble every project after completing it, you can economize on the cost of components by reusing them. The Minimum Shopping list assumes that you'll proceed on this basis. You will be recycling almost everything. See "Minimum Shopping: Experiments 1 Through 14" on page 333.
2. *Moderate Shopping.* If you don't need to keep the test experiments, but you may want to retain a few finished projects that can be fun to play with, the Moderate Shopping List contains the parts that I think you are most likely to require. See "Moderate Shopping: Experiments 1 Through 14" on page 336.
3. *Maximum Shopping.* This list contains every component required to build and keep every item in each of the thirty-six experiments. The list also contains additional spare parts of the types that are most easily damaged or likely to burn out. See "Maximum Shopping, Experiments 1 Through 14" on page 339.
4. *Incremental Shopping.* If you prefer to buy parts in small quantities, or if you just want to check that you have what you need for a particular project, the Incremental Shopping list itemizes components for one experiment at a time. See "Incremental Shopping" on page 343.

The first three options (Minimum, Moderate, and Maximum Shopping) are subdivided into separate summaries for experiments 1 through 14, experiments 15 through 25, and experiments 26 through 36.

The Kit Option

Component kits will be available for this book. They are compiled on the same basis as if you were doing Moderate Shopping as defined above. For information, see *http://www.makershed.com*.

Sources

For people who prefer to buy their own components rather than use a kit, there are two main sources:

Online retailers

The most important rule is, be willing to try more than one source. Personally, the places where I look for parts online are:

- *http://www.mouser.com*
- *http://www.radioshack.com*
- *http://www.jameco.com*

- *http://www.newark.com*
- *http://www.digikey.com*
- *http://www.alliedelec.com*
- *http://www.allelectronics.com*
- *http://www.sparkfun.com*

If one of those retailers has sold out of a component, a competitor will often have it.

Bear in mind that *http://www.allelectronics.com* is primarily a discount source for surplus components and doesn't have a huge, comprehensive stock. RadioShack, Jameco, and Sparkfun are hobbyist oriented, which means they are more likely to have the kinds of things we're interested in. Still, they don't have the amazingly diverse inventory of Mouser, Newark, or Digikey.

I realize that searching online takes time, even when you filter the results. This is why you should also have at least one paper catalog. The Mouser catalog, in particular, has a good index and is quicker to browse than the web site. The Jameco catalog is much smaller, but I find it useful because I often run across suggestions for parts that I didn't consider.

Both Mouser and Jameco offer their catalogs free to serious customers.

Buy it now

eBay is a great place to find parts that are obscure or obsolete. It's also a good source for state-of-the-art items, such as the latest LED lighting modules. And, it sells generic parts such as LEDs or resistors in bulk.

Don't be afraid to buy from Asian vendors that ship via international airmail. I have ordered from China, Cambodia, and Thailand without any problems. The descriptions are accurate, the prices are low, and the airmail service is usually reliable, although you will have to wait about two weeks for delivery.

Generic Components

In the shopping lists, you will find that I don't bother to specify precisely which type of LED to buy, or the brand of resistor, because they have become generic. Also I don't bother to specify the working voltage for each capacitor, because in this book, we don't need any capacitors rated above 16VDC.

Here's what you need to know to do your own buying.

Resistors

Any manufacturer is acceptable. Lead length is unimportant. Quarter-watt power rating (the most common value) is acceptable. Eighth-watt power rating allows a smaller component, but check each application to make sure the resistor will not be overloaded. Some people may find eighth-watt resistors too small to work with conveniently, while half-watt resistors take up an inconvenient amount of space on a breadboard.

A tolerance of 10% is acceptable, and the color bands on 10% resistors are easier to read than the bands on 5% or 1% resistors. However, you can buy 1% resistors if you wish.

Figure B-1 shows the value multipliers that are common in electronics for capacitors and resistors. For instance, resistor values of 1K or 1.5K are common, and so are 10K or 15K, and 100Ω or 150Ω. The mulipliers shown in black in the table are less common.

1.0	1.5	2.2	3.3	4.7	6.8
1.1	1.6	2.4	3.6	5.1	7.5
1.2	1.8	2.7	3.9	5.6	8.2
1.3	2.0	3.0	4.3	6.2	9.1

Figure B-1. *The traditional multipliers for resistor and capacitor values are shown in white along the top row. The additional black numbers show the full range of values for 5% resistors.*

Long ago, many resistors and almost all capacitors had an accuracy of plus-or-minus 20%, and therefore, a 1K resistor could have an actual resistance as high as 1 + 0.2 = 1.2K, while a 1.5K resistor could have a resistance as low as 1.5 – 0.3 = 1.2K. Therefore, it made no sense to have a 20% resistor with an intermediate value of, say, 1.4K, because its actual value could overlap with the actual value of a 1K resistor. Conversely, if there was a value of 1.7K instead of 1.5K, this could result in a gap in the range of values.

The six values in white type, in the top row of the table in Figure B-1, were the original multipliers for values of 20% components. They are still the most widely used values today, even though 5% resistors have become common. The additional 5% multipliers are shown in black type.

For most projects, you don't need all those 5% values. You can just stick to the six multipliers shown in white. I have used them exclusively throughout the book, so that you don't have to buy an unnecessarily large variety of resistors.

Figure B-2 shows the values that you might aim to acquire for a stockpile of resistors. The quantities in the table will be sufficient for all of the projects in this book, plus at least 50%. If you're wondering why the 220Ω and 10K values are so much more numerous than the rest, it's because 220Ω is commonly used as a series resistor with an LED, and 10K is commonly used for a pullup or pulldown resistor on an input pin of a logic chip.

- You may spend less money buying prepackaged assortments than if you try to buy small numbers of parts with specific values. The price of resistors drops radically when you buy them in bulk.

Resistor Value	Quantity	Resistor Value	Quantity	Resistor Value	Quantity	Resistor Value	Quantity
22	10	1K	40	10K	150	100K	60
47	10	1.5K	10	15K	10	150K	10
100	60	2.2K	20	22K	10	220K	10
220	150	3.3K	50	33K	10	330K	10
330	50	4.7K	30	47K	10	470K	10
470	60	6.8K	10	68K	20	680K	10
680	40					1M	10

Figure B-2. *The table shows quantities of resistors sufficient for all the projects in this book, plus at least 50%. All values are in ohms, except where otherwise indicated (K=kilohms; M=megohms).*

Of course, you don't have to buy the exact quantities shown in the table. You may be able to find a prepackaged assortment that matches the resistance values in the table and provides you with ten of each. You can then add the larger quantities separately.

Capacitors

Any manufacturer is acceptable. Radial leads are preferred. A working voltage of 16VDC is the minimum. Higher working voltages are acceptable. Multilayer ceramic capacitors are preferred. For capacitance values above 10µF, ceramics become more expensive and electrolytics are acceptable.

Figure B-3 suggests an assortment of capacitors to have in stock, sufficient for all of the projects in this book plus at least 50%. Try to use ceramic capacitors for the first three columns of values in the table. Purely for financial reasons, you'll probably want to use electrolytics for values above 10µF.

Capacitor Value (μF)	Quantity	Capacitor Value (μF)	Quantity	Capacitor Value (μF)	Quantity	Capacitor Value (μF)	Quantity
0.001	20	0.1	30	1	20	10	30
0.01	50	0.15	5	1.5	5	15	10
0.022	10	0.22	5	2.2	10	47	10
0.033	20	0.33	20	3.3	5	68	10
0.047	10	0.47	5	4.7	5	100	10
0.068	10	0.68	20	6.8	5	330	10

Figure B-3. *A recommended range of capacitor values to have on hand, sufficient for all the projects in this book plus at least 50%.*

Remember that 0.001μF = 1nF. I have avoided the nF (nanofarad) unit in the schematics in this book, because while it is common in Europe, it is less widely used in the United States.

Multilayer ceramic capacitors have become radically smaller and cheaper since the 1990s, and their durability makes them an attractive choice. Electrolytic capacitors are larger than the equivalent ceramic capacitors and may have a shorter shelf life, although this is a matter for debate.

Many authorities warn that electrolytics deteriorate with age because they have to be connected with a power source periodically to activate their internal chemistry. Yet I have electrolytics here which have been in storage for 15 years, and when I use them, they still seem to work. Should I trust the experts, or my own experience? I'm not sure, but with ceramic capacitors, the issue does not arise.

One problem with ceramic capacitors is that they seldom have any information printed on them. You can check their capacitance with your multimeter, if it offers this feature, but many meters cannot measure values higher than 20μF, and a meter cannot tell you the working voltage for which a capacitor is rated.

You have to be careful to label the storage containers for these components, and after you've taken one out of storage and used it in a project, you are unlikely to remember what its voltage rating is. For this reason, buying all your capacitors with the same voltage rating is helpful. A working voltage of 16VDC is the acceptable minimum, bearing in mind that capacitors should not generally be used at more than three-quarters of their rating.

LEDs

Any manufacturer is acceptable. LEDs come in a dizzying variety of shapes and sizes, but the type that is commonly described as "standard through hole" is what you want.

Often in this book you'll find that I use LEDs to test a circuit and verify the outputs. You'll find that 3mm diameter LEDs are useful for this purpose, as you can fit a series of them into adjacent rows on your breadboard.

- 3mm LEDs are also referred to as being T-1 size.

Choose your own color, intensity, viewing angle, clear or diffused. Maximum forward current of 10mA is preferred for driving LEDs from 74HC00 series chips. Forward voltage of 2VDC is typical.

LEDs fitting this specification are Kingbright WP132X*D, where a letter specifying the color is substituted for the asterisk—for example, WP132XGD is the green one. Vishay TLHK4200 is comparable. LEDs of this general type should cost 15 cents or less in the US. You can get much better prices if you buy in bulk.

As test indicators, low-power LEDs that use a forward current of 2mA can be useful to prolong battery life. The relatively low light output will not be an issue when you simply want to verify that something is working.

LEDs with internal resistor

This type of component contains an internal resistor with the correct value for a specified supply voltage so that an external resistor is not needed. These are extremely convenient for breadboarding work and are strongly recommended if you are willing to pay about twice the price of regular

LEDs. Vishay TLR*4420CU, where a letter specifying the color is substituted for the asterisk, is an example, rated for 12VDC but can be used with 9VDC or 5VDC. Chicago 4302F1-12V, 4302F3-12V, and 4302F5-12V (red, amber, and green, respectively) are acceptable but may be slightly more expensive.

Avago HLMP-1620 and HLMP-1640 series are rated for 5VDC. Personally I buy LEDs rated for 12VDC, because I'm not very interested in their brightness for testing purposes, and the 12VDC type can be used in both 9V and 5V circuits.

- Where I specify a "generic" LED in the shopping lists, this means it is the everyday type that does not have an internal resistor and must be protected with an external resistor.

Warning: Series Resistors

Most of the photographs of breadboarded circuits in this book show LEDs with internal resistors and no external series resistors. If you use a generic LED, you must remember to add your own series resistor.

If an LED does not have an internal series resistor, a 470Ω resistor in 9VDC circuits and a 220Ω resistor in 5VDC circuits should be appropriate. Quarter-watt series resistors can be used if we assume a maximum voltage drop across the resistor of 7V at 15mA, in a 9V circuit. This will entail a power dissipation of about 100mW, which is less than half the wattage for which the resistor is rated.

If there's a circuit that you want to build as a finished project, you may wish to use larger, brighter LEDs. Because specifications vary, it will be up to you to choose an appropriate series resistor to limit the current through the LED in accordance with the specification in the datasheet.

Chip Family Basics

Any manufacturer is acceptable. The "package" of a chip refers to its physical size, and this attribute should be checked carefully when ordering. All logic chips must be in a DIP package (meaning a dual-inline package with two rows of pins that have 0.1" spacing). This may also be referred to as PDIP (meaning a plastic dual-inline package). They are also described as "through hole." The DIP and PDIP descriptors may be appended with the number of pins, as in DIP-14 or PDIP-16. The numbers can be ignored.

Surface-mount chips will have packaging descriptors beginning with S, as in SOT or SSOP. Do not buy any chips with "S" type packages. You won't be able to use them.

The recommended chip family is HC (high-speed CMOS), as in 74HC00, 74HC08, and similar generic identifiers. These numbers will have additional letters or numbers added by individual manufacturers as prefixes or suffixes, as in SN74HC00DBR (a Texas Instruments chip) or MC74HC00ADG (from On Semiconductor). For our purposes, these versions are functionally identical. Look carefully, and you will see the 74HC00 generic number embedded in each proprietary number.

Sources often state that an HC chip can source or sink a maximum of 4mA to 6mA, but information from manufacturers (such as Application Note 313 from Fairchild Semiconductor) explicitly states that sourcing up to 25mA will not damage an HC series chip.

However, if the output of a logic chip is driving the input of another chip in addition to an LED, 10mA is a safer current value because the voltage on the output of a logic chip is pulled down by higher currents. If the voltage drops below 3.5VDC, another logic chip may fail to recognize this as a high state. A minimum of 4VDC is preferable.

Logic chips can be found in vendor sites by searching just for the generic number (e.g., 74HC86 for a quad two-input XOR chip). Generally speaking, the additional manufacturers' letters or numbers can be ignored, so long as the chip is specified to work with a 5VDC power supply. Often the specification will state "2V to 6V" for the HC series chips.

In some instances, the old 4000B series of logic chips may have functions that are unavailable in the 74HC00 family. For instance, a four-input OR gate is unavailable in the HC series, but the 4072B contains two four-input OR gates.

The 4000B series of logic chips can be substituted for 74HC00 series chips if their lower sourcing or sinking output current is acceptable. Most 4000B series logic chips are designed to interface with other chips, not with LEDs.

Figure B-4 shows the input and output voltage ranges that are interpreted as high and low logic states in the 4000B series and the 74HC00 series. You can see that these families of chips should have no problem understanding each other.

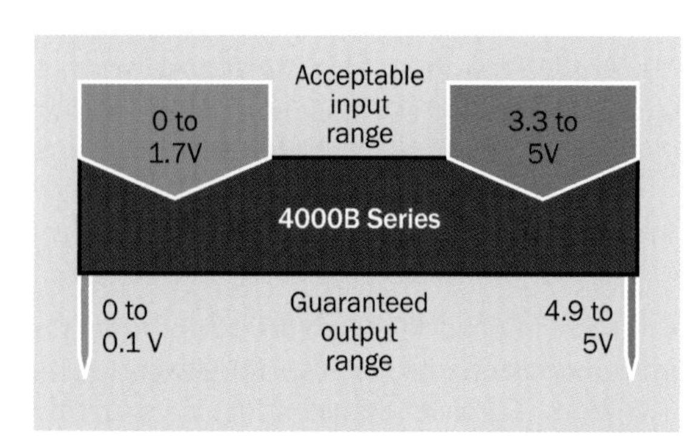

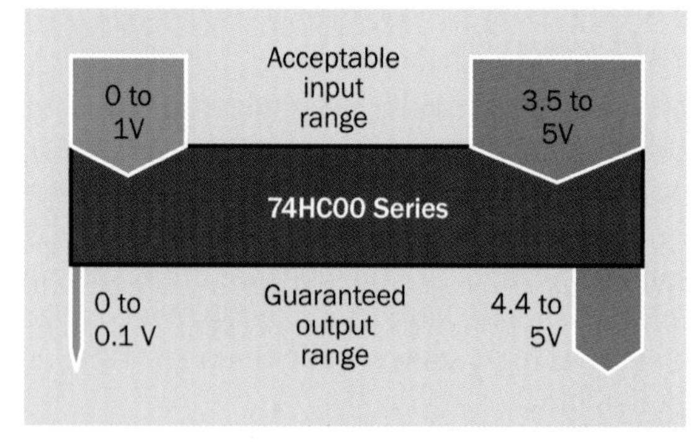

Figure B-4. *Acceptable input voltages and guaranteed output source-or-sink voltages for logic chips in the 4000B and 74HC00 families. The output voltages are specified assuming an output current of 4mA (for 74HC00 chips) and 0.5mA (for 4000B chips). Higher current values will pull down the voltages.*

TTL logic chips, such as the 74LS00 series, have significant compatibility issues. They are not used or recommended for any of the projects in this book.

Transistors

For convenience, only one type of transistor is used in this book: the 2N2222. This generic part number is often preceded by manufacturers with a P, as in PN2222, or PN2222A, or PN2222ATFR. The letters at the beginning of the part number make no difference.

Be careful not to buy a P2N2222, even if an online supplier's search algorithm suggests it as a substitute. In the case of this component, the letter and numeral (P2) make a big difference. The P2N2222 has pin functions that are reversed from all the other types. This can cause confusion, frustration, and a lot of wasted time.

Switches

Four types of switch are frequently used in the projects in this book: tactile switches, pushbutton switches, DIP switches, and toggle switches.

Tactile switches are very small pushbuttons that can be inserted in breadboards (or perforated boards, if you wish). The manufacturer and part number are generally irrelevant, so long as the switch is "through-hole" type, not surface mount. Its pins must fit a breadboard that has rows spaced at 0.1".

The most common tactile switches are 6mm x 6mm in size (about 1/4" square), often referred to as 6x6 in parts catalogues. They have four pins, but each pair of pins is typically joined together inside the switch so that this is a SPST device even though it looks as if it might have two poles.

For projects in this book, I prefer a half-size tactile switch, often described as being 3.5mm x 6mm in size. It has only two pins and occupies only one column on your breadboard. Examples are the Mountain Switch TS4311T series, with actual part numbers such as TS4311T5201 or TS4311T1601. These are different from each other only in the color of the button and the pressure that has to be applied to close the switch contacts.

Pushbutton switches of the type used in several projects here are DPDT, 4PDT, or 6PDT, depending on the application. They can be momentary switches that revert to their initial state when you let go of the button, or push-twice switches, also known as latching switches, where you press the button a second time to get back to the initial state.

This is really a matter of individual preference, and you can make up your mind when you read a description of a project and see how it is intended to be used. Any brand of pushbutton switch is acceptable, but they must be "through-hole" type with solder pins, and you should try to obtain them with 0.1" pin spacing (often listed as 2.54mm, although 2.5mm is acceptable). The Alps SPUJ series is typical. The E-Switch PBH series may also be used, such as the PBH4UOA-NAGX, where the number 4 identifies it as a four-pole switch.

Note that most pushbutton switches are sold without caps or buttons. You are expected to choose them separately. This allows you various options for color, shape, and size.

DIP switches are dual-inline pin format to fit across the center channel of your breadboard. The number of "positions" actually refers to the number of separate miniature switches in the component. Each should be SPST. The component must have a pin spacing of 2.54mm or 0.1", and must be described as "through-hole" or "PCB mount" or "solder pin" to fit your breadboard.

The BD series by C&K switches is typical, with the part number telling you how many "positions" (switches) each component has. Thus the BD02 has two positions, BD04 has four positions, and so on.

Toggle switches are used in this book only to switch the 5VDC or 9VDC power supply. You need an SPST or SPDT switch described as being "through hole," or having solder pins, or being suitable for PC board mounting, so long as the pin spacing is 0.1" (2.54mm) or 0.2" (5.08mm). Mountain Switch 108-2MS1T2B3M2QE-EVX is an example. The toggle switches in the projects described here will not be heavily used, so you can buy the smallest and cheapest you can find.

Power Supply, Breadboards, and Wiring

Each project requires either a 9VDC supply (which can be provided by a 9V battery) or a 5VDC regulated power supply. See Setup on page 25 for details.

If you expect to keep some projects in their final form, I leave it to you to decide how many power supplies you need. Don't forget to buy a snap-on connector for each 9V battery.

You will require between two and thirty breadboards, depending how many breadboarded projects you decide to keep. (A couple of projects are too big to fit on a single breadboard, so they require a pair; this is why the minimum number of breadboards is two.) The type of breadboard that I am recommending is very affordable if you buy them from Asian vendors through eBay. I have seen them at five for $10, with free shipping.

Setup also discusses hookup wire and jumpers.

So much for the procurement information. Now, here are the lists.

Minimum Shopping: Experiments 1 Through 14

See "Generic Components" on page 328 above for guidance about purchasing generic components including resistors, capacitors, LEDs, logic chips, transistors, and switches.

All quantities in the following lists are shown in parentheses. If no quantity is listed, the quantity is 1:

Power

- 1.5V battery (2)
- 9V battery and connector

- One 5VDC voltage supply requires an LM7805 voltage regulator, plus two capacitors, a resistor, and an LED, which are included below.
- AC adapter with 10VDC to 12VDC for long-term installation of Experiment 7

Resistors

- 22Ω, 47Ω, 100Ω, 220Ω (2), 470Ω (5), 1K (3), 1.5K, 2.2K (10), 3.3K, 4.7K (2), 10K (4), 33K, 68K (2), 100K (10), 150K (2), 220K (2), 1M (2)

Capacitors

- 0.01µF(2), 0.047µF, 0.068µF, 0.1µF (3), 0.33µF (3), 0.68µF (2), 1µF (5), 10µF (3), 15µF, 47µF, 100µF, 220µF, 330µF

Switches

- Toggle (SPST or SPDT)
- Tactile

LEDs

- 3mm generic (4)

Trimmer potentiometers

- 5K, 10K, 500K (2), 1M

Transistors

- 2N2222 or PN2222 (3)

Integrated circuit chips

- 555 timer, old-style bipolar type (2)
- LM339 comparator
- LM741 op-amp
- LM386 power amp

Sensors

- Phototransistor PT334-6C or similar, responsive to white light
- Microphone (electret, generic, two-terminal type)

Audio output

- Loudspeaker, 2" diameter, 50Ω or higher
- Beeper, 9VDC or 12VDC, 100mA max.

Other

- Alligator clip (2)
- Patch cord with alligator at each end (3)
- Corrugated cardboard, minimum 6" x 12" (1 piece)
- Elmer's glue or similar white glue (minimal quantity)
- LM7805 voltage regulator
- LM7806 voltage regulator
- UA78M33 voltage regulator
- 1N4001 small rectifier diode (2)
- Latching relay DS1E-SL2-DC3V or similar in SPDT or DPDT format with 3VDC coil to switch up to 2A
- Battery-powered digital alarm clock must use two 1.5V batteries

Minimum Shopping: Experiments 15 Through 25

See "Generic Components" on page 328 above for guidance about purchasing generic components including resistors, capacitors, LEDs, logic chips, transistors, and switches.

All quantities in the following lists are shown in parentheses. If no quantity is listed, the quantity is 1:

- I assume that you have previously bought the components listed for Minimum Shopping: Experiments 1 through 14, and you have those components available to you for reuse in experiments 15 through 25. The quantities listed below are therefore *additional* to those which you already have.

Resistors

- 100Ω (9), 220Ω (8), 330Ω (6), 470Ω (2), 3.3K, 10K (15)

Capacitors

- 0.001µF (2)

Switches

- Tactile (5)
- Pushbutton SPDT
- Pushbutton DPDT (3)
- Pushbutton 4PDT (6)
- Pushbutton 6PDT (2)
- Cap for pushbutton (9)
- DIP 4-position (2)
- DIP 8-position (2)

LEDs

- 3mm generic (16)

Trimmer potentiometers

- 1K, 50K

Integrated circuit chips

- 74HC08 quad two-input AND (2)
- 74HC32 quad two-input OR
- 74HC02 quad two-input NOR
- 74HC86 quad two-input XOR (2)
- 74HC4075 triple three-input OR
- 74HC4002 dual four-input NOR
- 74HC4514 decoder or 4514B decoder
- 74HC237 decoder (2)
- 4067B multiplexer
- 4520B counter
- 74HC148 encoder (2)
- 74HC11 triple three-input AND

Sensor

- Phototransistor PT334-6C or similar, responsive to white light

Audio output

- Buzzer or beeper, 9VDC or 12VDC (2)
- Loudspeaker 2" diameter, 50Ω or higher

Other

- Plain perforated board with no copper traces (6" x 6" piece minimum)

Optional extras

- Multicolored ribbon cable (2 feet)
- Pushbutton 3PDT
- Pushbutton DPDT (3)
- Pushbutton SPDT (3)
- Caps for pushbuttons (7)

Minimum Shopping: Experiments 26 Through 36

See "Generic Components" on page 328 above for guidance about purchasing generic components, including resistors, capacitors, LEDs, logic chips, transistors, and switches.

All quantities are shown in parentheses. If no quantity is listed, the quantity is 1:

- I assume that you have previously bought the components listed for Minimum Shopping experiments 1 through 14 and 15 through 25. Those components should be available to you for reuse in experiments 26 through 36. The quantities listed below are therefore *additional* to those which you already have.

Resistors

- 220Ω (14), 100K (2)

Capacitors

- 47pF, 68pF, 100pF, 0.0001µF, 0.01µF (2), 0.033µF, 2.2µF

LEDs

- 3mm generic (25), or (10) if light bars (36) are used in Experiment 28
- 5mm generic (16)

Trimmer Potentiometers

- 1K, 2K, 100K

Integrated circuit chips

- 7555 timer (3)
- 74HC4017 counter (3)
- 74HC164 shift register (3)

- 4078B single eight-input OR/NOR
- 74HC7266 quad two-input XNOR
- 74HC30 single eight-input NAND

Sensors

- SPST reed switch, any type
- Bipolar Hall-effect sensor, ATS177 or similar, and unipolar, or linear, or omnipolar Hall-effect sensor (optional)
- Transmissive infrared sensor Everlight ITR9606 or similar
- Thermistor 100K
- Bourns ECW1J-B24-BC0024L rotational encoder or similar with 24PPR and 24 detents and quadrature output (2)

Other

- Flexible jumper wires (35)
- Small iron bar magnet approximately 1/4" x 1/4" x 1.5" or very small neodymium bar magnet approximately 1/4" x 1/16" x 1/2"
- Lead sinkers (2) and galvanized wire (1 foot) as described in Experiment 33

Optional extras

- Header pins (33)
- Header sockets (33)
- Light bars (36) and ULN2003 Darlington array (3) if making a finished version of Experiment 28
- Ball magnet and aluminum tube, and ring magnets (4) as described in Experiment 29
- 330Ω (16), 680Ω (16), 3.3K (16), 5mm LED (16), Transmissive infrared sensor Everlight ITR9606 or similar (16), and ULN2003 Darlington array (3) if making a finished version of Experiment 31

Moderate Shopping: Experiments 1 Through 14

See "Generic Components" on page 328 above for guidance about purchasing generic components including resistors, capacitors, LEDs, logic chips, transistors, and switches.

All quantities in the following lists are shown in parentheses. If no quantity is listed, the quantity is 1:

Power

- 1.5V battery (2)
- 9V battery and connectors (5 max)
- AC adapter with 10VDC to 12VDC fixed output at 1A minimum
- 5VDC voltage supply; requires the following (multiply by the number of projects for which you will want to use a separate supply):
  - — LM7805 voltage regulator
  - — 2.2K resistor
  - — 0.33μF, 0.1μF capacitor
  - — Toggle switch SPST or SPDT
  - — Generic 3mm LED

Resistors

- 22Ω (5), 47Ω (5), 100Ω (5), 220Ω (5), 470Ω (10), 1K (10), 1.5K (5), 2.2K (10), 3.3K (5), 4.7K (5), 10K (15), 33K (5), 68K (5), 100K (20), 150K (5), 220K (5), 1M (5)

  Because resistors are very inexpensive, the minimum quantity in this list is 5.

Capacitors

- 0.01μF (10), 0.047μF (5), 0.068μF (5), 0.1μF (5), 0.33μF (5), 0.68μF (5), 1μF (5), 10μF (10), 15μF, 47μF, 100μF (2), 220μF, 330μF

  Because ceramic capacitors with values below 10μF are very inexpensive, their minimum quantity in this list is 5.

Switches

- Toggle
- Tactile

LEDs

- 3mm generic (10)

Trimmer potentiometers

- 5K, 10K, 500K (4), 1M (3)

Transistors

- 2N2222 or PN2222 (7)

Integrated circuit chips

- 555 timer, old-style bipolar type (7)
- LM339 comparator (3)
- LM741 op-amp (3)
- LM386 power amp

Sensors

- Phototransistor PT334-6C or similar, responsive to white light (3)
- Microphone, electret, generic, two-terminal type (4)

Audio output

- Loudspeaker, 2" diameter, 50Ω or higher (3)
- Buzzer or beeper, 9VDC or 12VDC, 100mA max

Other

- Alligator clip (2)
- Patch cord with alligator at each end (3)
- Corrugated cardboard, minimum 6" x 12" (1 piece)
- Elmer's glue or similar white glue (minimal quantity)
- LM7806 voltage regulator
- UA78M33 voltage regulator
- 1N4001 small rectifier diode (2)
- Latching relay DS1E-SL2-DC3V or similar in SPDT or DPDT format with 3VDC coil to switch up to 2A
- Battery-powered digital alarm clock; must use two 1.5V batteries

Optional extras

- Distilled or deionized water
- Table salt

Moderate Shopping: Experiments 15 Through 25

See "Generic Components" on page 328 above for guidance about purchasing generic components including resistors, capacitors, LEDs, logic chips, transistors, and switches.

All quantities in the following lists are shown in parentheses. If no quantity is listed, the quantity is 1. This list is entirely self-contained and does not require any parts that may have been purchased for experiments 1 through 14:

Power

- 9V battery and connectors
- 5VDC voltage supply; requires the following (multiply by the number of projects for which you will want to use a separate supply):
  - — LM7805 voltage regulator
  - — 2.2K resistor
  - — 0.33µF, 0.1µF capacitor
  - — Toggle switch SPST or SPDT
  - — Generic 3mm LED

Resistors

- 100Ω (15), 220Ω (15), 330Ω (10), 470Ω (15), 1K (5), 3.3K, 4.7K, 10K (20), 33K (5)

  Because resistors are very inexpensive, the minimum quantity in this list is 5.

Capacitors

- 0.001µF (5), 0.01µF (5), 0.1µF (5), 100µF

  Because ceramic capacitors with values below 10µF are very inexpensive, their minimum quantity in this list is 5.

Switches

- Tactile (10)
- Pushbutton DPDT (4)
- Pushbutton 4PDT (12)
- Cap for pushbutton (16)
- DIP 4-position (2)
- DIP 8-position (2)

LEDs

- 3mm with internal resistors (35)
- 5mm generic (10)

Trimmer potentiometers

- 50K

Transistors

- 2N2222 or PN2222

Integrated circuit chips

- 555 timer old-style bipolar type (3)
- 74HC08 quad two-input AND (4)
- 74HC32 quad two-input OR (2)
- 74HC02 quad two-input NOR
- 74HC86 quad two-input XOR (4)
- 74HC4075 triple three-input OR (2)
- 74HC4002 dual four-input NOR
- 74HC11 triple three-input AND
- 74HC4514 decoder or 4514B decoder (2)
- 74HC237 decoder (2)
- 4067B multiplexer
- 4520B counter
- 74HC148 encoder (2)

Sensors

- Phototransistor PT334-6C or similar, responsive to white light

Audio output

- Buzzer or beeper 9VDC or 12VDC (3)
- Loudspeaker 2" diameter, 50Ω or higher

Other

- Plain perforated board with no copper traces (6" x 6" piece minimum)

Optional extras

- Multicolored ribbon cable (2 feet)
- Pushbutton 3PDT
- Pushbutton DPDT (3)
- Pushbutton SPDT (3)
- Caps for pushbuttons (7)

Moderate Shopping: Experiments 26 Through 36

See "Generic Components" on page 328 above for guidance about purchasing generic components, including resistors, capacitors, LEDs, logic chips, transistors, and switches.

All quantities in the following lists are shown in parentheses. If no quantity is listed, the quantity is 1. This list is entirely self-contained and does not require any parts that may have been purchased for experiments 1 through 25:

Power

- 9V battery and connector
- 5VDC voltage supply; requires the following (multiply by the number of projects for which you will want to use a separate supply):
  - — LM7805 voltage regulator
  - — 2.2K resistor
  - — 0.33μF, 0.1μF capacitor
  - — Toggle switch SPST or SPDT
  - — Generic 3mm LED

Resistors

- 100Ω (5), 220Ω (50), 330Ω (20), 470Ω (5), 680Ω (20), 1K (5), 3.3K (20), 4.7K (5), 10K (20), 100K (5), 1M (5)

  Because resistors are very inexpensive, the minimum quantity in this list is 5.

Capacitors

- 47pF (5), 68pF (5), 0.0001µF (5), 0.001µF (5), 0.01µF (10), 0.033µF (5), 0.1µF (5), 0.33µF (5), 1µF (5), 2.2µF (5), 10µF (5), 100µF

  Because ceramic capacitors with values below 10µF are very inexpensive, their minimum quantity in this list is 5.

Switches

- Tactile (2)
- Pushbutton DPDT (4)
- Cap for pushbutton (4)

LEDs

- 3mm generic (40) or (4) if light bars (36) are used in Experiment 28
- 3mm with internal resistor (40)
- 5mm generic (20)

Trimmer potentiometers

- 1K, 2K, 100K

Transistors

- 2N2222 or PN2222 (2)

Integrated circuit chips

- 7555 timer (8)
- 74HC4017 counter (3)
- 74HC08 quad two-input AND (2)
- 74HC32 quad two-input OR (2)
- 74HC86 quad two-input XOR (2)
- 74HC7266 quad two-input XNOR (2)
- 4078B single eight-input OR/NOR
- 4520B binary counter
- 74HC30 single eight-input NAND
- 74HC164 shift register (5)
- 74HC4514 decoder or 4514B decoder
- 74HC4017 counter
- ULN2003 Darlington array (2), or (5) if light bars are used in Experiment 28

Sensors

- SPST reed switch, any type
- Bipolar Hall-effect sensor, ATS177 or similar (2), and unipolar, or linear, or omnipolar Hall-effect sensor (optional)
- Transmissive infrared sensor Everlight ITR9606 or similar
- Thermistor 100K
- Bourns ECW1J-B24-BC0024L rotational encoder or similar with 24PPR and 24 detents and quadrature output (2)

Other

- Flexible jumper wires (50)
- Small iron bar magnet 1/4" x 1/4" x 1.5" approx. or very small neodymium bar magnet 1/4" x 1/16" x 1/2" approx.
- Lead sinkers (2) and galvanized wire (1 foot) as described in Experiment 33

Optional extras

- Multicolor ribbon cable (2 feet)
- Header pins (33)
- Header sockets (33)
- Light bars (36) and ULN2003 Darlington array (3) if making a finished version of Experiment 28
- 330Ω (16), 680Ω (16), 3.3K (16), 5mm LED (16), Transmissive infrared sensor Everlight ITR9606 or similar (16), and ULN2003 Darlington array (3) if making a finished version of Experiment 31
- Ball magnet and aluminum tube, and ring magnets (4) as described in Experiment 29

Maximum Shopping, Experiments 1 Through 14

See "Generic Components" on page 328 above for guidance about purchasing generic components, including resistors, capacitors, LEDs, logic chips, transistors, and switches.

All quantities in the following lists are shown in parentheses. If no quantity is listed, the quantity is 1:

Power

- 9V battery and connector (8 max)
- AC adapter with 10VDC to 12VDC fixed output at 1A minimum
- 5VDC voltage supply; requires the following (multiply by the number of projects for which you will want to use a separate supply):
  - — LM7805 voltage regulator
  - — 2.2K resistor
  - — 0.33μF, 0.1μF capacitor
  - — Toggle switch SPST or SPDT
  - — Generic 3mm LED

Resistors

- 22Ω (5), 47Ω (5), 100Ω (5), 220Ω (5), 470Ω (10), 1K (15), 1.5K (5), 2.2K (10), 3.3K (10), 4.7K (10), 10K (20), 33K (5), 68K (10), 100K (40), 150K (5), 220K (5), 1M (5)

  Because resistors are very inexpensive, the minimum quantity in this list is 5.

Capacitors

- 0.01μF (10), 0.047μF (5), 0.068μF (5), 0.1μF (5), 0.33μF (5), 0.68μF (10), 1μF (10), 10μF (15), 15μF (2), 47μF (3), 100μF (4), 220μF (2), 330μF (3)

  Because ceramic capacitors with values below 10μF are very inexpensive, their minimum quantity in this list is 5.

Switches

- Toggle
- Tactile

LEDs

- 3mm generic (10)

Trimmer potentiometers

- 5K, 10K, 500K (4), 1M (3)

Transistors

- 2N2222 or PN2222 (10)

Integrated circuit chips

- 555 timer, old-style bipolar type (10)
- LM339 comparator (5)
- LM741 op-amp (7)
- LM386 power amp

Sensors

- Phototransistor PT334-6C or similar, responsive to white light (5)
- Microphone, electret, generic, two-terminal type (5)

Audio output

- Loudspeaker, 2″ diameter, 50Ω or higher (5)
- Buzzer or beeper, 9VDC or 12VDC, 100mA max.

Other

- Alligator clip (2)
- Patch cord with alligator at each end (3)
- Corrugated cardboard, minimum 6″ x 12″ (1 piece)
- Elmer's glue or similar white glue (minimal quantity)
- LM7806 voltage regulator
- UA78M33 voltage regulator
- 1N4001 small rectifier diode (2)
- Latching relay DS1E-SL2-DC3V or similar in SPDT or DPDT format with 3VDC coil to switch up to 2A
- Battery-powered digital alarm clock; must use two 1.5V batteries

Optional extras

- Distilled or deionized water
- Table salt, extra transistor
- Ammeter (50 microamperes)
- Ammeter (10 milliamperes)

Maximum Shopping: Experiments 15 Through 25

See "Generic Components" on page 328 above for guidance about purchasing generic components, including resistors, capacitors, LEDs, logic chips, transistors, and switches.

All quantities in the following lists are shown in parentheses. If no quantity is listed, the quantity is 1. This list is entirely self-contained and does not require any parts that may have been purchased for experiments 1 through 14:

Power

- 9V battery and connector (5 max)
- 5VDC voltage supply; requires the following (multiply by the number of projects for which you will want to use a separate supply):
  - — LM7805 voltage regulator
  - — 2.2K resistor
  - — 0.33µF and 0.1µF capacitor
  - — Toggle switch SPST or SPDT
  - — Generic 3mm LED

Resistors

- 100Ω (20), 220Ω (40), 330Ω (10), 470Ω (15), 1K (5), 3.3K (5), 4.7K (5), 10K (60), 33K (5)

  Because resistors are very inexpensive, the minimum quantity in this list is 5.

Capacitors

- 0.001µF (5), 0.01µF (5), 0.1µF (5), 100µF (2)

  Because ceramic capacitors with values below 10µF are very inexpensive, their minimum quantity in this list is 5.

Switches

- Tactile (16)
- Pushbutton DPDT (4)
- Pushbutton 4PDT (18)
- Pushbutton 6PDT (2)
- Cap for pushbutton (25)
- DIP 4-position (3)
- DIP 8-position (4)

LEDs

- 3mm generic (10)
- 3mm with internal resistors (50)
- 5mm generic (15)

Trimmer potentiometers

- 50K

Transistors

- 2N2222 or PN2222

Integrated circuit chips

- 555 timer old-style bipolar type (3)
- 74HC08 quad two-input AND (8)
- 74HC32 quad two-input OR (4)
- 74HC02 quad two-input NOR (2)
- 74HC86 quad two-input XOR (6)
- 74HC4075 triple three-input OR (3)
- 74HC4002 dual four-input NOR (2)
- 74HC11 triple three-input AND (2)
- 74HC4514 decoder or 4514B decoder (3)
- 74HC237 decoder (3)
- 4067B multiplexer (2)
- 4520B counter (2)
- 74HC148 encoder (3)

Sensors

- Phototransistor PT334-6C or similar, responsive to white light

Audio output

- Loudspeaker 2" diameter, 50Ω or higher
- Buzzer or beeper, 9VDC or 12VDC (3)

Other

- Plain perforated board with no copper traces (6" x 6" piece minimum)

Optional extras

- Multicolored ribbon cable (2 feet)
- Pushbutton 3PDT
- Pushbutton DPDT (3)
- Pushbutton SPDT (3)
- Caps for pushbuttons (7)

Maximum Shopping: Experiments 26 Through 36

See "Generic Components" on page 328 above for guidance about purchasing generic components, including resistors, capacitors, LEDs, logic chips, transistors, and switches.

All quantities in the following lists are shown in parentheses. If no quantity is listed, the quantity is 1.

This list is entirely self-contained and does not require any parts that may have been purchased for experiments 1 through 25:

Power

- 9V battery and connector (3 max)
- 5VDC voltage supply; requires the following (multiply by the number of projects for which you will want to use a separate supply):
  - — LM7805 voltage regulator
  - — 2.2K resistor
  - — 0.33μF, 0.1μF capacitor
  - — Toggle switch SPST or SPDT
  - — Generic 3mm LED

Resistors

- 100Ω (5), 220Ω (70), 330Ω (20), 470Ω (5), 680Ω (20), 1K (5), 3.3K (20), 4.7K (5), 10K (25), 100K (10), 1M (5)

  Because resistors are very inexpensive, the minimum quantity in this list is 5.

Capacitors

- 47pF (5), 68pF (5), 0.0001μF (5), 0.001μF (5),, 0.01μF (15), 0.033μF (10), 0.1μF (5), 0.33μF (5), 1μF (5), 2.2μF (5), 10μF (5), 100μF (3)

  Because ceramic capacitors with values below 10μF are very inexpensive, their minimum quantity in this list is 5.

Switches

- Tactile (4)
- Pushbutton DPDT (7)
- Cap for pushbutton (7)

LEDs

- 3mm generic (40) or (4) if light bars (36) are used in Experiment 28
- 3mm with internal resistor (60)
- 5mm generic (20)

Trimmer potentiometers

- 1K, 2K, 100K

Transistors

- 2N2222 or PN2222

Integrated circuit chips

- 7555 timer (11)
- 74HC4017 counter (3)
- 74HC08 quad two-input AND (2)
- 74HC32 quad two-input OR (2)
- 74HC86 quad two-input XOR (3)
- 74HC7266 quad two-input XNOR (2)
- 74HC30 single eight-input NAND
- 4078B single eight-input OR/NOR
- 4520B binary counter (2)
- 74HC164 shift register (6)
- 74HC4017 counter (3)
- 74HC4514 decoder or 4514B decoder
- ULN2003 Darlington array (2) or (5) if light bars are used in Experiment 28

Sensors

- SPST reed switch, any type (2)
- Bipolar Hall-effect sensor, ATS177 or similar (2), and unipolar, or linear, or omnipolar Hall-effect sensor (optional)
- Transmissive infrared sensor Everlight ITR9606 (20)
- Thermistor 100K (2)
- Bourns ECW1J-B24-BC0024L rotational encoder or similar with 24PPR and 24 detents and quadrature output (3)

Other

- Flexible jumper wires (50)
- Small iron bar magnet approximately 1/4" x 1/4" x 1.5" (2) or very small neodymium bar magnet approximately 1/4" x 1/16" x 1/2" (2)
- Lead sinkers (2) and galvanized wire (1 foot) as described in Experiment 33

Optional extras

- 3300Ω (16), 6800Ω (16), 3.3K (16), 5mm LED (16), Transmissive infrared sensor Everlight ITR9606 or similar (16), and ULN2003 Darlington array (3) if making a finished version of Experiment 31
- Multicolor ribbon cable (2 feet)
- Header pins (40) and header sockets (40)
- Light bars (36) and ULN2003 Darlington array (3) if making a finished version of Experiment 28
- Ball magnet and aluminum tube, and ring magnets (4) as described in Experiment 29

Incremental Shopping

See "Generic Components" on page 328 above for guidance about purchasing generic components, including resistors, capacitors, LEDs, logic chips, transistors, and switches.

The following lists show the precise quantity of each component required for each experiment. The quantities are in parentheses. If no quantity is listed, the quantity is 1:

Experiment 1

Power
: 9V battery and connector

Resistor
: 470Ω

Transistor
: 2N2222 or PN2222

LED
: Any type

Other
: Alligator clip (2), patch cord with alligator at each end (3), corrugated cardboard, minimum 6" x 12" (1 piece), Elmer's glue or similar white glue (minimal quantity)

Optional extras
: Distilled or deionized water, table salt, extra transistor

Experiment 2

Power
: 5VDC regulated

Resistor
: 220Ω, 470Ω(5), 1K, 1.5K

Trimmer potentiometer
: 1M

Transistor
: 2N2222 or PN2222

Experiment 3

Power
: 5VDC regulated

Resistor
: 100Ω, 3.3K, 10K, 33K

Capacitor
: 0.01μF, 10μF

Integrated circuit chip
555 timer, old-style bipolar type

Sensor
Phototransistor PT334-6C or similar, responsive to white light

Audio output
Loudspeaker, 2" diameter, 50Ω or higher

Experiment 4

Power
5VDC regulated

Resistor
3.3K

Sensor
Phototransistor PT334-6C or similar, responsive to white light

Experiment 5

Power
5VDC regulated

Resistor
100Ω, 3.3K, 10K (2), 33K, 150K

Capacitor
0.01µF, 1µF, 10µF (2), 47µF

Integrated circuit chip
555 timer old-style bipolar type (2)

Sensor
Phototransistor PT334-6C or similar, responsive to white light

Audio output
Loudspeaker 2" diameter, 50Ω or higher

Experiment 6

Power
5VDC regulated

Resistor
470Ω, 3.3K, 100K

LED
Generic 3mm

Trimmer potentiometer
500K (2)

Integrated circuit chip
LM339 comparator

Sensor
Phototransistor PT334-6C or similar, responsive to white light

Experiment 7

Power
9V battery and connector for circuit testing, 1.5 battery to fit clock (2), AC adapter with 10VDC to 12VDC for long-term installation

Resistor
47Ω, 220Ω (2), 1K (2), 3.3K, 10K (4), 100K (2), 220K (2), 1M (2)

Capacitor
0.01µF (2), 0.1µF (2), 0.33µF (2), 1µF (5), 100µF

Switch
Tactile

LED
Generic 3mm (2)

Trimmer potentiometer
500K (2)

Transistor
2N2222 or PN2222 (2)

Integrated circuit chip
LM339 comparator, 555 timer old-style bipolar type (2)

Sensor
Phototransistor PT334-6C or similar, responsive to white light

Other
LM7806 voltage regulator, UA78M33 voltage regulator, 1N4001 small rectifier diode (2), latching relay DS1E-SL2-DC3V or similar in SPDT or DPDT format with 3VDC coil to switch up to 2A, battery-powered digital alarm clock (see text for details), must use two 1.5V batteries

Experiment 8

Power
9V battery and connector

Resistor
4.7K

Sensor
Microphone, electret, generic, two-terminal type

Experiment 9

Power
9V battery and connector

Resistor
4.7K, 100K (10)

Capacitor
0.68μF (2)

Integrated circuit chip
LM741 op amp

Sensor
Microphone, electret, generic, two-terminal type

Experiment 10

Power
9V battery and connector

Resistor
470Ω, 1K (2), 4.7K, 10K, 100K (10)

Capacitor
0.68μF (2)

LED
Generic 3mm

Transistor
2N2222 or PN2222

Integrated circuit chip
LM741 op amp

Sensor
Microphone, electret, generic, two-terminal type

Experiment 11

Power
9V battery and connector

Resistor
2.2K (10), 68K (2), 10K, 100K (10), 220K, 1M

Capacitor
1μF, 10μF (2)

Trimmer potentiometer
5K

Integrated circuit chip
LM741 op amp

Experiment 12

Power
9V battery and connector

Resistor
22Ω, 100Ω, 1K, 3.3K, 4.7K, 10K, 68K (2), 100K, 150K

Capacitor
0.047μF, 0.1μF, 0.68μF, 10μF (2), 330μF

Trimmer potentiometer
10K

Integrated circuit chip
LM741 op amp, LM386 power amp

Sensor
Microphone, electret, generic, two-terminal type

Audio output
Loudspeaker 2" diameter, 50Ω or higher

Experiment 13

Power
9V battery and connector

Resistor
100Ω, 1K (2), 3.3K, 4.7K, 10K (4), 33K, 68K (2)

Capacitor
0.01μF, 0.068μF, 0.68μF, 10μF (3), 47μF, 100μF, 330μF

Trimmer potentiometer
1M

Transistor
2N2222 or PN2222

Integrated circuit chip
LM741 op amp, 555 timer old-style bipolar type

Sensor
Microphone, electret, generic, two-terminal type

Audio output
Loudspeaker 2″ diameter, 50Ω or higher

Experiment 14

Power
9V battery and connector

Resistor
220Ω (2), 470Ω, 1K (3), 3.3K, 4.7K (2), 10K (4), 68K (2), 100K, 150K (2)

Capacitor
0.01μF (2), 0.1μF, 0.68μF, 10μF (3), 15μF, 100μF, 220μF

LED
Generic 3mm (2)

Trimmer potentiometer
1M

Transistor
2N2222 or PN2222 (2)

Integrated circuit chip
LM741 op amp, 555 timer old-style bipolar type (2)

Sensor
Microphone, electret, generic, two-terminal type

Audio output
Buzzer or beeper, 9VDC or 12VDC, 100mA max

Experiment 15

Power
5VDC regulated

Resistor
220Ω, 10K (4)

Switch
Tactile (4)

LED
Generic 3mm

Integrated circuit chip
74HC08 quad two-input AND, 74HC32 quad two-input OR

Experiment 16

Power
5VDC regulated

Resistor
10K (4), and 220Ω (6) if required by LEDs

LED
3mm with internal resistors for testing (6), or add series resistors shown above, or 5mm for final (6)

Integrated circuit chip
74HC08 quad two-input AND, 74HC86 quad two-input XOR (2), 74HC11 triple three-input AND, 74HC02 quad two-input NOR

Switch
Tactile (4)

Experiment 17

- No parts required

Experiment 18

Power
9V battery and connector

Resistor
100Ω (6), 220Ω (6), 470Ω (2), 330Ω (6)

Switch
4PDT pushbutton (6) with caps (6)

LED
5mm with internal resistors (8), or add resistors shown above

Audio output
9VDC or 12VDC buzzer or beeper (3)

Other
Plain perforated board with no copper traces (6″ x 6″ piece minimum), optional multicolored ribbon cable (2 feet)

Experiment 19

Power
5VDC regulated

Resistor
10K (4), 220Ω (7) if needed for LEDs

Switch
Tactile (4), DIP 4-position

LED
3mm with internal resistors (7), or add resistors shown above

Integrated circuit chip
74HC4514 or 4514B decoder, 74HC32 quad two-input OR, 74HC4075 triple three-input OR

Experiment 20

Power
5VDC regulated

Resistor
10K (6), and 100Ω(10), 220Ω(10) if required by LEDs

Switch
4PDT pushbutton (6) with caps (6)

LED
3mm with internal resistors (10), or generic (10) with resistors shown above

Integrated circuit chip
74HC237 decoder (2), 74HC08 quad two-input AND (2), 74HC4075 triple three-input OR, 74HC4002 dual four-input NOR

Experiment 21

Power
9V battery and connector

Resistor
470Ω, 1K, 10K (8)

Capacitor
0.001µF, 0.01µF, 0.1µF, 100µF

Trimmer potentiometer
1K

Switch
Tactile (4), DPDT pushbutton, toggle, DIP 8-position (2)

LED
3mm with internal resistor, or add resistor shown above

Integrated circuit chip
4067B multiplexer, 4520B counter, 555 timer old-style bipolar type

Experiment 22

Power
5VDC regulated

Resistor
100, 1K, 3.3K (2), 4.7K, 10K (4), 33K

Capacitor
0.01µF (2)

Trimmer potentiometer
50K

Transistor
2N2222 or PN2222

Integrated circuit chip
555 timer old-style bipolar type (2), 74HC86 quad two-input XOR

Sensor
Phototransistor PT334-6C or similar, responsive to white light (2)

Audio output
Loudspeaker 2" diameter, 50Ω or higher

Experiment 23

Power
9VDC battery and connector

Resistor
470Ω (7) if required by LEDs

Switch
4PDT pushbutton (6)

LED
3mm with internal resistor (7), or add resistors shown above

Experiment 24

Power
: 5VDC regulated

Resistor
: 10K (6), and 220Ω (4) if required by LEDs

Switch
: DIP 4-position (2)

LED
: 3mm with internal resistor (4), or add resistors shown above

Integrated circuit chip
: 74HC86 quad two-input XOR (2), 74HC08 quad two-input AND (2), 74HC32 quad two-input OR

Experiment 25

Power
: 5VDC regulated

Resistor
: 10K (19), 220Ω (4) if required by LEDs

Switch
: Pushbutton 6PDT (2), DIP 8-position (2)

LED
: 3mm with internal resistor (18), or add resistor shown above

Integrated circuit chip
: 74HC4514 decoder, 74HC148 encoder (2)

Optional extras
: Pushbutton 3PDT, pushbutton DPDT (3), pushbutton SPDT (3)

Experiment 26

Power
: 5VDC regulated

Resistor
: 10K (4), 100K, and 220Ω (3) if required by LEDs

Capacitor
: 0.01μF (2), 0.033μF, 0.1μF (2), 1μF (2)

Switch
: DPDT pushbutton, and tactile

LED
: 3mm with internal resistor (30) or add resistors shown above

Trimmer potentiometer
: 100K

Integrated circuit chip
: 7555 timer (2), 74HC4017 counter (3), 74HC08 quad two-input AND

Other
: Optional multicolor ribbon cable (2 feet), optional header pins (33), optional header sockets (33), or use flexible jumpers (33)

Experiment 27

Power
: 5VDC regulated

Resistor
: 10K (2), 100K, 220Ω (9) if required by LEDs

Capacitor
: 0.01μF, 0.033μF, 0.1μF, 2.2μF

Switch
: DPDT pushbutton, and tactile

LED
: 3mm with internal resistor (9) or add resistors shown above

Integrated circuit chip
: 7555 timer, 74HC164 shift register

Experiment 28

Power
: 5VDC regulated

Resistor
: 220Ω (24), 3.3K, 4.7K, 10K (3), 1M

Capacitor
: 0.001μF, 0.033μF, 0.01μF (2), 0.1μF (2), 0.33μF, 33μF, 100μF

Switch
: Tactile

LED
: 3mm generic for demonstration (36), or light bars (36) as described in text

Transistor
    2N2222 or PN2222

Integrated circuit chip
    7555 timer (2), 74HC164 shift register (3), 74HC4514 decoder or 4514B decoder, 4078B single eight-input OR/NOR, 4520B binary counter, optional ULN2003 Darlington array (3) if light bars are used

Experiment 29

Power
    9V battery and connector

Resistor
    470Ω

LED
    3mm generic

Sensor
    SPST reed switch, any type

Other
    Small iron bar magnet (approximately 1/4″ x 1/4″ x 1.5″) or very small neodymium bar magnet (approximately 1/4″ x 1/16″ x 1/2″), optional ring magnets, optional ball magnet and aluminum tube described in text

Experiment 30

Power
    9V battery and connector

Resistor
    1K

LED
    Generic 3mm

Sensor
    Bipolar Hall-effect sensor, ATS177 or similar, and unipolar, or linear, or omnipolar Hall-effect sensor (optional)

Other
    Small iron bar magnet (approximately 1/4″ x 1/4″ x 1.5″) or very small neodymium bar magnet (approximately 1/4″ x 1/16″ x 1/2″)

Experiment 31

Power
    5VDC regulated

Resistor
    100Ω, 220Ω, 1K for demo—or 330Ω (16), 680Ω (16), 3.3K (16) for finished project

LED
    Generic 5mm for demo, or generic 5mm (16) for finished project

Trimmer potentiometer
    1K, 2K

Integrated circuit chip
    74HC32 quad two-input OR, and ULN2003 Darlington array (2) for finished project

Sensor
    Transmissive infrared sensor Everlight ITR9606 or similar, 1 for demo or 16 for finished project

Other
    Flexible jumper wires with plugs at the ends (9)

Experiment 32

- No parts required
- Optional extra: reed switch SPST (18) and appropriate activation magnets

Experiment 33

Power
    5VDC regulated

Resistor
    220Ω (2), 470Ω (2), 10K (4)

Switch
    Tactile

LED
    3mm generic (3)

Integrated circuit chip
    74HC86 quad two-input XOR

Sensor
Bourns ECW1J-B24-BC0024L rotational encoder or similar with 24PPR and 24 detents and quadrature output (2)

Other
Lead sinkers (2) described in text, galvanized wire (1 foot) described in text

Experiment 34

Power
5VDC regulated

Resistor
10K (4), 100K (2), and 220Ω (9) if required by LEDs

Capacitor
0.0001µF, 0.01µF (2), 0.033µF, 1µF, 10µF

Switch
DPDT pushbutton

LED
3mm with internal resistor (9), or generic with resistors shown above

Integrated circuit chip
7555 timer (2), 74HC4017 counter

Sensor
Thermistor 100K

Experiment 35

Power
5VDC regulated

Resistor
10K, 100K, and 220Ω (9) if required by LEDs

Capacitor
0.01µF, 0.033µF, 0.1µF, 2.2µF, 100µF

Switch
DPDT pushbutton

LED
3mm with internal resistor (9), or generic with resistors shown above

Integrated circuit chip
7555 timer, 74HC164 shift register, 74HC86 quad two-input XOR, 74HC7266 quad two-input XNOR

Experiment 36

Power
5VDC regulated

Resistor
10K (5), 100K (2), 220Ω (8) if required by LEDs

Capacitor
47pF, 68pF, 100pF, 0.01µF (4), 10µF (2), 100µF

Switch
DPDT pushbutton (3)

LED
3mm with internal resistor (4), or generic with resistors shown above

Integrated circuit chip
7555 timer (3), 74HC164 shift register, 74HC86 quad two-input XOR, 74HC7266 quad two-input XNOR, 74HC30 single eight-input NAND, 74HC08 quad two-input AND, 74HC32 quad two-input OR

Index

Symbols

A

B

C

D

E

F

G

H

I

J

K

L

M

N

O

P

T

U

V

W

X

Y

About the Author

Charles Platt is a contributing editor and regular columnist for *Make* magazine, where he writes about electronics. Following the success of his book *Make: Electronics*, he began writing *Encyclopedia of Electronic Components*, which will appear in three volumes. Volume One is currently available.

Platt was a senior writer for *Wired* magazine and has written various computer books. As a prototype designer, he created semiautomated rapid cooling devices with medical applications and air-deployable equipment for first responders. He was the sole author of four mathematical-graphics software packages and has been fascinated by electronics since he put together a telephone answering machine from a tape recorder and military-surplus relays at age 15. He lives in a Northern Arizona wilderness area, where he has his own workshop for prototype fabrication and projects that he writes about for *Make* magazine.

Colophon

The cover photograph is by Charles Platt. The cover and body font is Benton Sans, the heading font is Serifa, and the code font is Bitstream Vera Sans Mono.